The Geometry of Total Curvature on Complete Open Surfaces

This is a self-contained account of how some modern ideas in differential geometry can be used to tackle and extend classical results in integral geometry. The authors investigate the influence of the total curvature on the metric structure of complete noncompact Riemannian 2-manifolds, though their work, much of which has never appeared in book form before, can be extended to more general spaces.

Many classical results are introduced and then extended by the authors. The compactification of complete open surfaces is discussed, as are Busemann functions for rays. Open problems are provided in each chapter, and the text is plentifully illustrated with figures designed to improve the reader's intuitive understanding of the subject matter.

The treatment is self-contained, assuming only a basic knowledge of manifold theory, and so is suitable for graduate students and nonspecialists who seek an introduction to this modern area of differential geometry.

The Geometry of Total Curvature on Complete Open Surfaces

KATSUHIRO SHIOHAMA
Department of Mathematics, Saga University, Japan

TAKASHI SHIOYA
Mathematical Institute, Tohoku University, Japan

MINORU TANAKA
Department of Mathematics, Tokai University, Japan

CAMBRIDGE
UNIVERSITY PRESS

PUBLISHED BY THE PRESS SYNDICATE OF THE UNIVERSITY OF CAMBRIDGE
The Pitt Building, Trumpington Street, Cambridge, United Kingdom

CAMBRIDGE UNIVERSITY PRESS
The Edinburgh Building, Cambridge CB2 2RU, UK
40 West 20th Street, New York, NY 10011–4211, USA
477 Williamstown Road, Port Melbourne, VIC 3207, Australia
Ruiz de Alarcón 13, 28014 Madrid, Spain
Dock House, The Waterfront, Cape Town 8001, South Africa

http://www.cambridge.org

First published 2003

Printed in the United Kingdom at the University Press, Cambridge

Typeface Times 10/13 pt.   *System* LaTeX 2_ε [TB]

A catalog record for this book is available from the British Library

Library of Congress Cataloging in Publication data

Shiohama, K. (Katsuhiro), 1940–
The geometry of total curvature on complete open surfaces / Katsuhiro Shiohama,
Takashi Shioya, Minoru Tanaka.
  p.   cm. (Cambridge tracts in mathematics; 159)
Includes bibliographical references and index.
ISBN 0 521 45054 3
1. Riemannian manifolds.   2. Curves on surfaces.   3. Global differential geometry.
  I. Shioya, Takashi, 1963–   II. Tanaka, Minoru, 1949–   III. Title.   IV. Series.
QA670.S48   2003
516.3′52–dc21   2003041955

ISBN 0 521 45054 3 hardback

Contents

v

Preface

The study of the curvature and topology of Riemannian manifolds is mainstream in differential geometry. Many of the important contributions in this topic go back to the pioneering works by Cohn-Vossen in 1935–6, [19] and [20]. In fact the study of total curvature on complete noncompact Riemannian manifolds made by him contains many fruitful ideas. Many hints in his thoughts lead us to the study of the curvature and topology of Riemannian manifolds.

The well-known Gauss–Bonnet theorem states that the total curvature of a compact Riemannian 2-manifold is a topological invariant. Cohn-Vossen first proved that the total curvature of a finitely connected complete noncompact Riemannian 2-manifold M is bounded above by $2\pi \chi(M)$, where $\chi(M)$ is the Euler characteristic of M. Among many beautiful consequences of this result, he proved the splitting theorem for complete open Riemannian 2-manifolds of nonnegative Gaussian curvature admitting a straight line. The structure theorem for such 2-manifolds was also established by him. He investigated the global behavior of complete geodesics on these 2-manifolds and this gave rise to the study of poles. The Bonnesen-type isoperimetric problem for complete open surfaces admitting a total curvature was first investigated by Fiala [26] for the analytic case and then by Hartman [34] for the C^2 case. Here the Cohn-Vossen theorem plays an essential role. The total curvature of infinitely connected complete open surfaces was discussed by Huber from the point of view of complex analysis. Busemann considered the notion of total curvature on a G-surface X (see Section 43, [12]), in which he suggested that Cohn-Vossen's results would follow on Busemann G-surfaces when the total curvature was replaced by the Busemann total excess of X.

It took more than thirty years to obtain higher-dimensional extensions of Cohn-Vossen's results. They are the Toponogov splitting theorem [103] and the structure theorem for complete noncompact Riemannian manifolds of positive sectional curvature [30] and of nonnegative sectional curvature [17]. The total curvature of higher-dimensional complete noncompact Riemannian manifolds of nonnegative sectional curvature was discussed in [69].

The purpose of this book is to study the geometric significance of the total curvature of complete noncompact Riemannian 2-manifolds. The total curvature $c(M)$ of such a manifold M is not a topological invariant but is dependent upon the choice of Riemannian metric. Therefore we may consider that $c(M)$ describes certain geometric properties of M. These phenomena are seen in the asymptotic behavior of the mass of rays on M and that of the isoperimetric inequalities for metric balls and their boundaries. Moreover, the size of the ideal boundary equipped with the Tits metric is determined by $c(M)$ and the topology of M. The global behavior of complete geodesics on Riemannian planes is controlled by $c(M)$. It is expected that many results will be extended to complete noncompact Alexandrov surfaces with the Busemann total excess.

This book is written as a self-contained text including many examples, figures and exercises. First-year graduate students will find this book very useful. The reader will quickly acquire the tools necessary for the study of Riemannian geometry.

In Chapter 1, the basic tools in Riemannian geometry are prepared. We first use only local coordinates and introduce the Levi–Civita connection and curvature tensor. We then use vector field notation to simplify the working. We want to acknowledge two books which have been very useful in writing this chapter. The discussion in Sections 1.4 to 1.7 is based on the book by Gromoll, Klingenberg and Meyer [29]. The discussion on the Sasaki metric is based on Sakai's book [73].

In Chapter 2, the classical results by Cohn-Vossen and Huber on the total curvature of complete open surfaces are introduced. All the ideas employed by Cohn-Vossen in [19] and [20] are explained here. We deal with the Gauss–Bonnet theorem for compact simplicial complexes on surfaces in such a way that the Gauss–Bonnet theorem can be extended to them.

In Chapter 3, the ideal boundary $M(\infty)$ of a complete noncompact Riemannian 2-manifold M with total curvature is obtained by using the idea of Ballmann, Gromov and Schroeder [7], which they discussed using Hadamard manifolds. New ideas are introduced. We establish the Gauss–Bonnet theorem for the compactification $M \cup M(\infty)$ of M by attaching an ideal boundary equipped with the Tits metric. A new triangle comparison theorem is established for special triangles bounding domains having small total absolute curvature. Furthermore, we prove that the scaling limit of M with finite total curvature is the union of flat cones generated by $M(\infty)$ all having a common vertex. The behavior of Busemann functions is discussed.

In Chapter 4, the structure of the cut loci of circles on complete open surfaces with (or without) total curvature is discussed. The classical Hartman theorem on geodesic parallel circles is introduced. The topological structure of metric circles is discussed in detail.

In Chapter 5, the isoperimetric inequalities for metric circles and for balls around a smooth Jordan circle are discussed. The classical Fiala–Hartman theorem is extended. The infinitely connected case is also considered.

In Chapter 6, the mass of rays emanating from a point on M is discussed. Integral formulae for the mass of rays are treated in connection with the isoperimetric inequalities.

In Chapter 7, the classical result due to von Mangoldt is presented. The set of poles on a surface of revolution homeomorphic to a plane is determined explicitly: it consists of a unique trivial pole or forms a closed ball centered at the vertex. We find a necessary and sufficient condition for a surface of revolution to have many poles. The cut locus of a standard surface of revolution, such as a two-sheeted hyperboloid, is determined explicitly.

In Chapter 8, the global behavior of complete geodesics on a Riemannian plane M having a total curvature is discussed. The number of self-intersections of complete geodesics away from a compact set (near the ideal boundary) is estimated explicitly in terms of the total curvature of M. This involves the Whitney regular homotopy of curves and the rotation numbers.

The authors would like to express their thanks to Takao Yamaguchi, Kunio Sugahara, Qing Ming Cheng, Kazuyuki Enomoto and Yoshiko Kubo for reading and criticizing the first draft of this book. We would like to thank Manabu Ohura for typing the first draft.

1

Riemannian geometry

We introduce the basic tools of Riemannian geometry that we shall need assuming that readers already know basic facts about manifolds. Only smooth manifolds are discussed unless otherwise mentioned. The use of a local chart (local coordinates) will be convenient for readers at the beginning. Tensor calculus is used to introduce geodesics, parallelism, covariant derivative and the Riemannian curvature tensor. In Sections 1.1 to 1.3 the Einstein convention will be adopted without further mention. However, it is not convenient for later discussion, for example, the second variation formula or Jacobi fields. To avoid confusion we shall use vector fields and connection forms to discuss such matters as Jacobi fields and conjugate points.

1.1 The Riemannian metric

Let M be an n-dimensional, connected and smooth manifold and (U, x) local coordinates around a point $p \in M$. A point $q \in U$ is expressed as $x(q) = (x^1(q), \ldots, x^n(q)) \in x(U) \subset \mathbf{R}^n$. The tangent space to M at p is denoted by $T_p M$ or M_p and $TM := \bigcup_{p \in M} T_p M$ denotes the tangent bundle over M. We denote by $\pi : TM \to M$ the projection map. Let $\mathcal{X}(M)$ and $\mathcal{X}(U)$ be the spaces of all smooth vector fields over M and U respectively and $C^\infty(M)$ the space of all smooth functions over M. A positive definite smooth symmetric bilinear form $g : \mathcal{X}(M) \times \mathcal{X}(M) \to C^\infty(M)$ is by definition a *Riemannian metric* over M. The metric g is locally expressed in (U, x) as follows. Let $X_i \in \mathcal{X}(U)$ be the ith basis-vector field tangent to the ith coordinate curve, i.e., $X_i := \mathrm{d}(x^{-1})(\partial/\partial x^i)$, where $\partial/\partial x^i$ is the canonical vector field in $x(U) \subset \mathbf{R}^n$ parallel to the ith coordinate axis of $\mathbf{R}^n$. Then $T_q M$ for every $q \in U$ is spanned by $X_1(q), \ldots, X_n(q)$. If $X, Y : U \to TU$ are local vector fields expressed as

1

$X = \sum_i \phi^i X_i$ and $Y = \sum_j \psi^j X_j$ and if $g_{ij} := g(X_i, X_j)$ then

$$g(X, Y) = \sum_{i,j=1}^n \phi^i \psi^j g_{ij}, \qquad (1.1.1)$$

where $g_{ij} = g_{ji}$ and the g_{ij} are smooth functions on U. Thus the length (or norm) $\|v\|$ of a vector $v \in T_p M$ is defined by $\|v\| := g_p(v, v)^{1/2} = (\sum g_{ij}(p) v^i v^j)^{1/2}$, where $v := \sum v^i X_i(p)$. The *angle* $\angle(u, v)$ between two vectors u and v in $T_p M$ is thus defined by

$$\cos \angle(u, v) := \frac{\langle u, v \rangle}{\|u\| \, \|v\|}. \qquad (1.1.2)$$

Here the angle $\angle(u, v)$ takes values in $[0, \pi]$ and $\langle u, v \rangle = g_p(u, v)$.

The volume of a parallel n-cube in $T_p M$ spanned by $X_1(p), \ldots, X_n(p)$ is given by $|X_1(p) \wedge \cdots \wedge X_n(p)| = (\det g_{ij}(p))^{1/2}$. Thus the *volume element* dM of M is expressed as follows:

$$dM = (\det g_{ij})^{1/2} \, dx^1 \wedge \cdots \wedge dx^n. \qquad (1.1.3)$$

The manifold M equipped with a Riemannian metric g is called a *Riemannian manifold* and denoted by (M, g) or simply by M.

By a smooth curve $c : [\alpha, \beta] \to M$ we always mean that there exists an open interval $I \supset [\alpha, \beta]$ such that c is defined over I and is regular at all points on I. A piecewise-smooth curve $c : [\alpha, \beta] \to M$ is a continuous map consisting of finitely many smooth curves. Namely, there are finitely many points $t_0 = \alpha < t_1 < \cdots < t_k = \beta$ such that $c|_{[t_i, t_{i+1}]}$, for every $i = 0, \ldots, k - 1$, is a smooth curve. A piecewise-smooth vector field X along a curve $c : [\alpha, \beta] \to M$ is by definition a piecewise-smooth map $X : [\alpha, \beta] \to T_c M$ such that $\pi \circ X(t) = c(t)$ for all $t \in [\alpha, \beta]$, where $T_c M := \bigcup_{t \in [\alpha, \beta]} T_{c(t)} M$. A curve $c : [\alpha, \beta] \to U$ in a coordinate neighborhood is expressed by $x \circ c(t) = (x^1(t), \ldots, x^n(t))$, for $t \in [\alpha, \beta]$. Its velocity vector field is

$$\dot{c}(t) = \sum_{i=1}^n \frac{dx^i}{dt} X_i \circ c(t).$$

The length $L(c)$ of a curve $c : [\alpha, \beta] \to U$ is given by

$$L(c) = \int_\alpha^\beta \|\dot{c}\| \, dt = \int_\alpha^\beta \sqrt{\sum_{i,j=1}^n g_{ij}(c(t)) \frac{dx^i}{dt} \frac{dx^j}{dt}} \, dt. \qquad (1.1.4)$$

If $s(t)$ for $t \in [\alpha, \beta]$ is the length of the subarc $c|_{[\alpha, t]}$ of c then $ds(t)/dt = \|\dot{c}(t)\| > 0$, and hence $s(t)$ has an inverse, $t = t(s)$. Thus c can be parameterized by its arc length $s \in [0, L]$, where $L = L(c)$ is the total length of c. From the relation $ds(t) = \|\dot{c}\| \, dt$ we derive a quadratic differential form

$$ds^2 = \sum_{i,j=1}^{n} g_{ij} dx^i \, dx^j,$$

which we call the *line element* of M.

1.2 Geodesics

From now on the Einstein convention is used. Once the length of a curve in M has been established, we discuss a curve with the special property of having a local minimum in length among the neighboring curves with the same endpoints. This is referred to as the locally minimizing property. Such a curve, a *geodesic*, is obtained as the solution of a nonlinear second-order ordinary differential equation, (1.2.1) below, whose coefficients depend only on the g_{ij} and their partial derivatives. Thus geodesics are defined as the solutions of (1.2.1). The set of all solutions with a fixed starting point p corresponds to a domain in $T_p M$ that is star-shaped with respect to the origin. Thus the exponential map and its injectivity radius at that point is introduced here.

Definition 1.2.1. A unit-speed curve $c : [\alpha, \beta] \to M$ is said to have the *locally minimizing property* iff there exists for every $s \in [\alpha, \beta]$ a positive number δ with $[s - \delta, s + \delta] \subset I$ and a neighborhood $\mathcal{N}$ around $c([s - \delta, s + \delta])$ such that $c([s - \delta, s + \delta])$ has the minimum length among all the curves in $\mathcal{N}$ joining $c(s - \delta)$ to $c(s + \delta)$.

We define the *Christoffel symbols*

$$\Gamma^i_{jk} := \tfrac{1}{2} g^{i\ell} (\partial_j g_{\ell k} + \partial_k g_{j\ell} - \partial_\ell g_{jk}),$$

where $\partial_j g_{\ell k} := \partial g_{\ell k}/\partial x^j$ and (g^{ij}) is the inverse matrix of (g_{ij}), i.e.,

$$g^{i\ell} g_{\ell k} := \delta^i_k.$$

Here δ^i_j is the *Kronecker delta*, i.e., $\delta^i_k = 1$ for $i = k$ and $\delta^i_k = 0$ for $i \neq k$.

With this notation we prove

Theorem 1.2.1. *If a unit-speed curve $c : [\alpha, \beta] \to M$ has the locally minimizing property, then the local expression*

$$x \circ c(s) = (x^1(s), \ldots, x^n(s))$$

of c in a coordinate neighborhood U satisfies

$$\frac{d^2 x^i}{ds^2} + \Gamma^i_{jk} \frac{dx^j}{ds} \frac{dx^k}{ds} = 0. \tag{1.2.1}$$

Proof. Since the discussion is local, we may restrict ourselves to the case where $c([\alpha, \beta])$ is contained entirely in a coordinate neighborhood U and write $x \circ c(s) = (x^1(s), \dots, x^n(s))$. A *variation along* $c : [\alpha, \beta] \to U$ is by definition a (piecewise-smooth) map $V : (-\varepsilon_0, \varepsilon_0) \times [\alpha, \beta] \to U$ such that

$$V(0, s) = c(s) \qquad \text{for all } s \in [\alpha, \beta]$$

and $V_\varepsilon(s) := V(\varepsilon, s)$ for every $\varepsilon \in (-\varepsilon_0, \varepsilon_0)$ is a curve $V_\varepsilon : [\alpha, \beta] \to U$. If $x \circ V(\varepsilon, s) = (x^1(\varepsilon, s), \dots, x^n(\varepsilon, s))$ for $(\varepsilon, s) \in (-\varepsilon_0, \varepsilon_0) \times [\alpha, \beta]$ then the variational vector field $Y : [\alpha, \beta] \to T_c M$ associated with V is given by

$$Y(s) = dV_{(0,s)} \left(\frac{\partial}{\partial \varepsilon} \right) = \frac{\partial x^i}{\partial \varepsilon}(0, s) \, X_i \circ c(s).$$

The length $L(\varepsilon)$ of each variation curve V_ε is given by

$$L(\varepsilon) := L(V_\varepsilon) = \int_\alpha^\beta \sqrt{g_{ij} \frac{\partial x^i(\varepsilon, s)}{\partial s} \frac{\partial x^j(\varepsilon, s)}{\partial s}} \, ds.$$

Thus we have, by taking the arc length parameter $0 \le s \le L =: L(c)$,

$$\begin{aligned}
\frac{dL}{d\varepsilon}\bigg|_{\varepsilon=0} &= \int_0^L \frac{\partial}{\partial \varepsilon} \sqrt{g_{ij} \frac{\partial x^i(\varepsilon, s)}{\partial s} \frac{\partial x^j(\varepsilon, s)}{\partial s}} \bigg|_{\varepsilon=0} ds \\
&= \frac{1}{2} \int_0^L \left((\partial_k g_{ij}) \frac{\partial x^k}{\partial \varepsilon} \frac{\partial x^i}{\partial s} \frac{\partial x^j}{\partial s} + 2 g_{ij} \frac{\partial^2 x^i}{\partial s \partial \varepsilon} \frac{\partial x^j}{\partial s} \right) \bigg|_{\varepsilon=0} ds.
\end{aligned}$$

Since the second term in the integrand can be expressed as

$$\frac{\partial}{\partial s} \left(g_{ij} \frac{\partial x^i}{\partial \varepsilon} \frac{\partial x^j}{\partial s} \right) - \frac{\partial}{\partial s} \left(g_{kj} \frac{\partial x^j}{\partial s} \right) \frac{\partial x^k}{\partial \varepsilon}$$

we have

$$\begin{aligned}
\frac{dL(0)}{d\varepsilon} &= g_{ij} \frac{\partial x^i}{\partial \varepsilon} \frac{\partial x^j}{\partial s}(0, s) \bigg|_0^L \\
&\quad - \int_0^L \left(\partial_\ell g_{kj} \frac{\partial x^\ell}{\partial s} \frac{\partial x^j}{\partial s} + g_{kj} \frac{\partial^2 x^j}{\partial s^2} - \frac{1}{2} \partial_k g_{ij} \frac{\partial x^i}{\partial s} \frac{\partial x^j}{\partial s} \right) \frac{\partial x^k}{\partial \varepsilon} \bigg|_{(0,s)} ds.
\end{aligned}$$

By setting $2[ij; k] := \partial_i g_{jk} + \partial_j g_{ik} - \partial_k g_{ij}$, we see that $\Gamma^k_{ij} g_{k\ell} = [ij; \ell]$, and then the integrand can be rewritten as

$$\left(\frac{1}{2} (\partial_\ell g_{kj} + \partial_j g_{k\ell} - \partial_k g_{\ell j}) \frac{\mathrm{d}x^\ell}{\mathrm{d}s} \frac{\mathrm{d}x^j}{\mathrm{d}s} + g_{kj} \frac{\mathrm{d}^2 x^j}{\mathrm{d}s^2} \right) \frac{\partial x^k}{\partial \varepsilon}$$

$$= g_{kj} \left(\frac{\mathrm{d}^2 x^j}{\mathrm{d}s^2} + \Gamma^j_{\ell m} \frac{\mathrm{d}x^\ell}{\mathrm{d}s} \frac{\mathrm{d}x^m}{\mathrm{d}s} \right) \frac{\partial x^k}{\partial \varepsilon}.$$

Therefore we obtain

$$L'(0) = g(\dot{c}, Y)|_0^L - \int_0^L g_{kj} \left(\frac{\mathrm{d}^2 x^j}{\mathrm{d}s^2} + \Gamma^j_{\ell m} \frac{\mathrm{d}x^\ell}{\mathrm{d}s} \frac{\mathrm{d}x^m}{\mathrm{d}s} \right) Y^k \, \mathrm{d}s. \quad (1.2.2)$$

The locally minimizing property of c then implies that $L'(0) = 0$ for every variation V with $V(\varepsilon, 0) = c(0)$ and $V(\varepsilon, L) = c(L)$ for all $\varepsilon \in (-\varepsilon_0, \varepsilon_0)$. Therefore $g(\dot{c}, Y)|_0^L = 0$ follows from $Y(0) = Y(L) = 0$. The proof is concluded since the variation vector field Y can be taken as arbitrary. $\square$

Now the differential equation (1.2.1) will be discussed. Changing the parameter via $s = at$ for a constant $a > 0$, we observe that (1.2.1) becomes

$$\frac{\mathrm{d}^2 x^i}{\mathrm{d}t^2} + \Gamma^i_{jk} \frac{\mathrm{d}x^j}{\mathrm{d}t} \frac{\mathrm{d}x^k}{\mathrm{d}t} = 0.$$

Geodesics are always parameterized proportionally to arc lengths. The equation (1.2.1) is equivalent to the following system of first-order differential equations:

$$v^i(s) = \frac{\mathrm{d}x^i}{\mathrm{d}s}, \qquad \frac{\mathrm{d}v^i}{\mathrm{d}s} + \Gamma^i_{jk} v^j v^k = 0. \quad (1.2.3)$$

Because the Γ^i_{jk} are smooth functions, the above differential equations satisfy the Lipschitz condition and hence have a unique solution for given initial conditions. Let $p \in U$ and $\xi \in T_p M$ be expressed as $x(p) = (p^1, \dots, p^n)$ and $\xi = \xi^i X_i(p)$. Then (1.2.1) has locally a unique solution for the initial conditions $x^i(0) = p^i$ and $\mathrm{d}x^i/\mathrm{d}s(0) = \xi^i$ for $i = 1, \dots, n$. If $\gamma(s) := \gamma(p, \xi; s)$ for $s \in [0, s_0)$ is the maximal solution of (1.2.1) with $\gamma(0) := \gamma(p, \xi; 0) = p$ and $\dot{\gamma}(0) = \xi$ then $\gamma(p, \xi; t) = \gamma(p, a\xi; t/a) = \gamma(p, t\xi; 1)$. If we set

$$\widetilde{M}_p := \{ u \in T_p M; \gamma(p, u; 1) \text{ makes sense} \}$$

then $\widetilde{M}_p$ is a domain in $T_p M$ that is star-shaped with respect to the origin of $T_p M$.

Definition 1.2.2. The *exponential map* at $p \in M$ is a map defined on $\widetilde{M}_p$ such that

$$\exp_p u := \gamma(p, u; 1).$$

Clearly $\exp_p$ is a smooth map.

Theorem 1.2.2. *There exists an open set $U_p \subset \tilde{M}_p$ around the origin such that* $\exp_p |_{U_p} : U_p \to M$ *is an embedding. In particular, there exists an open set* $V_p \subset U_p$ *such that any two points in* $\exp_p V_p$ *can be joined by a geodesic.*

Proof. It is clear from the definition of the exponential map at p that

$$d(\exp_p)|_o = E_n,$$

where E_n is the $n \times n$ identity matrix and $u = (u^1, \ldots, u^n)$. Therefore we can find a small neighborhood U_p around the origin of T_pM as desired. Let $\widetilde{TM} := \bigcup_{p \in M} \tilde{M}_p$ and $\phi := (\pi, \exp) : \widetilde{TM} \to M \times M$. Then the above discussion shows that, at each zero section $o \in \widetilde{TM}$,

$$d\phi|_o = \begin{pmatrix} E_n & E_n \\ 0 & E_n \end{pmatrix}.$$

Therefore we can find an open set $W \subset \widetilde{TM}$ around the set of zero sections such that $\phi|_W$ is an embedding. Thus there exists an open set $V_p \subset U_p$ around p such that $V_p \times V_p \subset \phi(W)$. This proves Theorem 1.2.2. $\qquad\square$

Lemma 1.2.1 (The Gauss lemma). *If $u \in \tilde{M}_p$ and if $A \in T_u T_p M$ is orthogonal to u then*

$$\langle d(\exp_p)_u A, d(\exp_p)_u u \rangle = 0.$$

Proof. The conclusion is obvious if $d(\exp_p)_u A = 0$. Assume that $d(\exp_p)_u A \neq 0$. We then choose a geodesic variation $V : (-\varepsilon_0, \varepsilon_0) \times [0, \ell] \to M$ along the geodesic $\gamma : [0, \ell] \to M$ with $\gamma(0) = p$, $\dot{\gamma}(0) = u/\|u\|$ and $\ell = \|u\|$ such that

$$V(\varepsilon, t) := \exp_p t(u/\|u\| + \varepsilon A).$$

Then $V_\varepsilon : [0, \ell] \to M$ for every $\varepsilon \in (-\varepsilon_0, \varepsilon_0)$ is a geodesic with length $\ell\sqrt{1 + \varepsilon^2 \|A\|^2}$. If $Y(t) := dV_{(0,t)}(\partial/\partial\varepsilon)$ is the variational vector field associated with V then $Y(0) = 0$, $Y(\ell) = d(\exp_p)_u A$ and (1.2.2) implies that

$$L'(0) = \langle Y(t), \dot{\gamma}(t) \rangle |_0^\ell = 0. \qquad\square$$

A *geodesic polar coordinate system around a point* p is defined by the embedding $\exp_p |_{U_p} : U_p \to M$. Let $B(0, r) := \{u \in \mathbf{R}^n; \|u\| < r\}$ and $\mathbf{S}^{n-1} := \{u \in \mathbf{R}^n; \|u\| = 1\}$. They are placed in T_pM by a trivial identification. If $(\theta^1, \ldots, \theta^{n-1})$ is a local coordinate system of $\mathbf{S}^{n-1}$ around a point $u \in \mathbf{S}^{n-1}$ then $\exp_p |_{U_p}$ is expressed locally by $(\exp_p |_{U_p})^{-1}(q) = (r(q), \theta^1(q), \ldots, \theta^{n-1}$

$(q)) \in U_p$. By setting $u^1 := r, u^2 := \theta^1, \ldots, u^n := \theta^{n-1}$, we see from the Gauss lemma that the metric g can be expressed as

$$g_{ij}\, du^i\, du^j = dr^2 + h_{ab}\, d\theta^a\, d\theta^b, \tag{1.2.4}$$

where (h_{ab}) is a positive definite symmetric $(n-1) \times (n-1)$ matrix.

Definition 1.2.3. The *injectivity radius* of $\exp_p$ at p is defined by

$$i(p) := \sup\{r > 0\,;\, \exp_p\,|_{B(0,r)} \text{ is an embedding}\}.$$

Lemma 1.2.2. *If $r < i(p)$ then every point $q \in \exp_p B(0, r)$ is joined to p by a unique geodesic whose length attains the infimum of all the lengths of curves joining p to q. In particular every geodesic has the locally minimizing property.*

Proof. It follows from Theorem 1.2.2 that in $\exp_p B(0, i(p))$ there exists a unique geodesic joining p to q with length $r(q)$. Let $c : [0, 1] \to \exp_p B(0, i(p))$ be a (piecewise-smooth) curve with $c(0) = p$ and $c(1) = q$. There exists a lift $\psi : [0, 1] \to B(0, i(p)) \subset T_pM$ of c such that $c(t) = \exp_p \circ \psi(t)$ for all $t \in [0, 1]$. Then the lift can be expressed by $\psi(t) = (r(t), \theta^1(t), \ldots, \theta^{n-1}(t))$, and hence

$$\dot{c}(t) = d(\exp_p)_{\psi(t)}\dot{\psi}(t) = d(\exp_p)_{\psi(t)}\{\dot{r}\psi(t)/\|\psi\| + rA(t)\},$$

where $r(t) = \|\psi(t)\|$, $\psi(t)/\|\psi(t)\| = (\theta^1(t), \ldots, \theta^{n-1}(t)) \in S_p^{n-1}$ and $A(t)$ is the component of $\dot{\psi}(t)$ tangential to S_p^{n-1} at $\psi(t)/\|\psi(t)\|$. Then we have

$$L(c) = \lim_{\varepsilon \to 0} \int_\varepsilon^1 \|\dot{c}\|\, dt = \lim_{\varepsilon \to 0} \int_\varepsilon^1 \sqrt{g_{ij}\dot{u}^i\dot{u}^j}\, dt$$

$$= \lim_{\varepsilon \to 0} \int_\varepsilon^1 \sqrt{\dot{r}^2 + h_{ab}\dot{\theta}^a\dot{\theta}^b}\, dt \geq \int_\varepsilon^1 \left|\frac{dr}{dt}\right|\, dt$$

$$\geq \int_0^1 dr = r(q) - r(p) = r(q).$$

We now consider a (piecewise-smooth) curve $c : [0, 1] \to M$ joining p to q that is not contained in $\exp_p B(0, i(p))$. Then there exists a point $c(t_0)$ such that the subarc $c|_{[0,t_0)}$ is contained entirely in $\exp_p B(0, i(p))$. The previous discussion now implies that $L(c|_{[0,t_0]}) \geq i(p) > r(q)$. This means that such a curve has length greater than $i(p) > r(q)$. $\square$

1.3 The Riemannian curvature tensor

Newton's first law states that the motion of a particle is in a straight line at constant speed if no force acts upon it. Geodesics on a Riemannian manifold are understood to be subject to Newton's first law: it is considered that the straightness of a geodesic is equivalent to the existence of a parallel velocity vector field along it. In view of equation (1.2.1), this idea leads us to the definition of parallel fields along a curve.

Lemma 1.3.1. *Let $Z(t) = \xi^i(t)X_i \circ \gamma(t)$ be a vector field along a curve γ, where $\gamma(t) = (x^1(t), \ldots, x^n(t))$ in a coordinate neighborhood U. Then the map*

$$t \mapsto \left(\frac{d\xi^i}{dt} + \Gamma^i_{kj} \xi^k \frac{dx^j}{dt} \right) X_i \circ \gamma(t)$$

is independent of the choice of local coordinates, and hence is a vector field along γ.

Proof. Let (V, y) be another set of local coordinates such that $U \cap V$ contains a subarc of γ, which is expressed by $x \circ \gamma(t) = (x^1(t), \ldots, x^n(t))$ and $y \circ \gamma(t) = (y^1(t), \ldots, y^n(t))$. The line element ds^2 is expressed in $U \cap V$ as $ds^2 = g_{ij} dx^i dx^j = h_{ab} dy^a dy^b$. We denote by $(A^a_i) = (\partial y^a / \partial x^i)$ the Jacobian matrix and by $(B^i_a) = (\partial x^i / \partial y^a)$ its inverse matrix. Clearly we have $Z(t) = \xi^i X_i \circ \gamma(t) = \eta^a Y_a \circ \gamma(t)$, where $\xi^i = B^i_a \eta^a$. By differentiating the relation $g_{ij} = A^a_i A^b_j h_{ab}$ with respect to x^k, we obtain

$$\partial_k g_{ij} = A^a_i A^b_j A^c_k \frac{\partial h_{ab}}{\partial y^c} + \left(\frac{\partial^2 y^a}{\partial x^k \partial x^i} A^b_j + \frac{\partial^2 y^b}{\partial x^k \partial x^j} A^a_i \right) h_{ab}.$$

Also, by setting

$$\bar{\Gamma}^a_{bc} := \tfrac{1}{2} h^{ad} (\partial_b h_{dc} + \partial_c h_{bd} - \partial_d h_{bc}),$$
$$\overline{[bc, a]} := h_{da} \bar{\Gamma}^d_{bc},$$

we get

$$[ij, k] = A^b_i A^c_j A^a_k \overline{[bc, a]} + \frac{\partial^2 y^d}{\partial x^i \partial x^j} A^e_k h_{ed}.$$

It can be shown that

$$\Gamma^i_{jk} = A^b_j A^c_k B^i_a \bar{\Gamma}^a_{bc} + \frac{\partial^2 y^d}{\partial x^j \partial x^k} B^i_d.$$

Therefore we have

$$
\left(\frac{d\xi^i}{dt} + \Gamma^i_{jk}\xi^j \frac{dx^k}{dt} \right) X_i
$$

$$
= \left\{ \frac{d}{dt}\left(B^i_a \eta^a \right) + \left(A^b_j A^c_k B^i_a \bar{\Gamma}^a_{bc} + \frac{\partial^2 y^d}{\partial x^j \partial x^k} B^i_d \right) \xi^j \frac{dx^k}{dt} \right\} A^e_i Y_e
$$

$$
= \left\{ \left(\frac{\partial^2 x^i}{\partial y^a \partial y^c} \frac{dy^c}{dt} \eta^a A^e_i + B^i_a \frac{d\eta^a}{dt} A^e_i \right) + \eta^b \frac{dy^c}{dt} \delta^e_a \bar{\Gamma}^a_{bc} \right.
$$

$$
\left. + \frac{\partial^2 y^d}{\partial x^j \partial x^k} \delta^e_d B^j_a B^k_c \eta^a \frac{dy^c}{dt} \right\} Y_e
$$

$$
= \left\{ \frac{d\eta^e}{dt} + \eta^a \frac{dy^c}{dt} \left(\frac{\partial^2 x^i}{\partial y^a \partial y^c} A^e_i + \frac{\partial^2 y^e}{\partial x^j \partial x^k} B^j_a B^k_c + \bar{\Gamma}^e_{ac} \right) \right\} Y_e.
$$

It follows from

$$
0 = \frac{\partial}{\partial y^a} \left(\frac{\partial y^e}{\partial x^k} \frac{\partial x^k}{\partial y^c} \right) = \frac{\partial^2 y^e}{\partial x^k \partial x^j} B^j_a B^k_c + A^e_k \frac{\partial^2 x^k}{\partial y^c \partial y^a}
$$

that

$$
\left(\frac{d\xi^i}{dt} + \Gamma^i_{jk}\xi^j \frac{dx^k}{dt} \right) X_i \circ c(t) = \left(\frac{d\eta^e}{dt} + \bar{\Gamma}^e_{bc}\eta^b \frac{dy^c}{dt} \right) Y_e \circ c(t). \qquad \square
$$

The above reasoning shows that if Z is a vector field defined in a neighborhood around p and if its restriction to a curve c emanating from $p = c(0)$ with a given initial tangent vector $v = \dot{c}(0) \in M_p$ is expressed by $Z \circ c(t) = \xi^i(t) X_i \circ c(t)$ then the vector $(d\xi^i/dt + \Gamma^i_{jk}\xi^j dx^k/dt) X_i(p)$ is independent of the choice of c and depends only on v and Z. We write

$$
\nabla_v Z := \left(\frac{d\xi^i}{dt} + \Gamma^i_{jk}\xi^j \frac{dx^k}{dt} \right) X_i(p) \tag{1.3.1}
$$

and call $\nabla_v Z$ the *covariant derivative* of Z in v.

The covariant derivative also defines a map $\nabla : \mathcal{X}(U) \times \mathcal{X}(U) \to \mathcal{X}(U)$ as follows. If $Y, Z \in \mathcal{X}(U)$ then a vector field $\nabla_Y Z \in \mathcal{X}(U)$ is obtained by setting $Y = \phi^i X_i$ and $Z = \psi^i X_i$:

$$
\nabla_Y Z := \phi^i \left(\frac{\partial \psi^k}{\partial x^i} + \Gamma^k_{ij}\psi^j \right) X_k.
$$

Exercise 1.3.1. Show that the covariant derivative of a vector field has the following properties. If $Y, Z : U \to TU$ and α, β are functions defined on U then

$$\begin{cases} \nabla_{\alpha Y} Z = \alpha \nabla_Y Z, \\ \nabla_Y(\beta Z) = Y(\beta)Z + \beta \nabla_Y Z, \\ X g(Y, Z) = g(\nabla_X Y, Z) + g(Y, \nabla_X Z), \\ \nabla_X Y - \nabla_Y X = [X, Y]. \end{cases} \tag{1.3.2}$$

Here $[X, Y]$ is the vector field defined by $[X, Y]f := X(Yf) - Y(Xf)$ for a smooth function f.

A vector field $Z : [0, \ell] \to T_c M$ along a unit-speed curve $c : [0, \ell] \to U$ is by definition *parallel* iff $\nabla_{\dot c} Z = 0$. The *geodesic curvature vector* $\mathbf{k}(s)$ of c at $c(s)$ is defined by

$$\mathbf{k}(s) := \left(\frac{\mathrm{d}^2 x^i}{\mathrm{d}s^2} + \Gamma^i_{jk} \frac{\mathrm{d}x^j}{\mathrm{d}s} \frac{\mathrm{d}x^k}{\mathrm{d}s} \right) X_i \circ c(s),$$

where s is the arc length of c. The *geodesic curvature* $\kappa(s)$ of c is defined by

$$\kappa(s) := \|\mathbf{k}(s)\|.$$

Remark 1.3.1. The geodesic curvature vector $\mathbf{k}(s)$ at a point $c(s)$ of a unit-speed smooth curve c has the following property. Let $\gamma_\pm : [0, a_\pm] \to M$ for a sufficiently small $h > 0$ be a unit-speed minimizing geodesic with $\gamma_\pm(0) = c(s)$ and $\gamma_\pm(a_\pm) = c(s \pm h)$ and let $\tau : TM \to T_{c(s)}M$ be the parallel translation along the minimizing geodesics at $c(s)$. Then

$$\lim_{h \to 0} \frac{\tau \circ \dot c(s + h) - \dot c(s)}{h} = \mathbf{k}(s). \tag{1.3.3}$$

To see this we define *parallel fields* $\xi_\pm$ along $\gamma_\pm$ that are generated by $\xi_\pm(a_\pm) := \dot c(s \pm h)$. If $x \circ c(s) = (x^1(s), \dots, x^n(s))$ and $x \circ \gamma_\pm(t) = (x^1_\pm(t), \dots, x^n_\pm(t))$ and if $\xi_\pm(t) = \xi^i_\pm(t) X_i \circ \gamma_\pm(t)$ for $0 \le t \le a_\pm$ are local expressions then

$$\xi^i_\pm(a_\pm) - \xi^i_\pm(0) = a_\pm \frac{\mathrm{d}\xi^i_\pm}{\mathrm{d}t}(\tilde a^i_\pm), \qquad i = 1, \dots, n$$

for some $\tilde a^i_\pm \in (0, a_\pm)$. The ith component of the left-hand side of (1.3.3) is expressed as

$$\frac{1}{h} \left(\xi^i_\pm(0) - \frac{\mathrm{d}x^i(s)}{\mathrm{d}s} \right) = \frac{1}{h} \left(\frac{\mathrm{d}x^i(s \pm h)}{\mathrm{d}s} - \frac{\mathrm{d}x^i(s)}{\mathrm{d}s} \right) + \frac{1}{h} \left(a_\pm \Gamma^i_{jk} \xi^j_\pm \frac{\mathrm{d}x^k_\pm}{\mathrm{d}t} \right).$$

Taking the limit as $h \to 0$, we see that $\dot\gamma_\pm(a_\pm)$ converges to $\pm\dot c(s)$ and that $\lim_{h\to 0}(a_\pm/h) = 1$. This proves (1.3.3).

Theorem 1.3.1. *If Z and W are parallel vector fields along a curve c then the inner product g(Z, W) is constant. In particular, a curve is a geodesic iff it is auto-parallel (parallel to itself).*

Proof. Setting $Z(s) = \xi^i(s)X_i \circ c(s)$ and $W(s) = \eta^j(s)X_j \circ c(s)$, we see that

$$
\begin{aligned}
\frac{d}{ds}g(Z, W) \circ c(s) &= \partial_k g_{ij}\frac{dx^k}{ds}\xi^i\eta^j + g_{ij}\left(\frac{d\xi^i}{ds}\eta^j + \xi^i\frac{d\eta^j}{ds}\right)\\
&= \partial_k g_{ij}\frac{dx^k}{ds}\xi^i\eta^j - g_{ij}\left(\Gamma^i_{\ell m}\xi^\ell\frac{dx^m}{ds}\eta^j + \Gamma^i_{\ell m}\eta^\ell\frac{dx^m}{ds}\xi^j\right)\\
&= \frac{dx^k}{ds}\xi^i\xi^j\left(\partial_k g_{ij} - g_{pj}\Gamma^p_{ik} - g_{iq}\Gamma^q_{jk}\right)\\
&= 0.
\end{aligned}
$$

The rest is now clear. $\qquad\square$

Now the *Riemannian curvature tensor of M* is defined. Let $u, v, w \in T_pM$ and choose a local coordinate system (U, x) around p such that $x(p)$ is the origin of $\mathbf{R}^n$. At the origin in $x(U)$ consider a parallelogram P_{ab} spanned by $a(dx)_p u$ and $b(dx)_p v$ for sufficiently small numbers a, b. Let $\tau_{ab} : T_pM \to T_pM$ be the parallel translation along $x^{-1}(P_{ab})$ passing through p_1, p_3, p_2 in this order, where $x^{-1}(P_{ab})$ has corners at p, $p_1 := x^{-1}(au^1, \ldots, au^n)$, $p_2 := x^{-1}(bv^1, \ldots, bv^n)$ and $p_3 := x^{-1}(au^1 + bv^1, \ldots, au^n + bv^n)$. If $u := u^i X_i(p)$, $v := v^i X_i(p)$ and $w := w^i X_i(p)$ then $\tau_{ab}w - w$ is expressed as $w_1 - w_2$, where w_1 (resp. w_2) is obtained by the parallel translation of w along the subarc of $x^{-1}(P_{ab})$ from p to p_3 passing through p_1 (resp. p_2). By using the equation for a parallel field in Lemma 1.3.1 and a Taylor expansion, the ith component of $\tau_{ab}w - w$ is expressed as

$$
\begin{aligned}
(\tau_{ab}w - w)^i = \Big\{&a\big(\Gamma^i_{jk}(p_2) - \Gamma^i_{jk}(p)\big)u^k + b\big(\Gamma^i_{j\ell}(p) - \Gamma^i_{j\ell}(p_1)\big)v^\ell\\
&+ ab\big(\Gamma^i_{m\ell}(p_1)\Gamma^m_{jk}(p) - \Gamma^i_{mk}(p_2)\Gamma^m_{j\ell}(p)\big)u^k v^\ell\Big\}w^j,
\end{aligned}
$$

neglecting higher-order terms. By using the mean value theorem in the first term of the right-hand side of the above relation, we find points $p'_1 = x^{-1}(a'_i(dx)_p u)$ for some $a'_i \in (0, a)$ and $p'_2 = x^{-1}(b'_i(dx)_p v)$ for some $b'_i \in (0, b)$ such that

$$
\Gamma^i_{jk}(p_2) - \Gamma^i_{jk}(p) = b\partial_m\Gamma^i_{jk}(p'_2)v^m
$$

and also

$$
\Gamma^i_{j\ell}(p) - \Gamma^i_{j\ell}(p_1) = -a\partial_k\Gamma^i_{j\ell}(p'_1)u^k.
$$

Therefore, by setting

$$R^i{}_{jk\ell} := \partial_\ell \Gamma^i_{jk} - \partial_k \Gamma^i_{j\ell} + \Gamma^i_{a\ell}\Gamma^a_{jk} - \Gamma^i_{ak}\Gamma^a_{j\ell}, \tag{1.3.4}$$

we have

$$\lim_{a,b\to 0} \frac{\tau_{ab}w - w}{ab} = -R^i{}_{jk\ell}u^\ell v^k w^j X_i(p). \tag{1.3.5}$$

The $R^i{}_{jk\ell}$ are understood as follows. In view of the definition (1.3.1) of the covariant derivative, we see that $\nabla_{X_i} X_j = \Gamma^k_{ij} X_k$ and hence

$$R^i{}_{jk\ell} X_i = \nabla_{X_\ell} \nabla_{X_k} X_j - \nabla_{X_k} \nabla_{X_\ell} X_j.$$

If Z, V, W are vector fields defined in a coordinate neighborhood U such that $Z(p)=u$, $V(p)=v$, $W(p)=w$ then R defines a multilinear map $R : \mathcal{X}(U) \times \mathcal{X}(U) \times \mathcal{X}(U) \to \mathcal{X}(U)$ such that

$$R(Z, V)W := \nabla_Z \nabla_V W - \nabla_V \nabla_Z W - \nabla_{[Z,V]}W. \tag{1.3.6}$$

Thus we observe that $R(u, v)w = R(Z, V)W|_p = R^i{}_{jk\ell}u^\ell v^k w^j X_i(p)$ is independent of the choice of local coordinates around p. The quantity R is called the Riemannian curvature tensor of M and plays an essential role in the study of Riemannian geometry.

If $\tau_{ba} : T_pM \to T_pM$ is a parallel translation along the reversed orientation of $x^{-1}(P_{ab})$ then $\tau_{ab}^{-1} = \tau_{ba}$ and (1.3.5) implies that $R(u, v)w = -R(v, u)w$. From (1.3.4) we observe that $R^i{}_{jk\ell} = -R^i{}_{j\ell k}$. Other properties of R are summarized in the following.

Exercise 1.3.2. Prove that the Riemannian curvature tensor R has the following properties. If $g_{im} R^m{}_{jk\ell} = R_{ijk\ell}$ then

$$R_{ijk\ell} = \tfrac{1}{2}(\partial_j \partial_\ell g_{ki} + \partial_k \partial_i g_{j\ell} - \partial_i \partial_\ell g_{jk} - \partial_j \partial_k g_{j\ell})$$
$$+ g_{ab}\left(\Gamma^a_{ik}\Gamma^b_{j\ell} - \Gamma^a_{i\ell}\Gamma^b_{jk}\right),$$
$$R_{ijk\ell} = -R_{ij\ell k} = -R_{jik\ell} = R_{k\ell ij},$$
$$R_{ijk\ell} + R_{ik\ell j} + R_{i\ell jk} = 0,$$
$$R_{ijk\ell} = \langle R(X_\ell, X_k)X_j, X_i \rangle.$$

Example 1.3.1 (Geodesics on a surface of revolution in $\mathbf{R}^3$). For a positive smooth function $f : \mathbf{R} \to \mathbf{R}^+$ we define a surface of revolution with its profile curve $y = f(x)$ by $M := \{(f(u)\cos v, f(u)\sin v, u) \in \mathbf{R}^3; u \in \mathbf{R}, 0 < v \le 2\pi\}$. Then, setting $u^1 := u$, $u^2 := v$, we have $ds^2 = (1 + (f')^2)(du^1)^2 +$

$f^2(\mathrm{d}u^2)^2$ and

$$\Gamma^1_{12} = \Gamma^2_{22} = \Gamma^2_{11} = 0,$$

$$\Gamma^1_{11} = \frac{f' f''}{1 + (f')^2}, \qquad \Gamma^1_{22} = -\frac{f f'}{1 + (f')^2}, \qquad \Gamma^2_{12} = \frac{f'}{f}.$$

Computations show that $R^1_{212} = f f'' / (1 + (f')^2)^2$. If $\gamma(s) := (u(s), v(s))$ is a unit-speed geodesic on M then

$$\left(\frac{\mathrm{d}u}{\mathrm{d}s}\right)^2 (1 + (f')^2) + \left(\frac{\mathrm{d}v}{\mathrm{d}s}\right)^2 f^2 = 1 \tag{1.3.7}$$

and equation (1.2.1) is written as

$$\begin{cases} \dfrac{\mathrm{d}^2 u}{\mathrm{d}s^2} + \dfrac{f' f''}{1 + (f')^2} \left(\dfrac{\mathrm{d}u}{\mathrm{d}s}\right)^2 - \dfrac{f f'}{1 + (f')^2} \left(\dfrac{\mathrm{d}v}{\mathrm{d}s}\right)^2 = 0, \\[2mm] \dfrac{\mathrm{d}^2 v}{\mathrm{d}s^2} + 2\dfrac{f'}{f} \dfrac{\mathrm{d}u}{\mathrm{d}s} \dfrac{\mathrm{d}v}{\mathrm{d}s} = 0. \end{cases} \tag{1.3.8}$$

We observe from the above relations that every profile curve $v(s)$ equal to a constant is a geodesic. Therefore $\dot{\gamma}$ is not tangent to any profile curve if $\dot{\gamma}(s_0)$ for some s_0 is not a tangent. The second relation in (1.3.8) reduces to

$$\frac{\mathrm{d}}{\mathrm{d}s} \left(\frac{\mathrm{d}v}{\mathrm{d}s} f^2\right) = 0,$$

and hence $(\mathrm{d}v/\mathrm{d}s) f^2$ is constant. If $\theta(s)$ is the angle between $\dot{\gamma}(s)$ and the profile curve $v = v(s)$ passing through $\gamma(s)$ then $-\pi < \theta < \pi$ and $f \sin \theta(s)$ is constant. Thus we have proved

Theorem 1.3.2 (Clairaut's theorem). *If $\gamma(s) = (u(s), v(s))$ is a geodesic on a surface of revolution then the angle $\theta(s)$ between $\dot{\gamma}(s)$ and the profile curve passing through $\gamma(s)$ satisfies*

$$f(u(s)) \sin \theta(s) = \text{const.}$$

The general case of this theorem is proved later in Theorem 7.1.2.

Example 1.3.2 (The Sasaki metric over TM). We shall introduce the Sasaki metric over the tangent bundle TM of a Riemannian n-manifold (M, g). Let (U, φ) be a local set of coordinates. Let $\pi_M : TM \to M$ be the projection map. Then $\tilde{U} := \pi_M^{-1}(U)$ is a local coordinate neighborhood of TM and its coordinate map $\psi : \tilde{U} \to \mathbf{R}^{2n}$ is expressed as follows. For a point $(p, u) \in \tilde{U}$ we have $\varphi(p) = (x^1(p), \ldots, x^n(p))$ and $u = u^i (\partial/\partial x^i)(p)$, and hence $\psi(p, u) = (x^1(p), \ldots, x^n(p), u^1, \ldots, u^n) \in \mathbf{R}^{2n}$.

The tangent space $T_{(p,u)}TM$ to TM at $(p, u) \in \tilde{U}$ contains the n-dimensional linear subspace $T_u T_p M$, and this $T_u T_p M$ is naturally identified with $T_p M$ by parallel translation. Since $d(\pi_M)_{(p,u)} A = 0$ for every $A \in T_u T_p M$, we observe that the kernel of $d(\pi_M)_{(p,u)}$ is $T_u T_p M$. This kernel $T_u T_p M$ is called the vertical space of $T_{(p,u)}TM$ at (p, u) and denoted by V_u.

In order to find vectors in $T_{(p,u)}TM$ transversel to V_u we take a vector $X \in T_p M$ and a curve $c : (-\varepsilon, \varepsilon) \to M$ fitting X. In local expressions we write $\varphi \circ c(t) = (x^1(t), \ldots, x^n(t))$, $c(0) = p$ and $X = X^i (\partial/\partial x^i)(p)$. Let $\xi : (-\varepsilon, \varepsilon) \to TM$ be the parallel field along c generated by $u = \xi(0) \in T_p M$. Then $t \mapsto (c(t), \xi(t))$ is a curve in TM emanating from (p, u), and its velocity vector v at $t = 0$ is expressed as $v = (x^i, u^i, X^i, d\xi^i(0)/dt) = (x^i, u^i, X^i, -\Gamma^i_{jk} X^j \xi^k)$. This v is called the horizontal lift of X at (p, u). Clearly we have $d(\pi_M)_{(p,u)} v = X$, and the set of the horizontal lifts of all vectors in $T_p M$ is isomorphic to $T_p M$. It is called the horizontal space of $T_{(p,u)}TM$ at (p, u) and denoted H_u. Thus we have the decomposition $T_{(p,u)}TM = V_u \oplus H_u$.

If $\alpha = X^i (\partial/\partial x^i) + Z^i (\partial/\partial u^i) \in T_{(p,u)}TM$ then the decomposition of α into horizontal and vertical components $\alpha = \alpha_h + \alpha_v$ is expressed in local coordinates by $\alpha_h = (x^i, u^i, X^i, -\Gamma^i_{jk} u^j X^k)$ and $\alpha_v = (x^i, u^i, 0, Z^i + \Gamma^i_{jk} u^j X^k)$. The connection map $K : TTM \to TM$ is defined by $K(\alpha) := (p, Z^i + \Gamma^i_{jk} u^j X^k)$. Then $K|_{T_u T_p M} : T_u T_p M \to T_p M$ is clearly an isomorphism, by which V_u is identified with $T_p M$. Also H_u is identified with $T_p M$ by $d(\pi_M)_{(p,u)}$.

Making use of this decomposition we introduce the Sasaki metric G over TM as follows. If $\alpha, \beta \in T_{(p,u)}TM$ then

$$G(\alpha, \beta) := g(d(\pi_M)_{(p,u)} \alpha_h, \quad d(\pi_M)_{(p,u)} \beta_h) + g(K(\alpha_v), \quad K(\beta_v)).$$

The Sasaki metric is also defined on the unit-sphere bundle SM over M by the restriction of G.

1.4 The second fundamental form

Let (M, g) be a Riemannian n-manifold and N a smooth Riemannian m-manifold. If $f : N \to M$ is an immersion then the induced Riemannian metric h from g through f is defined as follows. If $X, Y \in \mathcal{X}(N)$ then

$$h(X, Y) := g(df(X), df(Y)).$$

To each point $q \in N$ an $(n-m)$-dimensional linear subspace $T_q N^\perp \subset T_{f(q)}M$ is assigned such that it is orthogonal to $df_q(T_q N)$ in $T_{f(q)}M$ with respect to g. We then have an n-dimensional vector bundle $f^* TM$ and an $(n - m)$-dimensional normal bundle $TN^\perp$ over N:

$$f^*TM := \bigcup_{q \in N} T_{f(q)}M, \qquad TN^\perp := \bigcup_{q \in N} T_q N^\perp.$$

The orthogonal decomposition of f^*TM is given as

$$f^*TM = df(TN) \oplus TN^\perp,$$

where $df(TN)$ is often identified with TN in localized contexts. A map $X : N \to f^*TM$ is by definition a *vector field along* f iff $\pi \circ X = f$. A map $\xi : N \to TN^\perp$ is called a *normal field along* f iff $\pi \circ \xi = f$. We denote by $\mathcal{X}_f$ and $\mathcal{X}_f^\perp$ the spaces of all vector fields and normal fields along f. Every vector field Y along f has a unique decomposition $Y = Y^\top + Y^\perp$, where $Y^\perp \in \mathcal{X}_f^\perp$ and $Y^\top$ may be considered as a vector field over N by the identification $df(Y^\top) \equiv Y^\top$.

The covariant derivatives of N and M are denoted by D and ∇ respectively. The covariant derivative along f is also denoted by ∇, where $\nabla : \mathcal{X}(N) \times \mathcal{X}_f \to \mathcal{X}_f$. For a point $q \in N$ and for vectors $x, y \in T_q N$ we take the (local) field extensions X, Y of x, y. By identifying X with $df(X)$, the covariant derivative along f of Y in x is expressed as

$$\nabla_x Y = (\nabla_x Y)^\top + (\nabla_x Y)^\perp.$$

Here the first term is identified with

$$(\nabla_x Y)^\top = (\nabla_x df(Y))^\top = df(D_x Y),$$

and the second term is written as

$$B(X, Y)_q := (\nabla_x Y)^\perp = \nabla_x Y - df(D_x Y). \qquad (1.4.1)$$

Thus $B : \mathcal{X}_f \times \mathcal{X}_f \to \mathcal{X}_f^\perp$ is a symmetric bilinear form. The symmetric property of B follows from that of the Christoffel symbols.

The tangential component of the covariant derivative along f of a normal field ξ along f is written as

$$A_{\xi(q)} X = \nabla_x \xi - (\nabla_x \xi)^\perp. \qquad (1.4.2)$$

The quantity $A_\xi : TN \to TN$ is linear and is called the *shape operator of* f. The symmetry property of A_ξ is observed from

$$g(B(X, Y), \xi) = -g(A_\xi X, Y) = -g(A_\xi Y, X). \qquad (1.4.3)$$

The normal component on the right-hand side of (1.4.2) induces a *normal connection along* f. That is, $\nabla^\perp : \mathcal{X}_f \times \mathcal{X}_f^\perp \to \mathcal{X}_f^\perp$ by the relation

$$\nabla_x^\perp \xi := \nabla_x \xi - A_{\xi(q)} X. \qquad (1.4.4)$$

It can be shown that if $\alpha, \beta \in C^\infty(N)$ then $\nabla^\perp_{\alpha X + \beta Y}\xi = \alpha\nabla^\perp_X\xi + \beta\nabla^\perp_Y\xi$ and $\nabla^\perp_X\alpha\xi = X(\alpha)\xi + \alpha\nabla^\perp_X\xi$.

Let R and S be the Riemannian curvature tensors of M and N respectively. The relation (1.3.6) then implies, by identifying $X, Y, Z \in \mathcal{X}(N)$ with the vector fields along f and by using (1.4.1),

$$
\begin{aligned}
R(X, Y)Z &= \nabla_X\nabla_Y Z - \nabla_Y\nabla_X Z - \nabla_{[X,Y]}Z \\
&= \nabla_X(D_Y Z + B(Y, Z)) - \nabla_Y(D_X Z + B(X, Z)) \\
&\quad - (D_{[X,Y]}Z + B([X, Y], Z)) \\
&= S(X, Y)Z + B(X, D_Y Z) + A_{B(Y,Z)}X + \nabla^\perp_X B(Y, Z) \\
&\quad - B(Y, D_X Z) - A_{B(X,Z)}Y - \nabla^\perp_Y B(X, Z) - B([X, Y], Z).
\end{aligned}
$$

If we set $R(X, Y)Z = R(X, Y)Z^\top + R(X, Y)Z^\perp$ then the *Gauss* and *Codazzi equations* are obtained as follows:

$$
\begin{cases}
R(X, Y)Z^\top = \mathrm{d}f(S(X, Y)Z) + A_{B(Y,Z)}X - A_{B(X,Z)}Y, \\
R(X, Y)Z^\perp = B(X, D_Y Z) - B(Y, D_X Z) - B([X, Y], Z) \\
\qquad\qquad\quad + \nabla^\perp_X B(Y, Z) - \nabla^\perp_Y B(X, Z).
\end{cases}
\tag{1.4.5}
$$

For a normal field $\xi : N \to TN^\perp$ we have

$$
\begin{aligned}
R(X, Y)\xi &= \nabla_X\nabla_Y\xi - \nabla_Y\nabla_X\xi - \nabla_{[X,Y]}\xi \\
&= \nabla_X\left(A_\xi Y + \nabla^\perp_Y\xi\right) - \nabla_Y\left(A_\xi X + \nabla^\perp_X\xi\right) \\
&\quad - \left(A_\xi[X, Y] + \nabla^\perp_{[X,Y]}\xi\right) \\
&= D_X A_\xi Y + B(X, A_\xi Y) + \left(A_{\nabla^\perp_Y\xi}X + \nabla^\perp_Y\nabla^\perp_X\xi\right) \\
&\quad - \left(D_Y A_\xi X + B(Y, A_\xi X) + A_{\nabla^\perp_X\xi}Y + \nabla^\perp_Y\nabla^\perp_X\xi\right) \\
&\quad - A_\xi[X, Y] - \nabla^\perp_{[X,Y]}\xi.
\end{aligned}
$$

If $R(X, Y)\xi = R(X, Y)\xi^\top + R(X, Y)\xi^\perp$ is the orthogonal decomposition, we then have

$$
R(X, Y)\xi^\top = \mathrm{d}f(D_X A_\xi Y - D_Y A_\xi X) - A_\xi[X, Y] + A_{\nabla^\perp_Y\xi}X - A_{\nabla^\perp_X\xi}Y
\tag{1.4.6}
$$

and

$$
R(X, Y)\xi^\perp = R^\perp(X, Y)\xi + B(X, A_\xi Y) - B(Y, A_\xi X).
\tag{1.4.7}
$$

Here $R^\perp$ is the curvature tensor for the normal connection $\nabla^\perp$ and is given by

$$
R^\perp(X, Y)\xi := \nabla^\perp_X\nabla^\perp_Y\xi - \nabla^\perp_Y\nabla^\perp_X\xi - \nabla^\perp_{[X,Y]}\xi.
\tag{1.4.8}
$$

Note that (1.4.4) and (1.4.6) are equivalent. The relation (1.4.8) is called the *Ricci equation*.

Definition 1.4.1. The *sectional curvature* $K_M(X, Y)$ of M with respect to the plane section $\sigma = \sigma(X, Y)$ spanned by X and Y is

$$K_{\sigma(X,Y)} = K_M(X, Y) := \frac{g(R(X, Y)Y, X)}{g(X, X)g(Y, Y) - g(X, Y)^2}.$$

Also, the *Ricci curvature* $Ric_M(X)$ with respect to a vector X at $p \in M$ is defined as follows. Let $\{e_1, \ldots, e_n\}$ be an orthonormal basis for $T_p M$ such that $X/\|X\| = e_n$. Then

$$Ric_M(X) := \sum_{i=1}^{n-1} K_M(e_i, e_n).$$

Moreover if $\dim M = 2$ then the sectional curvature is called the *Gaussian curvature* and denoted by G.

The Gauss equation, (1.4.5), implies that

$$\langle S(X, Y)Y, X \rangle = \langle R(X, Y)Y, X \rangle + B(X, X)B(Y, Y) - B(X, Y)^2.$$

If K_N and K_M are the sectional curvatures of N and M respectively then the above relation gives

$$K_N(X, Y) = K_M(X, Y) + \frac{B(X, X)B(Y, Y) - B(X, Y)^2}{g(X, X)g(Y, Y) - g(X, Y)^2}. \qquad (1.4.9)$$

1.5 The second variation formula and Jacobi fields

As discussed in Section 1.2, a curve having the locally minimizing property is a geodesic. We now discuss whether a geodesic $\gamma : [0, \ell] \to M$ possesses the minimizing property among curves near γ with the same endpoints. For this purpose we establish the second variation formula for a variation in γ. Namely, a symmetric bilinear form, called the index form of γ, is defined over the space of vector fields along γ by using the Riemannian curvature tensor. The second variation formula is expressed in index form. A Jacobi field along γ is a vector field along γ belonging to the kernel of the index form and is found as the solution of a second-order linear differential equation. We then prove the existence of convex balls around every point of M. Here we use the covariant differential along a map.

Let $V : (-\varepsilon_0, \varepsilon_0) \times [0, \ell] \to M$ be a variation along a unit-speed geodesic $\gamma : [0, \ell] \to M$. For the basis-vector fields $\partial/\partial\varepsilon$ and $\partial/\partial s$ of $\mathbf{R}^2$, the covariant differentials along V are

$$\nabla_{\partial/\partial\varepsilon} = \nabla_{dV(\partial/\partial\varepsilon)}, \qquad \nabla_{\partial/\partial s} = \nabla_{dV(\partial/\partial s)}.$$

In particular the following relation is important:

$$\nabla_{\partial/\partial\varepsilon}\, dV\left(\frac{\partial}{\partial s}\right) = \nabla_{\partial/\partial s}\, dV\left(\frac{\partial}{\partial\varepsilon}\right).$$

We denote by

$$T(\varepsilon, s) := dV_{(\varepsilon,s)}\left(\frac{\partial}{\partial s}\right), \qquad Y(\varepsilon, s) := dV_{(\varepsilon,s)}\left(\frac{\partial}{\partial\varepsilon}\right).$$

the velocity vector of V_ε and the variation vector field associated with V. The length $L(\varepsilon)$ of V_ε is given by

$$L(\varepsilon) = \int_0^\ell \|T(\varepsilon, s)\|\, ds.$$

Theorem 1.5.1. *Let* $V : (-\varepsilon_0, \varepsilon_0) \times [0, \ell] \to M$ *be a variation along a unit-speed geodesic* $\gamma : [0, \ell] \to M$. *We then have* the first variation formula

$$L'(0) = \langle Y, T\rangle(0, s)|_0^\ell.$$

If $Y_\perp := Y - \langle Y, T\rangle T$ *then we have* the second variation formula

$$L''(0) = \int_0^\ell \left(\langle\nabla_{\partial/\partial s}Y_\perp, \nabla_{\partial/\partial s}Y_\perp\rangle - \langle R(Y_\perp, \dot\gamma)\dot\gamma, Y_\perp\rangle\right)(0, s)\, ds$$

$$+ \langle\nabla_{\partial/\partial\varepsilon}Y_\perp, T\rangle(0, s)|_0^\ell.$$

In particular, if V *is proper, i.e.,* $V(\varepsilon, 0) = \gamma(0)$ *and* $V(\varepsilon, \ell) = \gamma(\ell)$ *for all* ε, *then*

$$L''(0) = \int_0^\ell \left(\langle\nabla_{\partial/\partial s}Y_\perp, \nabla_{\partial/\partial s}Y_\perp\rangle - \langle R(Y_\perp, \dot\gamma)\dot\gamma, Y_\perp\rangle\right)(0, s)\, ds.$$

Proof. We only need to prove the second variation formula. From

$$L'(\varepsilon) = \int_0^\ell \frac{\langle\nabla_{\partial/\partial\varepsilon}T, T\rangle(\varepsilon, s)}{\|T(\varepsilon, s)\|}\, ds$$

we have

$$L''(\varepsilon) = \int_0^\ell \frac{\partial}{\partial\varepsilon}\left(\frac{\langle\nabla_{\partial/\partial\varepsilon}T, T\rangle(\varepsilon, s)}{\|T(\varepsilon, s)\|}\right)\, ds.$$

Since $\partial/\partial\varepsilon$ and $\partial/\partial s$ are basis vector fields on $\mathbf{R}^2$, we see that

$$\nabla_{\partial/\partial\varepsilon} T = \nabla_{\partial/\partial\varepsilon}\left\{dV\left(\frac{\partial}{\partial s}\right)\right\} = \nabla_{\partial/\partial s}\left\{dV\left(\frac{\partial}{\partial\varepsilon}\right)\right\} = \nabla_{\partial/\partial s} Y = Y',$$

and the integrand in $L''(\varepsilon)$ can be rewritten as

$$\frac{\partial}{\partial\varepsilon}\left(\frac{\langle\nabla_{\partial/\partial\varepsilon} T, T\rangle}{\|T\|}\right) = \frac{\partial}{\partial\varepsilon}\left(\frac{\langle\nabla_{\partial/\partial s} Y, T\rangle}{\|T\|}\right)$$

$$= \frac{\langle\nabla_{\partial/\partial\varepsilon}\nabla_{\partial/\partial s} Y, T\rangle + \langle\nabla_{\partial/\partial s} Y, \nabla_{\partial/\partial\varepsilon} T\rangle}{\|T\|} - \frac{\langle\nabla_{\partial/\partial s} Y, T\rangle^2}{\|T\|^3}.$$

Insert $R(T, Y)Y = \nabla_T\nabla_Y Y - \nabla_Y\nabla_T Y$ and $\|T(0, s)\| = 1$ into the above relation to obtain

$$L''(0) = \int_0^\ell \left((\langle -R(T, Y)Y + \nabla_T\nabla_Y Y, T\rangle + \langle\nabla_T Y, \nabla_Y T\rangle\right.$$
$$\left. - \langle\nabla_T Y, T\rangle^2\right)(0, s)\,ds.$$

The skew-symmetric property of R implies that

$$\langle R(T, Y)Y, T\rangle = \langle R(T, Y_\perp)Y_\perp, T\rangle.$$

We also have $\langle\nabla_T\nabla_Y Y, T\rangle = T\langle\nabla_Y Y, T\rangle - \langle\nabla_Y Y, \nabla_T T\rangle$. By setting $\varepsilon = 0$ we get $\langle\nabla_T\nabla_Y Y, T\rangle(0, s) = (d/ds)\langle\nabla_Y Y, T\rangle(0, s)$. Finally we observe from $Y(0, s) = \{Y_\perp + \langle Y, T\rangle T\}(0, s)$ that the following relation holds at $(s, 0)$:

$$\nabla_T Y = \nabla_T Y_\perp + (T\langle Y, T\rangle)T + \langle Y, T\rangle\nabla_T T$$
$$= \nabla_{\partial/\partial s} Y_\perp + \frac{d}{ds}\langle Y, \dot\gamma\rangle\dot\gamma = \nabla_{\partial/\partial s} Y_\perp + \langle\nabla_{\partial/\partial s} Y, \dot\gamma\rangle\dot\gamma.$$

From $\langle Y_\perp, \dot\gamma\rangle = 0$ and the fact that γ is a geodesic we see that $\langle\nabla_{\partial/\partial s} Y_\perp, \dot\gamma\rangle = 0$. Therefore we have at $(0, s)$

$$\langle\nabla_T Y, \nabla_Y T\rangle - \langle\nabla_T Y, T\rangle^2$$
$$= \langle\nabla_T Y, \nabla_T Y\rangle - \langle\nabla_T Y, T\rangle^2$$
$$= \langle\nabla_{\partial/\partial s} Y_\perp + \langle\nabla_{\partial/\partial s} Y, \dot\gamma\rangle\dot\gamma, \nabla_{\partial/\partial s} Y_\perp + \langle\nabla_{\partial/\partial s} Y, \dot\gamma\rangle\dot\gamma\rangle - \langle\nabla_{\partial/\partial s} Y, \dot\gamma\rangle^2$$
$$= \langle\nabla_{\partial/\partial s} Y_\perp, \nabla_{\partial/\partial s} Y_\perp\rangle.$$

This proves Theorem 1.5.1. $\qquad\qquad\qquad\qquad\qquad\qquad\qquad\qquad\square$

Remark 1.5.1. For a piecewise-smooth variation $V : (-\varepsilon_0, \varepsilon_0) \times [0, \ell] \to M$ along γ we can also obtain the second variation formula as follows. Let $0 := s_0 < s_1 < \cdots < s_k := \ell$ be chosen such that $V_i := V|(-\varepsilon_0, \varepsilon_0) \times [s_i, s_{i+1}] \to$

M for every $i = 0, \ldots, k - 1$ is smooth, and set $Y_i(s) := \mathrm{d}V_i(\partial/\partial\varepsilon)|_{(0,s)}$ and $T_i(s) := \mathrm{d}V_i(\partial/\partial s)|_{(0,s)}$. The continuity of V implies that $Y_i(s_{i-1}) = Y_{i-1}(s_{i-1})$ is tangent to a curve $\varepsilon \mapsto V_{i-1}(\varepsilon, s_{i-1}) = V_i(\varepsilon, s_{i-1})$ at $\varepsilon = 0$ and also that $T_{i-1}(s_i) = T_i(s_i)$; hence

$$\langle \nabla_{\partial/\partial\varepsilon} Y_{i-1}, T_{i-1} \rangle (s_{i-1}) = \langle \nabla_{\partial/\partial\varepsilon} Y_i, T_i \rangle (s_{i-1}).$$

Thus we have

$$L''(0) = \sum_{i=0}^{k-1} \int_{s_i}^{s_{i+1}} \left(\langle \nabla_{\partial/\partial s}(Y_i)_\perp, \nabla_{\partial/\partial s}(Y_i)_\perp \rangle - \langle R((Y_i)_\perp, \dot\gamma)\dot\gamma, (Y_i)_\perp \rangle \right)(s)\, \mathrm{d}s$$
$$+ \left(\langle \nabla_{\partial/\partial\varepsilon}(Y_k)_\perp, T_k \rangle(\ell) - \langle \nabla_{\partial/\partial\varepsilon}(Y_0)_\perp, T_0 \rangle(0) \right).$$

If V is proper then

$$L''(0) = \sum_{i=0}^{k-1} \int_{s_i}^{s_{i+1}} \left(\langle \nabla_{\partial/\partial s}(Y_i)_\perp, \nabla_{\partial/\partial s}(Y_i)_\perp \rangle - \langle R((Y_i)_\perp, \dot\gamma)\dot\gamma, (Y_i)_\perp \rangle \right)(s)\, \mathrm{d}s.$$

Definition 1.5.1. A vector field J along a unit-speed geodesic γ is by definition a *Jacobi field along* γ iff J satisfies

$$J'' + R(J, \dot\gamma)\dot\gamma = 0, \tag{1.5.1}$$

where $J' = \nabla_{\partial/\partial s} J$ and $J'' = \nabla_{\partial/\partial s} \nabla_{\partial/\partial s} J$.

Because $\dot\gamma$ is parallel along itself and $R(Y, Z)W$ is skew symmetric in Y, Z, we see that $(as + b)\dot\gamma(s)$ for every pair of constants a, b is a Jacobi field along γ. If J_1 and J_2 are Jacobi fields along γ then $\langle J_1', J_2 \rangle - \langle J_1, J_2' \rangle = \mathrm{const}$. In particular $\langle J', \dot\gamma \rangle$ is constant for every Jacobi field J along γ. Let $\mathcal{J}_\gamma$ be the set of all Jacobi fields along γ. Then $\alpha J_1 + \beta J_2 \in \mathcal{J}_\gamma$ for all $\alpha, \beta \in \mathbf{R}$, and hence $\mathcal{J}_\gamma$ is a vector space over $\mathbf{R}$.

Lemma 1.5.1. *Let* $\gamma : [a, b] \to M$ *be a geodesic and* $t_0 \in [a, b]$. *Then there exists for given* $u, v \in T_{\gamma(t_0)}M$ *a unique Jacobi field* J *along* γ *such that* $J(0) = u$ *and* $J'(0) = v$.

Proof. Let $P_1, \ldots, P_n = \dot\gamma$ be an orthonormal parallel-frame field along γ, i.e., $g(P_i, P_j) = \delta_{ij}$ and $\nabla_{\dot\gamma} P_i = 0$. A vector field Y along γ is expressed as $Y = \eta^i P_i$, where $\eta^i : [a, b] \to \mathbf{R}$ for each i is smooth. Then $Y' = (\eta^i)' P_i$ and $Y'' = (\eta^i)'' P_i$. If Y satisfies (1.5.1) then

$$\frac{\mathrm{d}^2\eta^i}{\mathrm{d}t^2} + \eta^j R^i{}_{nnj} = 0 \qquad \text{for all } i = 1, \ldots, n.$$

The existence and uniqueness of the solution for given initial conditions is now clear. $\square$

Remark 1.5.2. From Lemma 1.5.1, $\mathcal{J}_\gamma$ forms a $2n$-dimensional vector space. The variation vector field associated with a geodesic variation is a Jacobi field. In fact, if V is such a geodesic variation and if $T := dV(\partial/\partial t)$ and $Y := dV(\partial/\partial \varepsilon)$ then $\nabla_T T = 0$. Therefore $Y'' = \nabla_T \nabla_T Y = \nabla_T \nabla_Y T = \nabla_Y \nabla_T T - R(Y, T)T$, and (1.5.1) holds for this Y. The converse is true: namely, for a given Jacobi field J along γ there exists a geodesic variation along γ whose variation vector field is J.

Lemma 1.5.2. *Let* $\gamma : [0, \ell] \to M$ *be a geodesic with* $p = \gamma(0)$ *and* $u = \dot{\gamma}(0) \in T_p M$. *For every vector* $v \in T_p M$ *let* $J(t) := d(\exp_p)_{tu} tv$. *Then* J *is a Jacobi field along* γ *with*

$$J(0) = 0, \qquad J'(0) = v.$$

Proof. Consider a variation $V : (-\varepsilon_0, \varepsilon_0) \times [0, \ell] \to M$ along γ such that $V(\varepsilon, t) := \exp_p t(u + \varepsilon v)$. Clearly each variational curve V_ε is a geodesic and we see from the construction of V that if $J(t) = dV(\partial/\partial \varepsilon)|_{(0,t)}$ then $J(0) = 0$ and $J'(0) = \nabla_{\partial/\partial t} dV(\partial/\partial \varepsilon)|_{(0,0)} = \nabla_{\partial/\partial \varepsilon} dV(\partial/\partial t)|_{(0,0)} = v$. $\qquad\square$

Definition 1.5.2. Let $\gamma : [0, \ell] \to M$ be a geodesic. A point $\gamma(t_0)$ is *conjugate* to $p = \gamma(0)$ iff there exists a nontrivial Jacobi field J along γ such that $J(0) = J(t_0) = 0$.

In view of Lemma 1.5.2, $\gamma(t_0)$ is conjugate to $\gamma(0)$ along γ iff $\exp_p$ is singular at $t_0 \dot{\gamma}(0)$. In other words, a point $\gamma(t_0)$ is *not conjugate* to $\gamma(0)$ along γ iff $d(\exp_p)$ has maximal rank at $t_0 \dot{\gamma}(0)$. Let $\mathcal{J}_\gamma^0(t_0)$ be the set of all Jacobi fields along γ vanishing at $\gamma(0)$ and $\gamma(t_0)$. The $\mathcal{J}_\gamma^0(t_0)$ form a linear subspace of $\mathcal{J}_\gamma$. If the dimension of $\mathcal{J}_\gamma^0(t_0)$ is positive then it is called the multiplicity of the conjugate point $\gamma(t_0)$. The rank of $d(\exp_p)_{t_0 \dot{\gamma}(0)}$ is equal to $n - \dim \mathcal{J}_\gamma^0(t_0)$.

Lemma 1.5.3. *Assume that a geodesic* $\gamma : [0, \ell] \to M$ *has no point conjugate to* $p = \gamma(0)$ *along* γ. *Then* γ *has the locally minimizing property.*

Proof. It follows from the assumption that $d(\exp_p)$ has maximal rank at $t\dot{\gamma}(0), t \in [0, \ell]$. Therefore there exists an open set $\Omega \subset \tilde{M}_p$ around the segment $\{t\dot{\gamma}(0); 0 \le t \le \ell\}$ such that $\exp_p |_\Omega : \Omega \to M$ is regular. If $V : (-\varepsilon_0, \varepsilon_0) \times [0, \ell] \to M$ is a variation of γ such that $V(\varepsilon, 0) = \gamma(0), V(\varepsilon, \ell) = \gamma(\ell)$ for all $\varepsilon \in (-\varepsilon_0, \varepsilon_0)$ then it is lifted into $\tilde{M}_p$ via $\exp_p |_\Omega$. Note that a geodesic polar coordinate system around p is not defined on Ω, for $\exp_p |_\Omega$ is not necessarily one-to-one. The regularity of $\exp_p |_\Omega$ then implies that γ has only finitely many self-intersections. Therefore Ω can be decomposed into finitely many domains on each of which geodesic polar coordinates around p make sense. Thus the Gauss lemma 1.2.1 applies, giving $L(\varepsilon) > L(0)$ for all $\varepsilon \in (-\varepsilon_0, \varepsilon_0) \setminus \{0\}$. $\square$

The geometric meaning of the sectional and Ricci curvatures are interpreted with the aid of Jacobi fields. We take unit vectors $u, v \in S_p(1)$ with $u \perp v$ and a geodesic γ with $\gamma(0) = p$, $\dot{\gamma}(0) = u$. If J is a Jacobi field along γ such that $J(0) = 0$, $J'(0) = v$ and $U := \dot{\gamma}$ and if V is the parallel field along γ generated by $V(0) = v$ then we have

$$J(t) = tV(t) - \frac{t^3}{3!} R(V(t), U(t))U(t) + o(t^3) \tag{1.5.2}$$

for sufficiently small t. If $S^1(t)$ is a circle with sufficiently small radius t centered at the origin of the plane σ spanned by u and v then $\exp_p S^1(t)$ has velocity vector $J(t)$. Therefore the length of $\exp_p S^1(t)$ is $2\pi \|J(t)\| = 2\pi t(1 - (t^2/3!)K_\sigma + o(t^2))$. Thus K_σ is interpreted as the divergence of the geodesics emanating from p.

Next, we take an orthonormal basis $e_1, \ldots, e_n = \dot{\gamma}(0)$ for T_pM. Let $d\sigma$ be the volume element of $S_p(1)$. Let Y_i for each $i = 1, \ldots, n-1$ be a Jacobi field along γ such that $Y_i(0) = 0$, $Y_i'(0) = e_i$. Then

$$(\exp_p)^*_{tu} \, dM = (\det d(\exp_p)_{tu}) \, d\sigma \wedge dt.$$

Thus we have

$$\det d(\exp_p)_{tu} = |Y_1 \wedge Y_2 \wedge \cdots \wedge Y_{n-1} \wedge \dot{\gamma}|(t).$$

Therefore $Ric_M(u)$ gives the limit as $t \to 0$ of the ratio of the area element of $S^{n-1}(t) := \exp_p tS_p(1)$ at the point $\exp_p tu$ and the area element of $S_p(1)$.

We shall provide examples of Riemannian n-manifolds of constant sectional (and Ricci) curvature.

Example 1.5.1 (Euclidean space). Let $M := \mathbf{R}^n$. The canonical metric ds^2 is expressed as $ds^2 := \delta_{ij} \, dx^i \, dx^j$ for $(x^1, \ldots, x^n) \in \mathbf{R}^n$. All the Christoffel symbols are zero and $R_{ijk\ell} = 0$ for all indices. Thus the sectional (and Ricci) curvature of $\mathbf{R}^n$ is zero.

Example 1.5.2 (Geographic coordinates on n-sphere). Let $M := S^n(r) \subset \mathbf{R}^{n+1}$ be the standard n-sphere with radius r. Set

$$D := \{x = (x^1, \ldots, x^n) \in \mathbf{R}^n; -\pi r/2 < x^1, \ldots, x^{n-1} < \pi r/2,$$
$$-\pi r < x^n < \pi r\}.$$

Let $\varphi : D \to \mathbf{S}^n(r) \subset \mathbf{R}^{n+1}$ be defined by setting $\varphi = (\varphi^1, \ldots, \varphi^n)$,

$$\varphi^k(x) := r \sin \frac{x^k}{r} \prod_{\ell=1}^{k-1} \cos \frac{x^\ell}{r} \qquad \text{for } k = 1, \ldots, n,$$

$$\varphi^{n+1}(x) := r \prod_{\ell=1}^{n} \cos \frac{x^\ell}{r}.$$

Then $x := \varphi^{-1}$ defines a coordinate map on a domain $U := \{x \in \mathbf{S}^n(r); \, x^n \neq 0, \, x^{n+1} > 0\}$. Since $X_i = d\varphi(\partial/\partial x^i)$ and

$$g_{ij} = d\varphi(\partial/\partial x^i) \, d\varphi(\partial/\partial x^j) \qquad \text{for } i, j = 1, \ldots, n,$$

we observe that

$$g_{11} = 1, \qquad g_{1i} = 0 \qquad \text{for all } i = 2, \ldots, n,$$

$$g_{ij} = \delta_{ij} \prod_{\ell=1}^{i-1} \cos^2 \frac{x^\ell}{r} \qquad \text{for all } i, j > 1$$

and also

$$g^{ij} = \delta^{ij} \prod_{\ell=1}^{i-1} \cos^{-2} \frac{x^\ell}{r}.$$

Further computations show that

$$\Gamma^i_{jk} = \Gamma^k_{kk} = 0 \qquad \text{for all } i \neq j \neq k \neq i,$$

$$\Gamma^k_{kj} = \Gamma^j_{kk} = 0 \qquad \text{for all } j \geq k,$$

$$\Gamma^k_{jj} = \frac{1}{r} \tan \frac{x^k}{r} \prod_{\ell=k}^{j-1} \cos^2 \frac{x^\ell}{r} \qquad \text{for all } k < j,$$

$$\Gamma^j_{kj} = -\frac{1}{r} \tan \frac{x^k}{r} \qquad \text{for all } k < j.$$

Making use of

$$R_{ijk\ell} = \tfrac{1}{2}(\partial_j \partial_\ell g_{ki} + \partial_k \partial_i g_{j\ell} - \partial_i \partial_\ell g_{jk} - \partial_j \partial_k g_{i\ell}) + g_{ab}\left(\Gamma^a_{ik} \Gamma^b_{j\ell} - \Gamma^a_{i\ell} \Gamma^b_{jk}\right),$$

we see that

$$R_{ijk\ell} = 0 \quad \text{and} \quad R_{ijki} = 0 \qquad \text{for all distinct indices.}$$

Then R_{ikik} for $i > k$ is expressed as

$$R_{ikik} = \tfrac{1}{2}\partial_k \partial_k g_{ii} + \sum_{a<k} g_{aa} \Gamma^a_{ii} \Gamma^a_{kk} - g_{ii}\left(\Gamma^i_{ik}\right)^2 = -\frac{g_{ii} g_{kk}}{r^2}.$$

Thus $R_{ijk\ell} = (1/r^2)(g_{ik}g_{j\ell} - g_{i\ell}g_{jk})$, and hence $R(X_\ell, X_k)X_j = (1/r^2) \times \{g_{jk} X_\ell - g_{j\ell}X_k\}$. Therefore we have, for $u, v, w \in T_pM$,

$$R(u, v)w = \frac{1}{r^2} (\langle v, w \rangle u - \langle u, w \rangle v).$$

In particular, the sectional and Ricci curvatures are constants, $1/r^2$ and $(n - 1)/r^2$ respectively.

Example 1.5.3 (Hyperbolic space). Let $H^n(-c^2)$ be a Riemannian manifold defined on an open half space $\{x = (x^1, \ldots, x^n) \in \mathbf{R}^n; x^n > 0\}$ of $\mathbf{R}^n$ with metric

$$ds^2 := \frac{\delta_{ij}dx^i dx^j}{c^2(x^n)^2}.$$

This is a complete Riemannian manifold. It can be shown that

$$\Gamma^i_{jk} = 0 \quad \text{for all distinct indices,}$$
$$\Gamma^i_{nn} = \Gamma^n_{ni} = \Gamma^i_{jj} = \Gamma^i_{ji} = \Gamma^i_{ii} = 0 \quad \text{for all } i, j = 1, \ldots, n - 1,$$
$$\Gamma^i_{ni} = \Gamma^n_{nn} = -\frac{1}{x^n} = -\Gamma^n_{ii} \quad \text{for all } i = 1, \ldots, n - 1.$$

We see from

$$R_{ijk\ell} = \tfrac{1}{2}(\partial_j\partial_\ell g_{ki} + \partial_k\partial_i g_{j\ell} - \partial_i\partial_\ell g_{jk} - \partial_j\partial_k g_{i\ell})$$
$$+ g_{ab} \left(\Gamma^a_{ik}\Gamma^b_{j\ell} - \Gamma^a_{i\ell}\Gamma^b_{jk}\right)$$

that

$$R_{ijk\ell} = 0 \quad \text{and} \quad R_{ijki} = 0 \quad \text{for all distinct indices,}$$
$$R_{ikik} = \tfrac{1}{2}\partial_i\partial_i g_{kk} + g_{nn}\Gamma^n_{ii}\Gamma^n_{kk} = c^2 g_{ii}g_{kk} \quad \text{for all } i > k.$$

Thus we have

$$R_{ijk\ell} = -c^2(g_{ik}g_{j\ell} - g_{i\ell}g_{jk}),$$

and hence $R(X_\ell, X_k)X_j = -c^2(g_{jk}X_\ell - g_{j\ell}X_k)$. The sectional and Ricci curvatures of $H^n(-c^2)$ are therefore $-c^2$ and $-(n - 1)c^2$ respectively.

1.6 Index form

We now discuss the locally minimizing property of a geodesic admitting a conjugate pair. The index form of a geodesic γ is introduced as a symmetric bilinear form on the space of vector fields along γ. If γ has no conjugate pair along it then Lemma 1.5.3 implies that it is locally minimizing. In this case the index form of γ is positive definite.

Definition 1.6.1. Let $\mathcal{X}$ be the set of all (piecewise-) smooth vector fields along a fixed geodesic $\gamma : [0, \ell] \to M$ and let $\mathcal{X}_\perp$ be the set of all (piecewise-) smooth vector fields along γ orthogonal to it. Define the *index form* $I : \mathcal{X} \times \mathcal{X} \to \mathbf{R}$ of γ as follows. Let $X, Y \in \mathcal{X}$ and $0 = t_0 < t_1 < \cdots , < t_k = \ell$ be such that $X|_{[t_{i-1}, t_i]}$ and $Y|_{[t_{i-1}, t_i]}$ for all $i = 1, \ldots, k - 1$ are smooth. Then

$$I(X, Y) := \sum_{i=1}^{k} \int_{t_{i-1}}^{t_i} (\langle X_i', Y_i' \rangle - \langle R(X_i, \dot\gamma)\dot\gamma, Y_i \rangle) \, dt.$$

Clearly I is symmetric and bilinear. If X and Y are smooth then

$$I(X, Y) = \langle X', Y \rangle|_0^\ell - \int_0^\ell (\langle X'' + R(X, \dot\gamma)\dot\gamma, Y \rangle) \, dt$$

$$= \langle X, Y' \rangle|_0^\ell - \int_0^\ell (\langle Y'' + R(Y, \dot\gamma)\dot\gamma, X \rangle) \, dt.$$

Therefore, if Y is a Jacobi field along γ then

$$I(X, Y) = \langle X, Y' \rangle|_0^\ell$$

for all $X \in \mathcal{X}$.

Lemma 1.6.1. *Let $\gamma : [0, \ell] \to M$ be a geodesic with a point $\gamma(t_0)$ conjugate to $p = \gamma(0)$ along it for $t_0 \in (0, \ell)$. Then there exists a piecewise-smooth variation $V : (-\varepsilon_0, \varepsilon_0) \times [0, \ell] \to M$ of γ such that $L(\varepsilon) < L(0)$ for $\varepsilon > 0$.*

Proof. From our assumption about γ we have a nontrivial Jacobi field J along it such that $J(0) = J(t_0) = 0$. Let $P \in \mathcal{X}_\perp$ be a parallel field along γ generated by $-J'(t_0) \neq 0$ and let $f : [0, \ell] \to [0, 1]$ be a smooth function such that $f(0) = f(\ell) = 0$ and $f(t_0) = 1$. For a small positive number h we define a vector field Y_h by

$$Y_h(t) := \begin{cases} J(t) + h(fP)(t) & \text{for } 0 \leq t \leq t_0, \\ h(fP)(t) & \text{for } t_0 \leq t \leq \ell. \end{cases}$$

Then $Y_h \in \mathcal{X}_\perp$ and $Y_h(0) = Y_h(\ell) = 0$. Therefore

$$I(Y_h, Y_h) = \langle J, J' \rangle(t_0) + 2h \langle J', fP \rangle(t_0) + h^2 I(fP, fP).$$

Because of $J'(t_0) = -P(t_0)$ we have

$$I(Y_h, Y_h) = -2h \langle J', J' \rangle(t_0) + h^2 I(fP, fP) < 0$$

for sufficiently small $h > 0$. $\qquad\square$

Now consider the set $\mathcal{X}'_\perp := \{Y \in \mathcal{X}_\perp; Y(0) = Y(\ell) = 0\}$. We will show that the space of all Jacobi fields in $\mathcal{X}'_\perp$ is characterized by the null space of I in $\mathcal{X}'_\perp$.

Lemma 1.6.2. *A vector field $J \in \mathcal{X}'_\perp$ is a Jacobi field iff $I(J, Y) = 0$ for all $Y \in \mathcal{X}'_\perp$.*

Proof. If $J \in \mathcal{X}'_\perp$ is a Jacobi field, it is clear that $I(J, Y) = 0$ for all $Y \in \mathcal{X}'_\perp$. If $J \in \mathcal{X}'_\perp$ satisfies $I(J, Y) = 0$ for all $Y \in \mathcal{X}'_\perp$, we then choose $0 = t_0, \ldots, t_k = \ell$ so that $J_i := J|_{[t_{i-1}, t_i]}$ for all $i = 1, \ldots, k$ is smooth. Let $f_i : [t_{i-1}, t_i] \to [0, 1]$ be a smooth function such that $f_i(t_{i-1}) = f_i(t_i) = 0$ and $f_i > 0$ on (t_{i-1}, t_i), and set $Z_i := f_i(J_i'' + R(J_i, \dot\gamma)\dot\gamma)$. Then setting $Z(t) := Z_i(t)$ for $t \in [t_{i-1}, t_i]$ we see that

$$0 = I(J, Z) = \sum_{i=1}^{k} \int_{t_{i-1}}^{t_i} f_i \, \|J_i'' + R(J_i, \dot\gamma)\dot\gamma\|^2 \, dt.$$

Thus J_i for each $i = 1, \ldots, k - 1$ is a Jacobi field along $\gamma|_{[t_{i-1}, t_i]}$. To prove the smoothness of J we choose a vector field $\widetilde{Z} \in \mathcal{X}'_\perp$ such that for each $i = 1, \ldots, k - 1$ we have $\widetilde{Z}(t_i) = J'_{i+1}(t_i) - J'_i(t_i)$. Then

$$0 = I(J, \widetilde{Z}) = \sum_{i=1}^{k} \langle J', \widetilde{Z} \rangle|_{t_{i-1}}^{t_i} = \sum_{i=1}^{k-1} \|J'_{i+1} - J'_i\|^2 (t_i). \qquad \square$$

Theorem 1.6.1. *Let $\gamma : [0, \ell] \to M$ be a geodesic without a point conjugate to $\gamma(0)$ along it. If J is a Jacobi field along γ with $\langle J, \dot\gamma \rangle = 0$ and if $X \in \mathcal{X}_\perp$ satisfies $X \neq J$, $X(0) = J(0)$ and $X(\ell) = J(\ell)$ then*

$$I(X, X) > I(J, J).$$

Proof. Since $X - J \in \mathcal{X}'_\perp$ and $X - J \neq 0$, we have from the above observation that $I(X - J, X - J) > 0$. Thus we conclude the proof with

$$\begin{aligned}
I(X - J, X - J) &= I(X, X) - 2I(X, J) + I(J, J) \\
&= I(X, X) - 2\langle X, J' \rangle|_0^\ell + \langle J, J' \rangle|_0^\ell \\
&= I(X, X) - I(J, J).
\end{aligned} \qquad \square$$

Remark 1.6.1. We observe from Theorem 1.6.1 that there exists for any given $0 \leq t_0 < t_1 \leq \ell$ and for any given vectors $u \in T_{\gamma(t_0)}M$ and $v \in T_{\gamma(t_1)}M$ a unique Jacobi field J such that $J(t_0) = u$ and $J(t_1) = v$.

The above discussion yields a refinement in the statement of Lemma 1.5.3.

Proposition 1.6.1. *The index form of a geodesic* $\gamma : [0, \ell] \to M$ *is positive definite on* $\mathcal{X}'_\perp$ *iff* γ *contains no conjugate pair.*

Proof. Assume first that I is positive definite on $\mathcal{X}'_\perp$. Suppose that $\gamma(t_0)$ is conjugate to $p := \gamma(0)$ along γ for some $t_0 \in (0, \ell)$. Lemma 1.6.1 then implies that there is a Jacobi field $Y \in \mathcal{X}'_\perp$ such that $I(Y, Y) < 0$, a contradiction.

Assume that γ contains no conjugate pair and set $q := \gamma(\ell)$. Then $\exp_p |_{[0,\ell]\dot\gamma(0)}$ and $\exp_q |_{[-\ell,0]\dot\gamma(\ell)}$ are regular. If $Y \in \mathcal{X}'_\perp$ satisfies $Y(t_0) \neq 0$ for some $t_0 \in (0, \ell)$, we then have a plane triangle in $\widetilde{M}_p$ with edges $\ell\gamma(0)$ and $[d(\exp_p)_{t_0\dot\gamma(0)}]^{-1}(Y(t_0))$ having a right angle at the corner $\ell\dot\gamma(0)$. Also we have a plane triangle in $\widetilde{M}_q$ with edges $-\ell\dot\gamma(\ell)$ and $[d(\exp_q)_{(t_0-\ell)\dot\gamma(\ell)}]^{-1} Y(t_0)$. As in the proof of the Gauss lemma 1.2.1, we can construct a broken geodesic proper variation $\widetilde{V} : (-\varepsilon, \varepsilon) \times [0, \ell] \to M$ along γ as follows. If $V : (-\varepsilon, \varepsilon) \times [0, \ell] \to M$ is a proper variation along γ associated with Y such that $V(s, t_0) = \widetilde{V}(s, t_0)$ for all $s \in (-\varepsilon, \varepsilon)$ then the Gauss lemma 1.2.1 implies that $L(V(s)) \geq L(\widetilde{V}(s))$ for all $s \in (-\varepsilon, \varepsilon)$ and that $L(V(0)) = L(\widetilde{V}(0)) = \ell$. If J is the variation vector field associated with $\widetilde{V}$ then J is a broken Jacobi field such that $J(0) = Y(0) = 0$, $J(\ell) = Y(\ell) = 0$ and $J(t_0) = Y(t_0)$. From Theorem 1.6.1 it follows that $I(Y, Y) \geq I(J, J) > 0$. Then $I(J, J) = L''(\widetilde{V}(0))$ follows from the construction of $\widetilde{V}$. $\qquad\square$

Finally we discuss the relation between the second fundamental form of geodesic spheres in M and the index form.

For an arbitrary fixed point $p \in M$ and a vector $u_0 \in \widetilde{M}_p$ such that $\exp_p$ is regular at u_0, we choose a domain $W \in \widetilde{M}_p$ such that $u_0 \in W$ and $\exp_p |_W : W \to M$ is an embedding. Setting $q := \exp_p u_0$ and $\ell := \|u_0\|$, we define a piece of smooth geodesic ℓ-sphere N around q by $N := (\exp_p |_W)(S_p(\ell) \cap W)$. For every point $x \in N$ there is a unique geodesic $\gamma_x : [0, \ell] \to M$ such that $\gamma_x(0) = p$, $\gamma_x(\ell) = x$ and $\ell\dot\gamma(0) \in S_p(\ell) \cap W$. A unit field ξ normal to N is defined by

$$\xi(x) := -\dot\gamma_x(\ell), \qquad x \in N.$$

For each $v \in T_q N$ and for a smooth curve $c : (-\varepsilon, \varepsilon) \to N$ such that $c(0) = q, \dot c(0) = v$ we define a geodesic variation $V : (-\varepsilon, \varepsilon) \times [0, \ell] \to M$ along $\gamma := \gamma_q$ by

$$V(s, t) := \gamma_{c(s)}(t), \qquad (s, t) \in (-\varepsilon, \varepsilon) \times [0, \ell].$$

If Y is the variation vector field associated with V then Y must be a Jacobi field along γ such that $Y(0) = 0$, $Y(\ell) = v$. The second variation formula then implies that

$$0 = L''(V_s)|_{s=0} = \int_0^\ell (\langle Y', Y'\rangle - \langle R(Y, \dot\gamma)\dot\gamma, Y\rangle)\,dt - \langle \nabla_{\partial/\partial\varepsilon} Y, \dot\gamma\rangle|_0^\ell.$$

Since $\dot\gamma(\ell) = -\xi(q)$ we see that

$$\langle \nabla_{\partial/\partial\varepsilon} Y, \dot\gamma\rangle(0, \ell) = -\langle A_\xi v, v\rangle.$$

Therefore we have

$$\langle A_\xi v, v\rangle = \langle Y, Y'\rangle(\ell) = I(Y, Y), \qquad v \in T_q N.$$

Thus the principal curvatures of A_ξ are obtained from the index forms of Jacobi fields along radial geodesics.

Example 1.6.1. Let $M = \mathbf{R}^n$ be as in Example 1.5.1. Then $Y(t) = (t/\ell)E(t)$, where E is the parallel field along γ and is generated by $v =: E(\ell)$. We then have $I(Y, Y) = \|v\|^2/\ell$ and $A_\xi = (1/\ell)E_{n-1}$. The principal curvatures of the ℓ-sphere are all equal to a constant, $1/\ell$.

Example 1.6.2. Let $M = \mathbf{S}^n(r)$ be a round r-sphere as in Example 1.5.2. Then

$$Y(t) = \frac{\sin(t/r)}{\sin(\ell/r)}E(t) \qquad \text{and} \qquad R(Y, \dot\gamma)\dot\gamma = \frac{1}{r^2}Y.$$

Thus we have $I(Y, Y) = \cot(\ell/r)\|v\|^2$ and

$$A_\xi = \begin{cases} \cot(\ell/r)E_{n-1} & r \neq \pi r/2, \\ 0 & r = \pi r/2. \end{cases}$$

Example 1.6.3. Let $M = H^n(-c^2)$ be as in Example 1.5.3. Then $Y(t) = (\sinh ct/\sinh c\ell)E(t)$ and $R(Y, \dot\gamma)\dot\gamma = -c^2 Y$. It can be shown that $I(Y, Y) = \|v\|^2 \coth c\ell$ and $A_\xi = \coth c\ell\, E_{n-1}$.

1.7 Complete Riemannian manifolds

Riemannian manifolds carry the structure of metric spaces. First we will prove the fundamental theorem in Riemannian geometry. Then, by using the index form and Jacobi fields, we will prove the Whitehead convexity theorem [108] and the completeness theorem due to Hopf, Rinow and de Rham. It will be shown that the topology of every Riemannian manifold M is equivalent to that of M as a metric space.

First of all the distance function d on M is defined, as follows. To a pair of points $p, q \in M$ a nonnegative number $d(p, q)$ is assigned according to

$$d(p, q) := \inf_c L(c),$$

where the infimum is taken over all piecewise-smooth curves joining p to q. In view of Lemma 1.2.2 it is not difficult to check that the above d defines a distance on M. Thus M carries the structure of a metric space with distance function d induced from g. A geodesic $\gamma : [0, \ell] \to M$ is said to be *minimizing* (or *minimal*) if its length realizes the distance between its endpoints.

We denote by $B(p, r) = \{x \in M; \ d(p, x) < r\}$ the open r-ball centered at p, and by $\overline{B(p, r)}$ its closure.

Definition 1.7.1

(1) An open set $A \subset M$ is said to be *convex* iff any points $p, q \in A$ can be joined by a minimizing geodesic whose image lies entirely in A.

(2) The *convexity radius* conv p of $p \in M$ is the maximal radius of open balls centered at p that are convex. The convexity radius conv A of a set $A \subset M$ is the infimum of conv over A.

(3) A set $A \subset M$ is said to be *locally convex* iff for every point $p \in \overline{A}$ there exists a small positive number r such that $\overline{A} \cap B(p, r)$ is convex.

(4) A function $f : M \to \mathbf{R}$ is said to be *convex* iff $f \circ \gamma$ is convex for every geodesic γ.

Theorem 1.7.1 (J. H. C. Whitehead). *There exists a (Lipschitz) continuous positive function* conv $: M \to \mathbf{R}$ *such that, for any point* $q \in B(p, \text{conv } p)$, *if* $B(q, s) \subset B(p, \text{conv } p)$ *then* $B(q, s)$ *is convex.*

Proof. To show the existence of a convex ball around each point $p \in M$, we set

$$K := \max_{\pi(\sigma) \in B(p, i(p))} |K_\sigma|.$$

From Theorem 1.2.2 and (1.4.5), there exists a constant $a(p) > 0$ with the property that if $J \in \mathcal{X}_\perp$ is a Jacobi field with $J(0) = 0$ along a geodesic $\gamma : [0, \ell] \to M$ such that $\gamma(0) = p$ and $\ell < a(p)$ then $\langle J, J' \rangle(\ell) > 0$. Set $\alpha(p) := \min\{i(p)/2, a(p)\}$. From Lemma 1.2.2 we see that every point $q \in B(p, \alpha(p))$ is joined to p by a unique minimizing geodesic. Let c be a geodesic with $c(0) \in B(p, \alpha(p))$ and let $V : (-\varepsilon_0, \varepsilon_0) \times [0, 1] \to B(p, \alpha(p))$ be a geodesic variation defined by $V(\varepsilon, t) := \exp_p t(\exp_p |_{U_p})^{-1} \circ c(\varepsilon)$. Then $L''(\varepsilon) > 0$ for all ε with $c(\varepsilon) \in B(p, \alpha(p))$. This means that the distance function to p is convex on $B(a, \alpha(p))$. Therefore $B(p, \alpha(p))$ is convex. To conclude the proof we only need to define conv as follows. Let $\operatorname{conv} p$ be the supremum of all $\alpha(p)$ that have the following property: if $B(q, s) \subset B(p, \alpha(p))$ then $B(q, s)$ is convex. It follows from the definition of conv that $|\operatorname{conv} p - \operatorname{conv} q| \leq d(p, q)$, and hence $\operatorname{conv} p$ is Lipschitz continuous with Lipschitz constant 1. $\qquad\square$

The fundamental theorem in Riemannian geometry is stated as follows:

Theorem 1.7.2. *The topology of a Riemannian manifold M is equivalent to that of M as a metric space.*

Exercise 1.7.1. Prove Theorem 1.7.2.

The completeness theorem due to Hopf, Rinow and de Rham is summarized as follows. A Riemannian manifold M is called *complete* iff (M, d) is a complete metric space.

Theorem 1.7.3 (Hopf–Rinow–de Rham). *Let M be a Riemannian manifold with distance function d induced from g. Then the following, (1)–(3), are equivalent.*

(1) *M is complete.*
(2) *$\widetilde{M}_p = T_p M$ for all points $p \in M$.*
(3) *$\widetilde{M}_p = T_p M$ for some point $p \in M$.*

In particular, if M is complete then any two points can be joined by a minimizing geodesic.

Proof. (1) implies (2) and hence (3). To see this suppose that there is a vector $u \in T_p M \setminus \widetilde{M}_p$ for some point $p \in M$. Then there is a $t_0 > 0$ such that $tu \in \widetilde{M}_p$ for all $t \in (0, t_0)$ and $t_0 u \notin \widetilde{M}_p$. If $\gamma_u(t) := \exp_p tu$ then we can find a Cauchy sequence $\{\exp_p t_j u\}$ for which $\lim_{j \to \infty} t_j = t_0$. The completeness of M then implies that $\gamma_u(t_0) \in M$, a contradiction.

To prove the final statement, we set

$$C_\rho := \{x \in \overline{B(p, \rho)};\ x \text{ is joined to } p \text{ by a minimizing geodesic}\}.$$

It follows from Lemma 1.2.2 that $C_\rho = \overline{B(p, \rho)}$ for all $\rho \in (0, i(p))$. Setting $\rho_0 := \sup\{\rho > 0;\ C_\rho = \overline{B(p, \rho)}\}$, we assert that $C_{\rho_0} = \overline{B(p, \rho_0)}$ and that $\overline{B(p, \rho)}$ is compact for all $\rho \in (0, \rho_0]$. Suppose that there is a sequence $\{q_j\}$ of points in $B(p, \rho_0)$ such that $\lim_{j \to \infty} q_j = q$. Let $\gamma_j : [0, \ell_j] \to M$ be a minimizing geodesic with $\gamma_j(0) = p$, $\gamma_j(\ell_j) = q_j$ and $\ell_j < \rho_0$. Clearly both $\{\dot{\gamma}_j(0)\} \subset S_p(1)$ and $\{\ell_j\}$ have convergent subsequences, say, $\lim_{k \to \infty} \dot{\gamma}_k(0) = v$ and $\lim_{k \to \infty} \ell_j = \ell$. Since the solution for a geodesic depends continuously on the initial condition, we have a geodesic $\gamma(y) = \exp_p tv$, $t \in [0, \ell]$ such that

$\gamma(0) = p$ and $\gamma(\ell) = q$. The continuity of d now implies that γ is minimizing and that $q \in C_{\rho_0}$. The above discussion shows the sequential compactness of $\overline{B(p, \rho)}$ for all $\rho \in (0, \rho_0]$.

We next assert that $C_{\rho_0} = \overline{B(p, \rho_0)} = M$. Suppose that $M \setminus \overline{B(p, \rho_0)} \neq \emptyset$. Then there is a point such that the distance from that point to $\overline{B(p, \rho_0)}$ is $a > 0$. The set $\partial B(p, \rho_0) \neq \emptyset$ is compact. Theorem 1.7.1 implies the existence of a small positive number $h < a$ such that $h = \text{conv}\, \partial B(p, \rho_0)$. Let

$$W := \bigcup_{x \in \partial B(p, \rho_0)} B(x, h/4).$$

For any point $y \in W$ we can find an $x \in \partial B(p, \rho_0)$ such that $d(x, y) < h/4$. If $z \in \partial B(p, \rho_0)$ is a point such that $d(y, z) = d(y, \overline{B(p, \rho_0)})$ then $z \in B(y, h/2) \subset B(x, h)$. Theorem 1.7.1 implies that z is joined to y by a unique minimizing geodesic. Since $z \in \overline{B(p, \rho_0)} = C_{\rho_0}$, it is also joined to p by a minimizing geodesic. The union of these geodesics together forms a minimizing geodesic with length $d(p, y)$. Clearly $B(p, \rho_0 + h/4) \subset B(p, \rho_0) \cup W$, and hence we have $\overline{B(p, \rho_0 + h/4)} = C_{\rho_0 + h/4}$. This contradicts the choice of ρ_0, which proves the final statement, (3).

It is now clear that (3) implies (1). $\qquad\square$

In the study of the curvature and topology of Riemannian manifolds, the Alexandrov–Toponogov triangle comparison theorem is essential. We state it without proof.

Theorem 1.7.4 (The triangle comparison theorem). *Let M be a complete Riemannian manifold whose sectional curvature is bounded below by a constant k for all plane sections. If $\Delta = \Delta(\alpha\beta\gamma)$ is a triple of minimizing geodesics forming a triangle in M then a corresponding triangle $\tilde{\Delta} = \tilde{\Delta}(\tilde{\alpha}\tilde{\beta}\tilde{\gamma})$ having the same edge lengths as Δ can be drawn on the complete simply connected 2-manifold $M^2(k)$ of constant curvature k such that every angle of Δ is not less than the corresponding angle of $\tilde{\Delta}$.*

1.8 The short-cut principle

In this section let M be a complete Riemannian n-manifold. A basic tool employed throughout this book is the short-cut principle.

Lemma 1.8.1 (The short-cut principle). *Let $\gamma_i : [0, a_i] \to M$ for $i = 1, 2$ be (minimizing) geodesics, with $\gamma_1(0) = \gamma_2(0) = x$ such that γ_1 and γ_2 make an angle θ at x. If $\theta < \pi$ then there exists for a sufficiently small $h > 0$ a geodesic polygon P joining $\gamma_1(a_1)$ to $\gamma_2(a_2)$ such that $a_1 + a_2 - L(P) > h\cos(\theta/2)$.*

Proof. Let $\alpha : [0, r] \to M$ be a geodesic with $\alpha(0) = x$ such that $\dot{\alpha}(0)$ bisects the angle at x between $\dot{\gamma}_1(0)$ and $\dot{\gamma}_2(0)$. If $\dot{\alpha}^{\perp}(h) \subset T_{\alpha(h)}M$ for small $h > 0$ is the hyperplane orthogonal to $\dot{\alpha}(h)$ and if conv is the convexity radius function as in Theorem 1.7.1 then the point $\exp_{\alpha(h)}\{\dot{\alpha}^{\perp}(h) \cap B(0, \text{conv } \alpha(h))\}$ intersects $\gamma_i[0, a_i]$ at a point p_i. The distance function $s \mapsto d(p_i, \alpha(s))$ is smooth convex and takes its minimum at $\alpha(h)$, and the first variation formula implies that

$$\frac{\mathrm{d}}{\mathrm{d}s}d(p_i, \alpha(s))|_{s=0} = -\cos\frac{\theta}{2} \qquad \text{for } i = 1, 2.$$

Therefore we have

$$d(p_i, \alpha(h)) = d(p_i, x) - h\cos\frac{\theta}{2} + \frac{h^2}{2}\frac{\mathrm{d}^2}{\mathrm{d}s^2}d(p_i, \alpha(\eta_i h))$$

for some $\eta_i \in (0, 1)$. The last term on the right-hand side of the above equation is determined by the index form along geodesics in a convex ball. Thus we find a small $h_0 > 0$ depending only on local invariants on a compact set A such that, for every $h \in (0, h_0)$,

$$-h\cos\frac{\theta}{2} + \frac{h^2}{2}\frac{\mathrm{d}^2}{\mathrm{d}s^2}d(p_i, \alpha(\eta_i h)) \leq -\frac{h}{2}h\cos\frac{\theta}{2}.$$

If $P := \gamma_1(a_1)p_1 \cup p_1 p_2 \cup p_2\gamma_2(a_2)$ is a broken geodesic obtained by joining $\gamma_1(a_1)$, p_1, p_2 and $\gamma_2(a_2)$ in this order then

$$a_1 + a_2 - L(P) = \sum_{i=1}^{2}\{d(p_i, x) - d(p_i, \alpha(h))\} > h\cos\frac{\theta}{2}.$$

$\square$

Remark 1.8.1. Let $A \subset M$ be a closed set and $p \notin A$. If $\gamma : [0, \ell] \to M$ is a minimizing geodesic with $\gamma(0) = p$ and $\gamma(\ell) \in A$ such that $L(\gamma) = d(p, A)$, we call γ a *minimizing geodesic from p to A*. Then $\gamma|_{[s,\ell]}$ for every $s \in (0, \ell)$ is a *unique* minimizing geodesic from $\gamma(s)$ to A. Namely, two minimizing geodesics drawn to A cannot meet in the interior of either geodesic. This fact is a direct consequence of the short-cut principle.

Now let $c : [0, 1] \to M$ be a curve with $c(0) = p, c(1) = q$. By means of the short-cut principle we can construct a homotopy between c and a geodesic in the space of all curves joining p to q such that the length of the homotopy curves is decreasing.

Let A be a compact set containing the $L(c)$-ball around $c[0, 1]$ and set $2r_0 := \text{conv} A$. For a small fixed positive number η and for a large fixed number m we take a partition $0 = a_0 < a_1 < \cdots < a_m = 1$ such that

$$\eta < L(c|_{[a_j, a_{j+1}]}) < \tfrac{1}{2}r_0 \qquad \text{for all } j = 0, \ldots, m - 1.$$

Let $\gamma_{j,t} : [a_j, (1 - t)a_j + ta_{j+1}] \to M$ be a unique minimizing geodesic with $\gamma_{j,t}(a_j) = c(a_j)$ and $\gamma_{j,t}((1-t)a_j + ta_{j+1}) = c((1-t)a_j + ta_{j+1})$. Clearly $\gamma_{j,0}$ is a point curve and $\gamma_{j,1}$ is a minimizing geodesic joining $c(a_j)$ to $c(a_{j+1})$. Let $P := pc(a_1) \cup c(a_1)c(a_2) \cup \cdots \cup c(a_{m-1})q$ be a broken geodesic joining p to q. A homotopy $H : [0, 1] \times [0, 1] \to M$ between c and P is obtained as follows:

$$H(t, s) := \begin{cases} \gamma_{j,t}(s) & \text{for } s \in [a_j, (1 - t)a_j + ta_{j+1}), \\ c(s) & \text{for } s \in [(1 - t)a_j + ta_{j+1}, a_{j+1}). \end{cases}$$

From the construction of H we observe that $H(0, s) = c(s)$ and that the image of $H(1, \cdot)$ is P. Then the short-cut principle implies that $L(H(t, \cdot))$ is strictly decreasing in t unless c is a geodesic. This H will be called *the length-decreasing deformation of c*.

Lemma 1.8.2. *Let $c : [0, 1] \to M$ be a curve joining p to q. Then there exists a geodesic γ having the same extremal points as c such that γ is homotopic to c and $L(\gamma) \le L(c)$.*

Proof. Using the same notation as above, we set $P_0 := P$. The corners of P_0 are expressed as $p_{0,0} := p, p_{0,1}, \ldots, p_{0,m} := q$. Let $q_{0,i}$ for every $i = 1, \ldots, m - 1$ be the midpoint of an edge $p_{0,i-1}p_{0,i}$ of P_0, and let $q_{0,0} := p$, $q_{0,m} := q$. Similarly, let $p_{1,i}$ for every $i = 1, \ldots, m - 1$ be the midpoint of an edge $q_{0,i}q_{0,i+1}$ and set $p_{1,0} := p$, $p_{1,m} := q$. If P_1 is the broken geodesic $P_1 := pp_{1,0} \cup \cdots \cup p_{1,m-1}q$, we then observe that P_1 is homotopic to a broken geodesic $pq_{0,1} \cup \cdots \cup q_{0,m-1}q$ and that this curve is also homotopic to P via a length-decreasing deformation. Thus a sequence $\{P_j\}$ of broken geodesics joining p to q is obtained in a compact set A such that P_j is obtained by a length-decreasing deformation of P_{j-1} for all j. Moreover the corners $p_{j,0}, \ldots, p_{j,m}$ of P_j are in this order and $d(p_{j,i}, p_{j,i+1}) < r_0/2$ for all i and j. The Ascoli theorem implies that there is a subsequence $\{P_k\}$ of $\{P_j\}$ such that $\{p_{k,i}\}$ for every $i = 1, \ldots, m - 1$ converges to a point $p_{\infty,i}$. Let $P_\infty := pp_{\infty,1} \cup \cdots \cup p_{\infty,m-1}q$. To see that P_∞ is a geodesic joining p to q, suppose that the angle θ_i at some corner $p_{\infty,i}$ of P_∞ is less than π. By Lemma 1.8.1 we then find for a sufficiently small fixed $h > 0$ a large number k such that

$$d(p_{\infty,j}, p_{k,j}) < \frac{h}{4m} \cos \frac{\theta_i}{2} \qquad \text{for all } j = 1, \ldots, m - 1,$$

$$L(P_k) - L(P_{k+1}) > \frac{h}{2} \cos \frac{\theta_i}{2}.$$

Then

$$L(P_\infty) < L(P_{k+1}) < L(P_k) - \frac{h}{2} \cos \frac{\theta_i}{2}$$

and also

$$L(P_\infty) > L(P_k) - 2 \sum_{j=1}^{m-1} d(p_{\infty,j},\ p_{k,j}) > L(P_k) - \frac{2(m-1)h}{4m} \cos \frac{\theta_i}{2},$$

a contradiction.   □

1.9 The Gauss–Bonnet theorem

In this section we prove the well-known Gauss–Bonnet theorem without using
the Green–Stokes theorem. Our proof does not require the orientability of surfaces and domains. It is very interesting to generalize it for closed Alexandrov
surfaces and Busemann G-surfaces as well.

Let M be a connected, compact (not necessarily oriented) Riemannian
2-manifold, dM the Lebesgue measure and G the Gaussian curvature. Let $\Omega \subset M$ be a domain whose boundary $\partial\Omega$ consists of a finite union of piecewise-
smooth simple closed curves $c_1, \ldots, c_k$. Let $\mathbf{e}$ be the inward-pointing unit nor-
mal field along the smooth pieces of $\partial\Omega$. The curvature measure over Ω is
written as

$$c(\Omega) := \int_\Omega G \, dM \tag{1.9.1}$$

and is called *the total curvature* of Ω. Let $\omega_{i,1}, \ldots, \omega_{i,r_i}$ be the inner angles of
the corners of c_i and $\mathbf{k}_i$ its geodesic curvature vector. Let I be a measurable set
on c_i. The total geodesic curvature $\lambda(I)$ of I is given by

$$\lambda(I) := \int_I \langle \mathbf{k}_i, \mathbf{e} \rangle \, ds + \sum_I (\pi - \omega_{i,j}),$$

where the second term on the right-hand side is a sum over the corners lying
in I. The inner angles of $\partial\Omega$ are independent of the orientation of Ω. Thus we
observe that

$$\lambda(\partial\Omega) := \sum_{i=1}^k \lambda(c_i) = \sum_{i=1}^k \left\{ \oint_{c_i} \langle \mathbf{k}_i, \mathbf{e} \rangle \, ds + \sum_{j=1}^{r_i} (\pi - \omega_{i,j}) \right\} \tag{1.9.2}$$

is independent of the orientation of $\partial\Omega$; it is called the *total geodesic curvature
of* $\partial\Omega$. With this notations the *Gauss–Bonnet theorem* will be stated as follows.

The Gauss–Bonnet theorem. *If $\chi(\Omega)$ is the Euler characteristic of Ω then*

$$c(\Omega) + \lambda(\partial\Omega) = 2\pi\chi(\Omega). \qquad (1.9.3)$$

In particular if M has no boundary then

$$c(M) = 2\pi\chi(M). \qquad (1.9.4)$$

The following two lemmas are needed for the proof of the above theorem. These lemmas are independent of the orientation of M.

Lemma 1.9.1. *Let $\Delta = \Delta(pqr) \subset M$ be a geodesic triangle, with vertices at p, q and r, that is contained entirely in a convex ball. If A, B and C are the inner angles of Δ then*

$$c(\Delta) = A + B + C - \pi. \qquad (1.9.5)$$

Proof. If x and y are in a convex ball then there is a unique minimizing geodesic xy joining x to y. For a sufficiently large number N the edges of Δ are divided into N sufficiently small subarcs by taking points $p_{0,0}, \dots, p_{N,0}$ on qr, $q_{0,0}, \dots, q_{0,N}$ on rp and $r_{0,0}, \dots, r_{0,N}$ on qp such that $q = p_{0,0} = r_{0,0}$, $r = p_{N,0} = q_{0,0}$ and $p = r_{0,N} = q_{0,N}$. For each $i, j = 1, \dots, N-1$ we set

$$p_{i,j} := pp_{i,0} \cap r_{0,j}q_{0,j}.$$

Let $\square_{ij}$ be the oriented rectangle with vertices at $p_{i,j}$, $p_{i+1,j}$, $p_{i+1,j+1}$ and $p_{i,j+1}$. Let θ_{ij} be the oriented angle obtained by the parallel translation of vectors at $p_{i,j}$ along $\square_{ij}$ and let u_i and v_i be unit vectors at $p_{i,j}$ tangent to the geodesics $p_{i,j}p_{i+1,j}$ and $p_{i,j}p_{i,j+1}$ respectively. If α is the angle between u_i and v_i and if $a_i := d(p_{i,j}, p_{i+1,j})$ and $b_i := d(p_{i,j}, p_{i,j+1})$ then (1.3.5) implies (infinitesimally)

$$-R(u_i, v_i)v_i = \frac{\theta_{ij}}{a_i b_i}\left\{\cos\left(\frac{\pi}{2} + \alpha\right)u_i + \sin\left(\frac{\pi}{2} + \alpha\right)u_i^\perp\right\},$$

where $u_i^\perp$ is defined by $v_i := u_i\cos\alpha + u_i^\perp\sin\alpha$. Thus we see that

$$G(p_{i,j})\,\text{Area}\square_{ij} = \theta_{ij}.$$

If $\tau : T_qM \to T_qM$ is the parallel translation along Δ then summing up the above relation over all $i, j = 1, \dots, N$ and letting $N \to \infty$, we see that $c(\Delta)$ is the oriented angle obtained by τ. If $w \in T_qM$ is tangent to qr then $\angle(\tau(w), w) = A + B + C - \pi$. $\qquad\square$

Lemma 1.9.2. *Let $D \subset M$ be a disk domain contained entirely in a convex ball such that ∂D consists of two minimizing geodesics pq, pr and a unit-speed* *smooth curve* $c : [0, \ell] \to M$ *with* $c(0) = q$ *and* $c(\ell) = r$. *Let* **e** *be the inward-pointing unit normal vector field along* c *and* $\mathbf{k}(s)$ *the geodesic curvature vector at* $c(s)$. *Assume that all the interior points of* $pc(s)$ *for all* $s \in (0, \ell)$ *are in* D. *If* A, B *and* C *are the inner angles of* D *at* p, q *and* r *respectively then*

$$c(D) + \int_0^\ell \langle \mathbf{k}, \mathbf{e} \rangle(s)\, ds = A + B + C - \pi.$$

Proof. Let $0 = s_0 < s_1 < \cdots < s_N = \ell$ be a partition of $[0, \ell]$ and set $p_i := c(s_i)$. We approximate $c[0, \ell]$ by a broken geodesic $P_N := p_0 p_1 \cup p_1 p_2 \cup \cdots \cup p_{N-1, N} p_N$, where $s_{i+1} - s_i < \ell/N$. If D_N is the disk domain bounded by the geodesics pq, pr and P_N then $\lim_{N \to \infty} D_N = D$, and hence

$$c(D) = \lim_{N \to \infty} c(D_N).$$

If ω_i is the inner angle of D_N at p_i for $i = 1, \ldots, N - 1$ then Lemma 1.8.1 implies that

$$c(D_N) = A + B + C - \pi + \sum_{i=1}^{N-1} (\omega_i - \pi).$$

For each $i = 1, \ldots, N - 1$ we denote by $\sigma_i : [0, 1] \to M$ the minimizing geodesic with $\sigma_i(0) = p_i$ and $\sigma_i(1) = p_{i+1}$. Since c is smooth, there is a point $z_i = c(\tilde{s}_i)$ for $\tilde{s}_i \in (s_i, s_{i+1})$ and for $i = 1, \ldots, N - 1$ with the following properties. If $\gamma_\pm : [0, a_\pm] \to M$ are minimizing geodesics with $\gamma_\pm(0) = c(\tilde{s}_i)$, $\gamma_+(a_+) = c(s_{i+1})$ and $\gamma_-(a_-) = c(s_i)$ then

$$|\angle(\dot{\gamma}_+(a_+), \dot{\sigma}_i(1)) - \angle(\dot{\gamma}_+(0), \dot{c}(\tilde{s}_i))|$$
$$+ |\angle(\dot{c}(\tilde{s}_i), -\dot{\gamma}_-(0)) - \angle(\dot{\sigma}_i(0), -\dot{\gamma}_i(a_-))|$$
$$< \text{Area}\{\Delta(c(s_i) c(\tilde{s}_i) c(s_{i+1}))\} \sup_D |G| < C(s_{i+1} - s_i)^2.$$

Here C is a positive constant depending only on the set D. Let $\tau_x : TM|_{B(x, r(x))} \to T_x M$ be the parallel translation along minimizing geodesics from points on the convex ball $B(x, r(x))$ to $T_x M$. Then (1.3.3) implies that

$$\frac{\|\tau_{z_i}(\dot{c}(s_{i+1})) - \dot{c}(\tilde{s}_i)\|}{s_{i+1} - \tilde{s}_i} = \langle \mathbf{k}, \mathbf{e} \rangle(\tilde{s}_i) + C(s_{i+1} - \tilde{s}_i)$$

and

$$\frac{\|\dot{c}(\tilde{s}_i) - \tau_{z_i}(\dot{c}(s_i))\|}{\tilde{s}_i - s_i} = \langle \mathbf{k}, \mathbf{e} \rangle(\tilde{s}_i) + C(\tilde{s}_i - s_i).$$

Therefore, from

$$\pi - \omega_{i+1} = \angle(\dot{\sigma}_{i+1}(0), \dot{c}(s_{i+1})) + \angle(\dot{c}(s_{i+1}, \dot{\sigma}_i(1))$$

and letting $N \to \infty$, we have

$$\lim_{N \to \infty} \sum_{i=1}^{N-1} (\pi - \omega_i) = \int_0^\ell \langle \mathbf{k}, \mathbf{e} \rangle(s)\, ds.$$

This proves Lemma 1.9.2. $\qquad\qquad\qquad\qquad\qquad\qquad\qquad\square$

Remark 1.9.1. We observe that $M \setminus D$ is strictly convex in a small neighborhood around $c(s)$ iff $\langle \mathbf{k}, \mathbf{e} \rangle(s) < 0$. Also D is strictly convex iff $\langle \mathbf{k}, \mathbf{e} \rangle(s) > 0$.

Proof of the Gauss–Bonnet theorem. We divide $\Omega \subset M$ into finitely many geodesic triangles $\Delta_1, \ldots, \Delta_\ell$ and disk domains $D_1, \ldots, D_m$, as in Lemmas 1.9.1 and 1.9.2, in such a way that each break point of $\partial\Omega$ is a vertex of some disk domain and $\partial\Omega$ is simply covered by the union of all the nongeodesic edges of all disk domains. Such a division is obtained by covering the closure $\overline{\Omega}$ of Ω by finitely many small convex balls. The vertices consist of the centers of convex balls, the break points of $\partial\Omega$ and finitely many smooth points on $\partial\Omega$.

If k is the total number of components of $\partial\Omega$, and if v, e and f respectively are the total numbers of vertices, edges and faces of the division of Ω, then we have

$$v - e + f = \chi(\Omega).$$

Let v_b and e_b be the total numbers of vertices and edges respectively on $\partial\Omega$. Since each component of $\partial\Omega$ is a circle, we observe that $v_b = e_b$. Setting $v_i := v - v_b$ and $e_i := e - e_b$, we see that

$$3f = e_b + 2e_i = v_b + 2e_i. \tag{1.9.6}$$

If $\omega_1, \ldots, \omega_r$ are the inner angles at vertices on $\partial\Omega$ that are not equal to π then there are $v_b - r$ vertices on $\partial\Omega$ at which the inner angles are equal to π. By means of Lemmas 1.9.1 and 1.9.2, we have

$$c(\Omega) + \sum_{i=1}^k \oint_{c_i} \langle \mathbf{k}_i, \mathbf{e} \rangle\, ds = \sum_{i=1}^\ell c(\Delta_i) + \sum_{j=1}^m c(D_j) + \sum_{i=1}^k \oint_{c_i} \langle \mathbf{k}_i, \mathbf{e} \rangle\, ds$$

$$= 2\pi v_i + \pi(v_b - r) + \sum_{j=1}^r \omega_j - f\pi.$$

Making use of $v_i = v - v_b = v - e_b$ and (1.9.6), we have proved (1.9.3). If M has no boundary then (1.9.4) is direct from (1.9.3). Thus the proof is complete. $\qquad\qquad\qquad\qquad\qquad\qquad\qquad\qquad\qquad\qquad\square$

Finally, we discuss the Gauss–Bonnet theorem for special connected compact sets on M. Let $A \subset M$ be a connected and compact set with the following

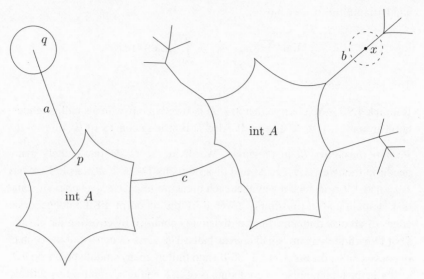

Figure 1.9.1

properties (see Figure 1.9.1):

(a) int A is a disjoint union of finitely many domains;
(b) $A \setminus$ int A is a finite union of piecewise-smooth curves;
(c) if $A_\varepsilon \subset M$ is an ε-ball around A then A_ε does not change its topology for all sufficiently small $\varepsilon > 0$.

The decreasing sequence $\{A_\varepsilon\}$ of domains has its limit A as $\varepsilon \downarrow 0$. The boundary ∂A of A is considered as the boundary of the completion of $M \setminus A$. If $x \in A$ satisfies $B(x, \varepsilon) \cap$ int $A = \emptyset$ for a small $\varepsilon > 0$ and if $B(x, \varepsilon) \setminus A$ has two components then such a point x on ∂A is understood as a double point and $B(x, \varepsilon) \cap \partial A$ as a double curve with reversed orientation. Such a ∂A is called the *fine boundary* of A. We say that such an A has a fine boundary.

Theorem 1.9.1 (The Gauss–Bonnet theorem). *If $A \subset M$ is a compact set with a fine boundary ∂A then*

$$c(A) + \lambda(\partial A) = 2\pi \chi(A). \tag{1.9.7}$$

Proof. We first note that $\lambda(\partial A)$ exists by the assumption for ∂A, that $\lambda(\partial A_\varepsilon)$ exists and that $c(A_\varepsilon) + \lambda(\partial A_\varepsilon) = 2\pi \chi(A_\varepsilon)$ for all sufficiently small $\varepsilon > 0$. If b is a curve in ∂A passing through a point $x \in A$ such that $B(x, \varepsilon) \setminus b$ has two components for a small $\varepsilon > 0$ then the total geodesic curvature of b near x is counted twice with opposite signs along ∂A_ε. If a is a curve in ∂A joining

p to q such that $B(q, \varepsilon) \setminus A$ is connected for all small $\varepsilon > 0$ and such that p is a regular point of $\partial \operatorname{int}(A)$ then the sum of two exterior angles of A_ε near p is $-\pi$ and the geodesic curvature integral of a semi-circle around q in ∂A_ε is π. If c is a curve in ∂A joining two points of $\partial \operatorname{int}(A)$ at which ∂A is regular in such a way that $\operatorname{int} c$ does not meet other curves in ∂A then the sum of the four exterior angles of A_ε near the endpoints of c is -2π. Summing up these cases and letting $\varepsilon \to 0$, we conclude the proof. $\qquad \square$

2

The classical results of Cohn-Vossen and Huber

We shall introduce the classical results obtained by Cohn-Vossen in [19] and [20] and Huber in [39] by exhibiting a simpler method under more general assumptions. For this purpose we need to consider the curvature measure over an unbounded domain, not necessarily oriented, with possibly noncompact boundary. The boundary curves may be divergent and hence certain conditions for them will be required. Our ideas will lead to the natural compactification of complete open surfaces for which the Gauss–Bonnet theorem is valid. Some examples on the total curvature of complete open surfaces in $\mathbf{R}^3$ are provided. From these examples, one can begin to see the geometric significance of the total curvature of complete open surfaces.

2.1 The total curvature of complete open surfaces

From now on let M be a connected, complete and noncompact Riemannian 2-manifold either with or without a piecewise-smooth boundary. In general we cannot expect the Gauss–Bonnet theorem to hold for such an M. However, it is still of interest to consider the Gauss–Bonnet theorem for noncompact surfaces. For this purpose we need to ascertain:

(1) how to define the Euler characteristic of M;
(2) how to make sense of the curvature integral over M.

For point (1) we are led to consider the notion of the *finite connectivity* of M.

Definition 2.1.1. M is by definition *finitely connected* iff there exist a compact 2-manifold N and finitely many points $p_1, \ldots, p_k \in N$ for $k \geq 1$ such that M is homeomorphic to $N \setminus \{p_1, \ldots, p_k\}$. M is by definition *infinitely connected* iff it is not finitely connected.

If M is homeomorphic to $N \setminus \{p_1, \ldots, p_k\}$, if $p_1, \ldots, p_\ell \in$ int N and if $p_{\ell+1}, \ldots, p_k \in \partial N$ then the Euler characteristic $\chi(M)$ of M is

$$\chi(M) := \chi(N) - \ell. \qquad (2.1.1)$$

In the above case we say that M has k *ends*. Furthermore we observe from $\chi(\mathbf{S}^1) = 0$ that

$$\chi(\partial M) = k - \ell,$$

where $\chi(\partial M)$ is the number of unbounded components of ∂M.

From now on let M be finitely connected. In view of the definition of the finite connectivity of M, with boundary or without, we can always choose a compact set $\Omega \subset M$ such that

(1) all compact components of ∂M are contained in int Ω;
(2) $\partial \Omega$ is a disjoint union of finitely many simple closed piecewise-smooth curves;
(3) $M \setminus \Omega$ has ℓ *tubes* (or half-cylinders) and $k - \ell$ half-planes.

Definition 2.1.2. A compact set $\Omega \subset M$ is called a *core* of M iff Ω satisfies (1), (2) and (3) above.

It is clear that M admits a monotone increasing sequence $\{C_i\}$ of cores exhausting M. Then all the C_i are homeomorphic to each other and

$$\chi(M) = \chi(C_i) \qquad \text{for all } i. \qquad (2.1.2)$$

Regarding point (2) at the start of this section, we have already defined the *total curvature of M* and the *total geodesic curvature of ∂M*. The total curvature of a complete open 2-manifold M is defined without assuming its finite connectivity, as follows.

Definition 2.1.3. The *total curvature* $c(M)$ of M is defined by taking an increasing sequence $\{C_i\}$ of cores exhausting M; then

$$c(M) := \lim_{i \to \infty} c(C_i). \qquad (2.1.3)$$

The total curvature $c(M)$ exists iff the improper integral $\int_M G \, dM$ exists in $[-\infty, \infty]$. Namely, $c(M)$ exists iff

$$\int_M G_+ \, dM < \infty \qquad \text{or} \qquad \int_M G_- \, dM < \infty,$$

where $G_+ := \max\{G, 0\}$ and $G_- := G_+ - G$. If $c(M)$ is finite then there exists for every $\varepsilon > 0$ a number $i(\varepsilon)$ such that

$$|c(M) - c(C_i)| < \varepsilon \qquad \text{for all } i > i(\varepsilon), \qquad (2.1.4)$$

and in particular

$$\int_{M \setminus C_i} |G| \, dM < \varepsilon \qquad \text{for all } i > i(\varepsilon). \qquad (2.1.5)$$

If $c(M) = -\infty$ then

$$\int_{M \setminus C_i} G_+ \, dM < \varepsilon \qquad \text{for all } i > i(\varepsilon). \qquad (2.1.6)$$

The boundary condition for M will be discussed now. The boundary of M may contain a divergent curve; here, a curve $c : [0, 1) \to M$ is by definition a *divergent* (or *proper*) *curve* iff for every compact set A there is a number $t_A \in (0, 1)$ such that $c(t_A, 1) \subset M \setminus A$. If $c : \mathbf{R} \to \partial M$ is a nonclosed boundary curve then both $c|_{(-\infty, 0]}$ and $c|_{[0, \infty)}$ are divergent. For a break point $x_i \in \partial M$ of ∂M, the inner angle at x_i of ∂M is denoted by $\omega_i \in (0, 2\pi)$. Let $\mathbf{e}$ be the inward-pointing unit normal to ∂M and $\mathbf{k}$ the geodesic curvature vector of ∂M. Let $\{I_j\}$ be a monotone increasing sequence of compact sets of ∂M such that $\bigcup_{j=1}^\infty I_j = \partial M$. With this notation the *total geodesic curvature* of ∂M is defined as follows. The total geodesic curvature $\lambda(\partial M) \in [-\infty, +\infty]$ of ∂M is well defined iff

$$\lambda(\partial M) := \lim_{j \to \infty} \lambda(I_j) \qquad (2.1.7)$$

exists and is independent of the choice of $\{I_j\}$.

Remark 2.1.1. If M has k ends as before then $M \setminus C_i$ has ℓ tubes and $k - \ell$ half-planes having divergent boundary curves $c_{\ell+1}, \ldots, c_k$. If $|\lambda(\partial M)| < \infty$ then so are all of $\lambda(c_{\ell+1}), \ldots, \lambda(c_k)$. If $\lambda(\partial M) = \infty$ then $\lambda(c_j) > -\infty$ for all $j = \ell+1, \ldots, k$, and if $\lambda(\partial M) = -\infty$ then $\lambda(c_j) < \infty$ for all $j = \ell+1, \ldots, k$.

Now some examples of total curvature are provided.

Let $M \subset \mathbf{R}^3$ be a complete noncompact surface without a boundary. The total curvature of M is obtained by the Gauss normal map $\nu : M \to \mathbf{S}^2(1)$ as follows. Let $D := \{(u^1, u^2) \in \mathbf{R}^2\}$ be a domain and $X : D \to M$ be a local expression for $M \subset \mathbf{R}^3$. By setting $X_i := dX(\partial/\partial u^i)$ we have $g_{ij} = X_i \cdot X_j$ for $i, j = 1, 2$, and the area element dM of M is given by

$$dM = \|X_1 \times X_2\| \, du^1 \, du^2.$$

Let $\mathbf{n}$ be the unit field normal to M in $\mathbf{R}^3$. Then $\nu(p)$ for $p \in M$ is defined by the parallel translation of $\mathbf{n}(p)$ to the origin of $\mathbf{R}^3$. The structure equation gives

$$X_{ij} = \Gamma_{ij}^k X_k + h_{ij}\mathbf{n} \qquad \text{for } i, j = 1, 2,$$
$$\mathbf{n}_i = -h_{ij}g^{jk}X_k \qquad \text{for } i = 1, 2,$$

where $X_{ij} = \partial^2 X/(\partial u^i \partial u^j)$, $\mathbf{n}_i = \partial\mathbf{n}/\partial u^i$ and the h_{ij} are the coefficients in the second fundamental tensor of M. We then have

$$\det d\nu = \|\mathbf{n}_1 \times \mathbf{n}_2\| = \frac{\det h_{ij}}{\det g_{ij}}\|X_1 \times X_2\|.$$

If $d\Sigma$ is the area element of $\mathbf{S}^2(1)$ then the above discussion implies that $\nu^* d\Sigma = K\,dM$. Therefore

$$c(M) = \int_M G\,dM = \int_{\nu(M)} d\Sigma. \qquad (2.1.8)$$

This relation is employed to find the total curvature of $M \subset \mathbf{R}^3$.

Example 2.1.1. For a plane $M = \mathbf{R}^2$ we have $c(M) = 0$.

Example 2.1.2. For a hyperbolic plane $M = H^2(-c^2)$ we have $c(M) = -\infty$.

Example 2.1.3 (Rotation surface of parabola). For a positive constant k, the surface of revolution generated by a parabola $y = kx^2/2$ is given as

$$M = \{(r\cos\theta, r\sin\theta, kr^2/2) \in \mathbf{R}^3; r > 0, 0 < \theta \leq 2\pi\}.$$

Setting $u^1 := r$ and $u^2 := \theta$, we have $ds^2 = (1 + k^2r^2)(du^1)^2 + r^2(du^2)^2$ and

$$\Gamma_{12}^1 = \Gamma_{11}^2 = \Gamma_{22}^2 = 0,$$

$$\Gamma_{22}^1 = \frac{-r}{1 + k^2r^2}, \qquad \Gamma_{11}^1 = \frac{k^2r}{1 + k^2r^2}, \qquad \Gamma_{12}^2 = \frac{1}{r}.$$

Therefore we get

$$R_{212}^1 = \frac{-k^2r^2}{(1 + k^2r^2)^2},$$

and hence

$$G = \frac{k^2}{(1 + k^2r^2)^2}.$$

Clearly $\nu(M)$ covers the northern hemisphere of $\mathbf{S}^2(1)$. Also, a direct computation shows that

$$c(M) = \lim_{r\to\infty} \int_0^{2\pi} \int_0^r \frac{k^2r}{(1 + k^2r^2)^{3/2}}\,dr\,d\theta = 2\pi.$$

Example 2.1.4 (Two-sheeted hyperboloid). For a component of the two-sheeted hyperboloid $M = \{(r\cos\theta, r\sin\theta, \sqrt{k^2r^2 + 1}) \in \mathbf{R}^3; r > 0, 0 < \theta \le 2\pi\}$, we have

$$ds^2 = \frac{k^2(k^2 + 1)r^2 + 1}{k^2r^2 + 1}(du^1)^2 + r^2(du^2)^2,$$

where $u^1 := r, u^2 := \theta$ and also

$$\Gamma^1_{12} = \Gamma^2_{11} = \Gamma^2_{22} = 0, \qquad \Gamma^2_{12} = \frac{1}{r},$$

$$\Gamma^1_{11} = \frac{k^4 r}{(k^2r^2 + 1)\{k^2(k^2 + 1)r^2 + 1\}}, \qquad \Gamma^1_{22} = \frac{-r(k^2r^2 + 1)}{k^2(k^2 + 1)r^2 + 1}.$$

Thus we have

$$R^1{}_{212} = \frac{-k^4 r^2}{(k^2(k^2 + 1)r^2 + 1)^2}$$

and

$$K = \frac{k^4}{(k^2(k^2 + 1)r^2 + 1)^2}.$$

If we set $\tan\varphi = k$ then $v(M)$ is the φ-ball around the north pole, and hence $c(M) = 2\pi(1 - \cos\varphi)$. This can be shown by direct evaluation also:

$$c(M) = \lim_{r\to\infty} \int_0^{2\pi} \int_0^r \frac{k^4 r}{(k^2r^2(k^2 + 1) + 1)^{3/2}(1 + k^2r^2)^{1/2}} \, dr \, d\theta$$

$$= 2\pi k^2 \sqrt{\frac{k^2r^2 + 1}{k^2r^2(k^2 + 1) + 1}} \Bigg|_0^\infty$$

$$= 2\pi \left(1 - \frac{1}{\sqrt{1 + k^2}}\right) = 2\pi(1 - \cos\varphi).$$

Example 2.1.5 (One-sheeted hyperboloid). By setting

$$f(t) := \sqrt{k^2t^2 + 1},$$

a one-sheeted hyperboloid is given as

$$M = \{(f(t)\cos\theta, f(t)\sin\theta, t) \in \mathbf{R}^3; t \in \mathbf{R}, 0 < \theta \le 2\pi\}.$$

Then

$$ds^2 = \frac{k^2(k^2 + 1)t^2 + 1}{k^2t^2 + 1}(du^1)^2 + (k^2t^2 + 1)(du^2)^2$$

and

$$\Gamma^1_{12} = \Gamma^2_{11} = \Gamma^2_{22} = 0,$$

$$\Gamma^1_{22} = \frac{-k^2 t (k^2 t^2 + 1)}{k^2 (k^2 + 1) t^2 + 1}, \qquad \Gamma^2_{12} = \frac{k^2 t}{k^2 t^2 + 1},$$

$$\Gamma^1_{11} = \frac{k^4 t}{(k^2 t^2 + 1)(k^2 (k^2 + 1) t^2 + 1)}.$$

This gives

$$R^1{}_{212} = \frac{k^2 (k^2 t^2 + 1)}{(k^2 (k^2 + 1) t^2 + 1)^2}$$

and

$$G = \frac{-k^2}{(k^2 (k^2 + 1) t^2 + 1)^2}.$$

Setting $\tan \varphi := k$ for $\varphi \in (0, \pi/2)$, we see that $v(M)$ omits $(\pi/2 - \varphi)$-balls around the north and south poles of $\mathbf{S}^2(1)$ and hence that $c(M) = -4\pi \sin \varphi$. Again, using direct evaluation,

$$c(M) = \lim_{t \to \infty} \int_0^{2\pi} \int_{-t}^t \frac{-k^2}{\left(k^2 (k^2 + 1) t^2 + 1 \right)^{3/2}} \, dt \, d\theta$$

$$= -4\pi \frac{k}{\sqrt{k^2 + 1}} = -4\pi \sin \varphi.$$

2.2 The classical theorems of Cohn-Vossen and Huber

The classical well-known theorem due to Cohn-Vossen will be stated under more general assumptions. We show here that the Huber theorem is derived directly from the Cohn–Vossen theorem.

Theorem 2.2.1 (compare Satz 6, [19], Theorem 10, [39] and Section 3.4, [93]). *Let M be a connected noncompact finitely connected complete Riemannian 2-manifold. If M admits a total curvature and ∂M a total geodesic curvature and if $c(M) = -\lambda(\partial M) = \pm\infty$ does not hold then*

$$2\pi \chi(M) - (c(M) + \lambda(\partial M)) \geq \pi \chi(\partial M). \tag{2.2.1}$$

Theorem 2.2.2 (Huber [39]). *If a connected, infinitely connected, complete Riemannian 2-manifold M without boundary admits a total curvature $c(M)$ then*

$$c(M) = -\infty. \tag{2.2.2}$$

The following definitions of special geodesics are used throughout.

Definition 2.2.1. A unit-speed geodesic $\gamma : [0, \infty) \to M$ is called a *ray from a compact set A* iff $d(\gamma(s), A) = s$ for all $s \geq 0$. A unit-speed geodesic $\gamma : \mathbf{R} \to M$ is called a *straight line* iff $d(\gamma(s), \gamma(t)) = |s - t|$ for all $s, t \in \mathbf{R}$.

Note that if $\partial M = \emptyset$ then there exists at least one ray emanating from every point of M and that if M has more than one end then M admits a straight line.

Let $H \subset M$ be a half-plane and let its boundary curve $c : \mathbf{R} \to \partial H$ admit a total geodesic curvature. The *inner distance* ρ_H of H is induced from that of M as follows:

$$\rho_H(x, y) := \inf \{L(c) : c \text{ is a curve in } H \text{ joining } x \text{ to } y\}.$$

It follows from the length-decreasing deformation that every pair of points $x, y \in H$ can be joined by a curve γ in H whose length realizes the inner distance $\rho_H(x, y)$. Such a curve will be called a *segment in H* joining x to y. A unit-speed curve $\gamma : [0, \infty) \to H$ is called a *ρ_H-ray* iff every subarc of it is a segment in H. A *ρ_H-straight line* is defined in a similar manner. The following proposition, 2.2.1, is a direct consequence of the short-cut principle. The proof is left to the reader.

Proposition 2.2.1. *Let $H \subset M$ be a half plane such that its boundary curve admits a total geodesic curvature. If $\gamma : [0, a] \to H$ is a unit-speed segment in H then the following statements are true.*

(1) *Each component of $\gamma[0, a] \cap \text{int } H$ is a geodesic in M.*
(2) *If the geodesic curvature vector $\mathbf{k}(s_0)$ of γ at an interior point $\gamma(s_0)$ of γ exists and if $\gamma(s_0) \in \partial H$ then*

$$\langle \mathbf{k}, \mathbf{e} \rangle(s_0) \leq 0.$$

(3) *If the geodesic curvature vector of γ at an interior point $\gamma(s_0)$ does not exist and if $\gamma(s_0) \in \partial H$ then the inner angle between $-\lim_{s \uparrow s_0} \dot{\gamma}(s)$ and $\lim_{s \downarrow s_0} \dot{\gamma}(s)$ is not less than π.*
(4) *Each segment, ρ_H-ray and ρ_H-straight line in H admits a total geodesic curvature.*

Let $c : \mathbf{R} \to \partial H$ be the boundary curve of H and let $\rho_c : [0, \infty) \to \mathbf{R}^+$ be defined by

$$\rho_c(s) := \rho_H(c(-s), c(s)), \qquad s \geq 0.$$

Clearly ρ_c is Lipschitz continuous, with Lipschitz constant 2, and hence is differentiable almost everywhere. Let γ_s for every $s \geq 0$ be a segment in H

joining $c(-s)$ to $c(s)$. Let $A_s \subset H$ for $s > 0$ be a compact set with fine boundary $\gamma_s \cup c[-s, s]$. If γ_s does not meet ∂H at its interior then A_s is a closed 2-disk domain. If γ_s intersects ∂H at its interior then int A_s may have at most countably many disk domains. Moreover, the ε-ball for sufficiently small ε around A_s in H is a disk domain. Then the Gauss–Bonnet theorem 1.9.1 implies that

$$c(A_s) + \lambda(\partial A_s) = 2\pi,$$

and if the total geodesic curvature of γ_s is measured with respect to $A_s \subset H$ then Proposition 2.2.1 implies that γ_s is convex with respect to A_s and that

$$\lambda(\gamma_s) \geq 0.$$

With this notation we prove

Lemma 2.2.1. *If $\alpha(s)$ and $\beta(s)$ for $s \geq 0$ are inner angles at $c(-s)$ and $c(s)$ of the compact set $A_s \subset H$ whose fine boundary is $\gamma_s \cup c[-s, s]$ then*

$$\liminf_{s \to \infty} (\alpha(s) + \beta(s)) \leq \pi, \tag{2.2.3}$$

and

$$\limsup_{s \to \infty} \rho_c'(s) \geq 0. \tag{2.2.4}$$

Proof. The first variation formula implies that if ρ_c is differentiable at s then

$$\frac{d\rho_c(s)}{ds} = \cos\alpha(s) + \cos\beta(s). \tag{2.2.5}$$

Suppose that (2.2.4) is false. Then there is an $\varepsilon > 0$ such that $\limsup_{s\to\infty} \rho_c'(s) \leq -\varepsilon$. We can find a sufficiently large number s_0 such that $\rho_c'(s) \leq -\varepsilon/2$ for all $s \geq s_0$. It then follows from

$$\rho_c(s_1) - \rho_c(s_0) = \int_{s_0}^{s_1} \rho_c'(s)\,ds \qquad \text{for } s_1 > s_0$$

that

$$\rho_c(s_0) \geq \rho_c(s_1) + \frac{\varepsilon}{2}(s_1 - s_0) > \frac{\varepsilon}{2}(s_1 - s_0).$$

Thus a contradiction is obtained for sufficiently large $s_1 > s_0$. The rest is now clear from (2.2.4) and (2.2.5). □

Lemma 2.2.2 (see Proposition 3.4.2, [93]). *Let $H \subset M$ be a half-plane admitting both $c(H)$ and $\lambda(\partial H)$. If $c(H) = -\lambda(\partial H) = \pm\infty$ does not hold then*

$$c(H) + \lambda(\partial H) \leq \pi\chi(\partial H) = \pi. \tag{2.2.6}$$

Proof. Let $c : \mathbf{R} \to \partial H$ be a unit-speed boundary curve. For a monotone divergent sequence $\{s_i\}$ of positive numbers with $\lim_{i \to \infty} s_i = \infty$ we choose a monotone increasing sequence $\{C_i\}$ of cores exhausting M, such that $\partial C_i \cap H$ for every i is a piecewise-smooth curve joining $c(-s_i)$ to $c(s_i)$. Let $\{\varepsilon_i\}$ be a strictly decreasing sequence of positive numbers converging to 0. From Lemma 2.2.1 we find for every $i = 1, \ldots,$ a large number $s_i' > s_i$ and a segment γ_i joining $c(-s_i')$ to $c(s_i')$ in $H \setminus C_i$ such that $c[-s_i', s_i'] \cup \gamma_i$ forms the fine boundary of a compact contractible set $A_i \subset H$. If α_i and β_i are inner angles at $c(-s_i')$ and $c(s_i')$ of A_i then

$$\alpha_i + \beta_i \leq \pi + \varepsilon_i \qquad \text{for all } i. \tag{2.2.7}$$

Further, the Gauss–Bonnet theorem 1.9.1 implies

$$c(A_i) + \lambda(\partial A_i) = 2\pi \qquad \text{for all } i.$$

Here we note that

$$\lambda(\partial A_i) = \lambda(c|_{[-s_i', s_i']}) + 2\pi - \alpha_i - \beta_i + \lambda(\gamma_i). \tag{2.2.8}$$

Since γ_i is convex with respect to A_i, we have $\lambda(\gamma_i) \geq 0$ and

$$c(A_i) + \lambda(c|_{[-s_i', s_i']}) \leq \alpha_i + \beta_i. \tag{2.2.9}$$

Because $\lim_{i \to \infty} c(A_i) = c(H)$ and $\lim_{i \to \infty} \lambda(c|_{[-s_i', s_i']}) = \lambda(\partial H)$ and because $c(H) = -\lambda(\partial H) = \pm\infty$ does not hold, we see that the limit as $i \to \infty$ on the left-hand side in (2.2.9) exists in $[-\infty, \pi]$. This proves Lemma 2.2.2. □

Corollary 2.2.1 (compare Satz 1 and Satz 2 in [20]). *In addition to the assumptions in Lemma 2.2.2, if ρ_c is bounded above then*

$$c(H) + \lambda(\partial H) = \pi \chi(\partial H). \tag{2.2.10}$$

Moreover, if $\rho_c(s) \geq 2s - L$ for a constant $L > 0$ and for all sufficiently large s then

$$c(H) + \lambda(\partial H) \leq 0. \tag{2.2.11}$$

Proof. If ρ_c is bounded above then there exists a monotone divergent sequence $\{s_j\}$ with $\lim_{j \to \infty} s_j = \infty$ such that $\lim_{j \to \infty} \rho_c'(s_j) = 0$. This proves (2.2.10). If $\rho_c(s) \geq 2s - L$ then $\limsup_{s \to \infty} \rho_c'(s) = 2$. In particular, there is a monotone divergent sequence $\{s_j\}$ with $\lim_{j \to \infty} s_j = \infty$ such that $\lim_{j \to \infty}(\alpha(s_j) + \beta(s_j)) = 0$. This proves (2.2.11). □

Lemma 2.2.3 (see Proposition 3.4.3, [93]). *Let* $U \subset M$ *be a half-cylinder whose boundary is a piecewise-smooth circle. If* $c(U)$ *exists then*

$$c(U) + \lambda(\partial U) \leq \pi \chi(\partial U) = 0.$$

Proof. Let $\tilde{U}$ be a half-plane obtained by cutting open U along a geodesic ray $\gamma : [0, \infty) \to U$ from ∂U, and let $\pi : \tilde{U} \to U$ be the Riemannian covering projection. Let $\tilde{\gamma}_1, \tilde{\gamma}_2 : [0, \infty) \to \partial\tilde{U}$ be the lifted images of γ. Then $\tilde{\gamma}_1[0, \infty) \cup \pi^{-1}(\partial U) \cup \tilde{\gamma}_2[0, \infty)$ forms the boundary of $\tilde{U}$ and $\lambda(\partial\tilde{U})$ is finite. Lemma 2.2.2 then implies that $c(\tilde{U}) + \lambda(\partial\tilde{U}) \leq \pi$. If α and β are the inner angles of $\tilde{U}$ at the corners $\tilde{\gamma}_1(0)$ and $\tilde{\gamma}_2(0)$, we can see from the construction of $\tilde{U}$ that

$$\lambda(\partial\tilde{U}) - \lambda(\partial U) = (\pi - \alpha) + (\pi - \beta) - (\pi - (\alpha + \beta)) = \pi.$$

This proves Lemma 2.2.3.   □

Example 2.2.1. Let $M \subset \mathbf{R}^3$ be a rotation surface generated by a parabola, as in Example 2.1.3, and let $U := \{p \in M; r(p) \geq 1\}$ and $D := \{p \in M; r(p) \leq 1\}$. The parallel r-circle $S(r)$ of M has a positive constant geodesic curvature κ_r such that $\lim_{r \to \infty} \kappa_r = 0$ and has length $2\pi r$. Since $c(M) = 2\pi$, we observe that $\lim_{r \to \infty} 2\pi r \kappa_r = 0$. The universal Riemannian covering H of U has the property that $c(H) = -\lambda(\partial H) = \infty$. Then the limit of the left-hand side of (2.2.9) as $i \to \infty$ depends on the choice of $\{A_i\}$. Let $\alpha > \pi$ be an arbitrary fixed number and $\{\theta_i\}$ a fixed monotone increasing sequence of positive numbers such that $\lim_{i \to \infty} \theta_i = \infty$. We then choose for these constants a sequence $\{r_i\}$ of increasing positive numbers such that the geodesic curvature of $S(r_i)$ is $(\alpha - \pi)/(2r_i\theta_i)$. Let $C_i \subset H$ for every i be a compact set such that $C_i := \{p \in H; 1 \leq r(p) \leq r_i, -\theta_i \leq \theta(p) \leq \theta_i\}$ and set $\gamma_i := \partial C_i \setminus \partial H$. Then $\{C_i\}$ is strictly increasing and $\bigcup_i C_i = H$. Moreover, γ_i for every i is convex with respect to C_i and $\alpha_i = \beta_i = \pi/2$. Therefore $\lambda(\gamma_i) = \alpha$ holds for all i. Thus the limit on the left-hand side of (2.2.9) does not exist.

Example 2.2.2. Let $M \subset \mathbf{R}^3$ be a two-sheeted hyperboloid as in Example 2.1.4. Then every parallel r-circle $S(r)$ has a positive constant geodesic curvature κ_r and κ_r satisfies

$$\lim_{r \to \infty} \kappa_r = 0 \qquad \text{and} \qquad \lim_{r \to \infty} 2\pi r \kappa_r = 2\pi \sin\theta.$$

If $U := \{p \in M; r(p) \geq 1\}$ and if H is the universal Riemannian covering of U then $c(H) = -\lambda(\partial H) = \infty$. If D_i is a monotone increasing sequence of locally convex disk domains exhausting H, then $\lim_{i \to \infty} \lambda(\partial D_i) = -\infty$. Thus we have $\lim_{i \to \infty}(c(D_i) + \lambda(\partial D_i)) = \infty$. We exclude the case where

$c(H) = -\lambda(\partial H) = \infty$, for the compactification discussed later is not obtained in general.

Proof of Theorem 2.2.1. Let $\Omega \subset M$ be a core of M such that $M \setminus \Omega$ consists of ℓ tubes and $k - \ell$ half-planes. Let $U_1, \ldots, U_\ell$ be all the tubes and $H_{\ell+1}, \ldots, H_k$ all the half-planes of $M \setminus \Omega$. We may assume without loss of generality that Ω is chosen such that if c_j for each $j = \ell + 1, \ldots, k$ is the component of $\partial\Omega$ touching ∂H_j then the two corners $c_j \cap \partial H_j \cap \partial M$ of c_j are smooth points of ∂M. If α_j and β_j are the inner angles at these corners with respect to Ω then the total geodesic curvature of c_j with respect to Ω satisfies

$$\lambda(c_j) = \lambda(c_j \cap \partial M) + (\pi - \alpha_j) + (\pi - \beta_j) + \lambda(c_j \cap \partial H_j),$$

where **e** is taken to be inward pointing with respect to Ω. Thus we get

$$\lambda(\partial\Omega) = \sum_{i=1}^{\ell} -\lambda(\partial U_i) + \sum_{j=\ell+1}^{k} (\lambda(c_j \cap \partial M) + \lambda(c_j \cap \partial H_j) + 2\pi - \alpha_j - \beta_j).$$

The Gauss–Bonnet theorem 1.9.1 and Lemmas 2.2.2 and 2.2.3 imply that

$$c(\Omega) + \lambda(\partial\Omega) = 2\pi \chi(M),$$

$$\sum_{i=1}^{\ell} (c(U_i) + \lambda(\partial U_i)) \leq 0,$$

$$\sum_{j=\ell+1}^{k} (c(H_j) + \lambda(\partial H_j)) \leq (k - \ell)\pi,$$

where

$$\lambda(\partial H_j) = -\lambda(c_j \cap \partial H_j) + \lambda(\partial H_j \cap \partial M) + \alpha_j + \beta_j.$$

Thus the proof of Theorem 2.2.1 is complete. $\qquad\square$

Remark 2.2.1. We note that, in Theorem 2.2.1, $2\pi \chi(M) - \{c(M) + \lambda(\partial M)\}$ is independent of a change in the Riemannian metric on any compact set. Let M be as in Theorem 2.2.1. By eliminating all the compact components of ∂M, we can construct from M a new surface M_1 such that M_1 coincides with M outside a compact set. Namely, ∂M_1 has $k - \ell$ components each of which is a divergent curve and M_1 has ℓ tubes. Such an M_1 is obtained by pasting a disk domain on-to each component of ∂M_0. Let g_1 be a complete Riemannian metric on M_1 such that $g = g_1$ on $M \setminus C_i$ for some $i \geq 1$. If follows from the first sentence of this remark that

$$2\pi \chi(M) - (c(M) + \lambda(\partial M)) = 2\pi \chi(M_1) - (c(M_1) + \lambda(\partial M_1))$$
$$\geq \pi \chi(\partial M_1), \qquad (2.2.12)$$

where $\chi(\partial M_1) = \chi(\partial M)$ is the number of (unbounded) components of ∂M_1. We see from this fact that we need only to discuss the noncompact components of ∂M.

Theorem 2.2.1 has many consequences, as obtained in [19] and [20]. They are stated as follows.

Corollary 2.2.2 (compare Satz 7, [19] and Theorem 11, [39]). *In addition to the assumptions in Theorem 2.2.1, if every half-plane H of $M \setminus \Omega$ has the property that $\rho_H : [0, \infty) \to \mathbf{R}^+$ is bounded above then*

$$c(M) + \lambda(\partial M) = 2\pi \chi(M),$$

where every tube is cut open along a ray from Ω lying with in it.

Corollary 2.2.3 (Satz 8, [19]). *If a complete Riemannian 2-manifold without boundary has everywhere positive Gaussian curvature then M is diffeomorphic to a sphere, a real projective plane or a plane.*

Corollary 2.2.4 (Satz 10, [19]). *If M is homeomorphic to a plane (such an M is called a* Riemannian plane*) and if $G > 0$ everywhere then M admits no closed geodesic (either with self-intersection or without).*

Corollary 2.2.5 (Satz 5, [20]). *If a Riemannian plane M admits a straight line and if $c(M)$ exists then*

$$c(M) \leq 0.$$

In particular, if $G \geq 0$ everywhere on M then M is isometric to a flat plane $\mathbf{R}^2$.

It was pointed out by Huber that the Gauss–Bonnet theorem holds for complete noncompact Riemannian 2-manifolds with finite total area.

Theorem 2.2.3 (see Theorem 12, [39] and Corollary, [82]). *Let M be a finitely connected complete noncompact Riemannian 2-manifold with finite total area. If M admits a total curvature then*

$$c(M) = 2\pi \chi(M).$$

Theorem 2.2.3 is given later by a Bonnesen-type isoperimetric inequality for sufficiently large metric balls (see Chapter 5).

Exercise 2.2.1. Prove Corollaries 2.2.2–2.2.5.

Generalizations to higher dimensions for Corollary 2.2.3 were obtained by Gromoll and Meyer in [**30**]. They proved that a complete noncompact Riemannian n-manifold without boundary of everywhere positive sectional curvature is diffeomorphic to $\mathbf{R}^n$. Also, Cheeger and Gromoll proved in [**17**] that a complete noncompact Riemannian n-manifold without boundary of nonnegative sectional curvature admits a compact totally geodesic submanifold S such that M is diffeomorphic to the normal bundle over S in M.

Corollary 2.2.5 was generalized by Toponogov in [**103**] as follows. If a complete Riemannian n-manifold without boundary of nonnegative sectional curvature admits k independent straight lines then M splits isometrically into a Riemannian product $N \times \mathbf{R}^k$, where N is a totally geodesic submanifold of nonnegative sectional curvature and admits no straight line. The Alexandrov–Toponogov triangle comparison theorem plays an important role in the proofs of these generalizations.

For the proof of Theorem 2.2.2 we need the following lemma on the existence of Morse exhaustion functions on noncompact manifolds.

Lemma 2.2.4. *Let M be an n-dimensional connected and noncompact manifold without boundary. There exists a Morse function $f : M \to \mathbf{R}$ such that $f^{-1}(-\infty, a]$ for every $a \in \mathbf{R}$ is compact and such that if $p_0, \ldots, p_m, \ldots,$ are all the critical points of f then $f(p_i) = i$ for all i.*

Proof. By means of the Whitney embedding theorem we can find a proper embedding $E : M \to \mathbf{R}^{2n+1}$ such that $E(M) \subset V := \{q \in \mathbf{R}^{2n+1}; \sum_{i=1}^{2n}(q^i)^2 < 1, q^{2n+1} \geq 0\}$. Let $v := (0, \ldots, 0, 1)$ and choose a monotone divergent sequence $\{t_j\}$ with $\lim_{j \to \infty} t_j = \infty$ such that t_j for every j is a regular value of the height function in v. Let $V_j := \{q \in V; t_{j-1} \leq q^{2n+1} \leq t_j\}$. The Sard theorem for the Gauss normal map of E implies that if $M_j \subset M$ is a submanifold with $E(M_j) = E(M) \cap V_j$ then there is a sequence $\{v_j\}$ of unit vectors for which each v_j lies sufficiently close to v that the height function h_j for v_j has no degenerate critical points on M_j. Clearly the gradient vector field of h_j is transverse to ∂M_j, and hence there is a sufficiently small positive η_j for every j such that h_j has no critical point on $O_j := \{p \in E(M); p^{2n+1} \in (t_{j-1} - \eta_j, t_{j-1} + \eta_j) \cup (t_j - \eta_j, t_j + \eta_j)\}$. Setting $W_j := V_j \cup O_j$ and $Y_j := V_j \setminus \overline{O}_j$, we take a partition of unity $\{\varphi_j\}$ subordinate to $\{W_j\}$ such that $\varphi_j = 1$ on Y_j and $\mathrm{supp}\,\varphi_j \subset W_j$. Then $h := \sum_{j=1}^{\infty} \varphi_j h_j$ is a well-defined

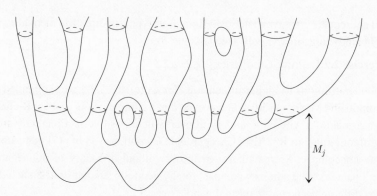

Figure 2.2.1 The submanifold M_j comprises the surfaces within the vertical region shown.

Morse function having no critical points on $\bigcup_{j=1}^{\infty} O_j$. It is not hard to construct the desired f from h. This proves Lemma 2.2.4.   □

Proof of Theorem 2.2.2. Let $f : M \to \mathbf{R}$ be a Morse exhaustion function as obtained in the previous lemma, and let $\{t_j\}$ be a monotone increasing sequence of regular values of f such that $\lim_{j \to \infty} t_j = \infty$. The sublevel set $\hat{M}_j :=$ $f^{-1}(-\infty, t_j]$ for every j has as its boundary a finite union of circles. Then $f^{-1}[t_j, \infty)$ consists of a finite union of compact surfaces, tubes and noncompact surfaces that are not tubes. Let M_j be the union of $\hat{M}_j$ and all the compact surfaces and tubes in $f^{-1}[t_j, \infty)$ (see Figure 2.2.1). We observe that M_j for every j is a connected, finitely connected, noncompact (or compact) surface with compact boundary. We then choose a subsequence $\{M_k\}$ of $\{M_j\}$ with the following properties (see Figure 2.2.1):

(1) $\{M_k\}$ is strictly increasing and $\bigcup_{k=1}^{\infty} M_k = M$;
(2) $\{\chi(M_k)\}$ is strictly decreasing;
(3) ∂M_k for every k is a finite union of circles.

For an arbitrary fixed $k > 1$, we can replace ∂M_k by simple closed geodesics as follows. Let $c_{k,1}, \ldots, c_{k,m}$ be since redundant the simple closed curves in ∂M_k. Since each $c_{k,i}$ is the boundary of an unbounded component of $f^{-1}[t_k, \infty)$ that is not a tube, there exists for each $i = 1, \ldots, m$ a closed nonnull homotopic curve $b_i \subset M$ such that if c is a closed curve freely homotopic to $c_{k,i}$ then $b_i \cap c \neq \emptyset$. A length-decreasing deformation can be applied to each $c_{k,i}$ to obtain a simple closed geodesic $\sigma_{k,i}$, whose length attains the minimum of all closed curves freely homotopic to $c_{k,i}$. Thus the boundary curves of M_k can be replaced by simple closed geodesics $\sigma_{k,1}, \ldots, \sigma_{k,m}$, and hence we may consider that $\partial M_k = \sigma_{k,1} \cup \cdots \cup \sigma_{k,m}$. Now apply Theorem 2.2.1 to each M_k to get $c(M_k) \leq 2\pi \chi(M_k)$. This proves Theorem 2.2.2.   □

Exercise 2.2.2. Let M be a connected, infinitely connected, complete Riemannian 2-manifold with nonempty boundary. If $c(M)$ and $\lambda(\partial M)$ exist such that $c(M) = -\lambda(\partial M) = \infty$ does not hold then does the Huber theorem hold?

2.3 Special properties of geodesics on Riemannian planes

We shall introduce some interesting results on the global behavior of geodesics that were proved in [20]. Throughout this section let M be a Riemannian plane. Namely, M is called a *Riemannian plane* iff it is homeomorphic to $\mathbf{R}^2$. Poles, simple points and special properties of geodesics on M will be discussed.

A geodesic is called a *complete geodesic* iff it is defined over the whole real line. A geodesic $\gamma : [a, b] \to M$ is called a *geodesic loop* iff $\gamma(a) = \gamma(b)$ and $\gamma|[a, b)$ is injective. The point $\gamma(a) = \gamma(b)$ is called the base point of the loop.

Definition 2.3.1. A point $p \in M$ is called a *simple point* iff it is not a base point of any geodesic loop. Define the following sets:

$S_0 := \{p \in M; p \text{ is a simple point of } M\}$;

$S_1 := \{q \in M; \text{ no geodesic loop passes through } q\}$;

$\mathcal{P} := \{q \in M; q \text{ is not a corner of any nontrivial geodesic biangle}\}$.

From the definition we observe that

$$\mathcal{P} \subset S_1 \subset S_0.$$

It is proved in Theorem 2.3.7 below that every point q on $\mathcal{P}$ has the special property that $\exp_q : T_q M \to M$ is injective. Such a point is called a *pole* of M. If a complete geodesic has a self-intersection then it contains a geodesic loop. Let $\gamma : [a, b] \to M$ be a geodesic loop, $D \subset M$ the disk domain bounded by $\gamma[a, b]$ and ω the inner angle of D at the base point $\gamma(a) = \gamma(b)$. Then the Gauss–Bonnet theorem implies that

$$c(D) = \pi + \omega$$

and hence that

$$\int_M G_+ \, dM > \pi.$$

Thus M does not admit any geodesic loop, or equivalently no complete geodesic of M has a self-intersection, if

$$\int_M G_+ dM \leq \pi.$$

Remark 2.3.1. Does there exist a geodesic loop on M if

$$\int_M G_+ dM > \pi?$$

No such loop does exist. In fact, a counter-example is constructed as follows. Remove a disk from a plane and place a *low hill D* to cover the hole such that $\int_D G_+ dM = \alpha > 0$, where α is a sufficiently small positive number. Place m such low hills, disjoint on the plane M, in such a way that $m\alpha > \pi$ and the distance between any two hills is sufficiently large. Then there exists no geodesic loop on M. In this case it has zero total curvature.

An example of the proof technique used often in this section is as follows.

Let $A \subset M$ be a closed disk bounded by a geodesic polygon. For a point $q \in M \setminus A$ and for a point $p \in \partial A$ let pq be a geodesic segment such that it does not meet int A. The set $\hat{A} := A \cup pq$ has as its fine boundary $pq \cup \partial A$. A is called *locally concave* iff all the inner angles of A are not less than π; also, $\hat{A}$ is called *locally concave* iff the inner angles of all the corners of its fine boundary are not less than π (see Figure 2.3.1). If $\hat{A}$ is locally concave then a length-decreasing deformation proceeds to its fine boundary. We then obtain a geodesic loop at p that has the minimum length among all closed curves in $M \setminus$ int A with base point at p and freely homotopic to $\partial(\text{int } \hat{A})$ in $M \setminus$ int A. If A is locally concave then we find for every point $p \notin A$ a geodesic loop at p that has the minimum length among all the loops in $M \setminus$ int A with base point at p and freely homotopic to ∂A.

The following theorem shows that $\mathcal{S}_0$ is a bounded nonempty set on a Riemannian plane with total curvature greater than π.

Theorem 2.3.1 (see Satz 6, [20]). *Assume that the total curvature of M is greater than π. For every bounded set $A \subset M$ there exists a bounded set $N \subset M$ containing A such that all points of $M \setminus N$ are the base point of some geodesic loop and the disk domain bounded by such a geodesic loop contains A in its interior.*

The following proposition is useful for the proof of Theorem 2.3.1.

Proposition 2.3.1. *Let $A \subset M$ be a closed disk in a Riemannian plane M such that ∂A consists of a geodesic polygon. Let c_p for every $p \in M \setminus$ int A be a*

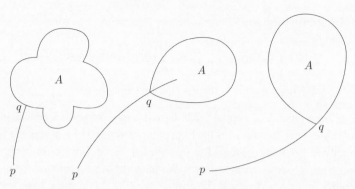

Figure 2.3.1 Each compact set with a fine boundary is locally concave.

closed curve with base point at p that has the minimum length among all the curves in M \ int A with base point at p and freely homotopic to ∂A. Then we have the following.

(1) *The function $p \to L(c_p)$ is Lipschitz continuous with Lipschitz constant 2.*
(2) *Let $\sigma : [0, \ell] \to M \setminus$ int A be a minimizing geodesic from A to a point $q \in M \setminus A$ and let $f_\sigma : [0, \ell] \to \mathbf{R}$ be defined by $f_\sigma(t) := L(c_{\sigma(t)})$. Then f is Lipschitz continuous with Lipschitz constant 2.*
(3) *Let $\gamma : [0, \infty) \to M \setminus$ int A be a ray from A such that $c_{\gamma(t)} \cap A \neq \emptyset$ for all sufficiently large t. Then there exists for any $\varepsilon > 0$ a monotone divergent sequence $\{t_j\}$ such that the angle at $\gamma(t_j)$ of the curve $c_{\gamma(t_j)}$ is less than ε.*

Proof. If pq for points $p, q \in M$ is a minimizing geodesic segment in $M \setminus A$ from p to q, then $pq \cup c_q \cup qp$ is freely homotopic to c_q in $M \setminus A$ and $L(c_p) \leq L(c_q) + 2d(p, q)$. This proves (1).

The proof of (2) is straightforward from (1).

We argue for the proof of (3) by deriving a contradiction. Let $\theta(t)$ be the angle at $\gamma(t)$ of the disk domain D_t bounded by $c_{\gamma(t)}$. By construction D_t contains A. Suppose (3) is not true. Then there exists a small $\varepsilon_0 > 0$ and a large number t_0 such that $\theta(t) \geq \varepsilon_0$ for all $t > t_0$.

The triangle inequality implies the asymptotic behavior of f_γ at infinity. Namely,

$$\left| f_\gamma(t) - 2t \right| \leq L(\partial A) \qquad \text{for all } t > 0.$$

Clearly D_t contains $\gamma[0, t)$ in its interior. Thus the angle $\theta(t)$ is divided by $-\dot{\gamma}(t)$ into $\alpha(t)$ and $\beta(t)$. If f_γ is differentiable at t then $f_\gamma'(t) = \cos\alpha(t) + \cos\beta(t) \leq$

$2\cos(\theta(t)/2)$. By means of (2) we have

$$f_\gamma(t_1) - f_\gamma(t_0) \leq \int_{t_0}^{t_1} 2\cos\frac{\theta(t)}{2}\,dt \leq 2(t_1 - t_0)\cos\frac{\varepsilon_0}{2}$$

for all $t_1 > t_0$. This contradicts the asymptotic behavior of f_γ.   □

Proof of Theorem 2.3.1. Suppose that there is a monotone divergent sequence $\{p_j\}$ of points such that no p_j is the base point of geodesic loops that are freely homotopic to ∂A in $M \setminus \text{int } A$. Let $\gamma_j : [0, \ell_j] \to M \setminus \text{int } A$ for every j be a minimizing geodesic from A to p_j. We then choose a subsequence of $\{\gamma_j\}$ converging to a ray γ from A. Because $p_j = \gamma_j(\ell_j)$ is not the base point of geodesic loops freely homotopic to ∂A in $M \setminus \text{int } A$, we can see that $c_{p_j} \cap A \neq \emptyset$. Clearly every point x on γ_j has the same property as p_j, i.e., $c_x \cap A \neq \emptyset$. This fact implies $c_{\gamma(t)} \cap A \neq \emptyset$, since $\gamma(t)$ is the limit of $\{\gamma_j(t)\}$.

Setting $a := c(M) - \pi > 0$, we can choose a sufficiently large disk domain B bounded by a geodesic polygon such that $B \supset A$,

$$c(B) > \pi + \frac{a}{2} \quad \text{and} \quad \int_{M \setminus B} |G|\,dM < \frac{a}{4}.$$

It follows from what we have supposed that there is a large number $t(B)$ such that p_j for every $j > t(B)$ is not the base point of any geodesic loop that is freely homotopic to ∂B in $M \setminus \text{int } B$. Therefore (3) in Proposition 2.3.1 applies to a ray $\tau : [0, \infty) \to M \setminus \text{int } B$ from B. We then obtain a convex disk domain D_t bounded by $c_{\tau(t)}$, for our choice of B and for t sufficiently large that $\theta(t) < a/4$. Then the convexity of D_t implies that $c(D_t) + (\pi - \theta(t)) \leq 2\pi$. However, it also follows from our choice of B and from $D_t \supset B$ that

$$c(D_t) = c(B) + c(D_t \setminus B) > c(B) - \int_{M \setminus B} |G|\,dM > c(B) - \frac{a}{4}.$$

Thus a contradiction is derived.   □

Example 2.3.1. Consider the rotation surface of the parabola defined in Example 2.1.3. If $\gamma(s) = (r(s), \theta(s))$ is a unit-speed geodesic then

$$r''(s) + \frac{k^2 r}{1 + k^2 r^2}(r'(s))^2 - \frac{r}{1 + k^2 r^2}(\theta'(s))^2 = 0,$$

$$\theta''(s) + \frac{2}{r}\theta'(s)r'(s) = 0.$$

For an arbitrary fixed positive number r_0, the initial conditions for γ are set as

$$\gamma(0) = (r_0, 0), \qquad \dot{\gamma}(0) = \frac{\partial}{\partial\theta}(r_0, 0)/\sqrt{g_{22}(r_0, 0)}.$$

The modified Clairaut theorem (Theorem 1.3.2) then implies that

$$\theta'(s)\,r^2(s) = r_0 \cos \angle \left(\dot{\gamma}(0), \frac{\partial}{\partial \theta}(r_0, 0) \right) = r_0.$$

Since γ is parameterized by arc length,

$$\|\dot{\gamma}(s)\|^2 = (r')^2(1 + k^2 r^2) + (\theta')^2 r^2 = 1,$$
$$\mathrm{d}\theta^2\, r^4 = \mathrm{d}s^2\, r_0^2.$$

Eliminating $\mathrm{d}s^2$ from the above relations, we have for $r > r_0$

$$\left(\frac{\mathrm{d}\theta}{\mathrm{d}r} \right)^2 = \frac{r_0^2(1 + k^2 r^2)}{r^2\left(r^2 - r_0^2\right)} > 0.$$

We observe that both $\theta(s)$ and $r(s)$ are strictly increasing in $s > 0$. Thus we have along γ

$$\frac{\mathrm{d}\theta}{\mathrm{d}r} = \frac{r_0 \sqrt{1 + k^2 r^2}}{r \sqrt{r^2 - r_0^2}} > 0 \qquad \text{for } s > 0.$$

By using the inequality

$$k^2 - \frac{1 + k^2 r^2}{r^2 - r_0^2} < 0,$$

we obtain

$$\frac{\mathrm{d}\theta}{\mathrm{d}r} > k \frac{r_0}{r}.$$

Integration gives

$$\theta(r_1) - \theta(r_0) > \int_{r_0}^{r_1} k \frac{r_0}{r}\, \mathrm{d}r = k \log \frac{r_1}{r_0},$$

and hence

$$\lim_{r_1 \to \infty} \theta(r_1) = +\infty.$$

Summing up, we see that the following are true.

(1) The image of γ has infinitely many self-intersections.
(2) There exists a subarc $\gamma|_{[-t_0, t_0]}$ that is a geodesic loop with the property that neither of the subarcs $\gamma|_{(t_0, \infty)}$ and $\gamma|_{(-\infty, t_0)}$ has a self-intersection or meets the loop $\gamma|_{[-t_0, t_0]}$.

Moreover, since $\theta(s)$ is monotone increasing in $s > 0$, there exists a unique number $R(r_0) > 0$ such that $\theta(R(r_0)) = \pi$ or, equivalently, $\gamma(R(r_0)) = (R(r_0), \pi)$. From $\sqrt{1 + k^2 r^2} < 1 + kr$, we get

$$\frac{d\theta}{dr} < \frac{r_0(1 + kr)}{r\sqrt{r^2 - r_0^2}} = \frac{kr}{\sqrt{r^2 - r_0^2}} + \frac{r_0}{r\sqrt{r^2 - r_0^2}}.$$

We then have

$$\pi = \theta(R(r_0)) < kr_0 \log \frac{2R(r_0)}{r_0} + \tan^{-1} \frac{\sqrt{R(r_0)^2 - r_0^2}}{r_0}.$$

Because $R(r_0)$ is bounded below by the convexity radius around the pole of the paraboloid, we see that

$$\lim_{r_0 \to 0} \tan^{-1} \frac{\sqrt{R(r_0)^2 - r_0^2}}{r_0} = \frac{\pi}{2}.$$

Also, from

$$\liminf_{r_0 \to 0} (kr_0 \log 2R(r_0)) \geq \frac{\pi}{2}$$

we see that $\liminf_{r_0 \to 0} R(r_0) = \infty$.

Note that $\inf_{r_0 > 0} R(r_0)$ is attained at some $r_* > 0$. Let γ_* be the geodesic with initial conditions

$$\gamma_*(0) = (r_*, 0), \qquad \dot{\gamma}_*(0) = \frac{\partial}{\partial \theta}(r_*, 0)/\sqrt{g_{22(r_*, 0)}}.$$

Then the point $\gamma_*(-R(r_*)) = \gamma_*(R(r_*))$ is conjugate to itself along γ_* (see Figure 2.3.2). Note also that $\mathcal{S}_0 = \{(r, \theta); r < R(r_*)\}$ and that the set $\mathcal{S}_1 = \mathcal{P}$ is simply the origin of $\mathbf{R}^3$.

Example 2.3.2. Consider one component of the two-sheeted hyperboloid defined in Example 2.1.4. Let $\gamma(s) = (r(s), \theta(s))$ be a unit-speed geodesic with initial conditions

$$\gamma(0) = (r_0, 0), \qquad \dot{\gamma}(0) = \frac{\partial}{\partial \theta}(r_0, 0)/\sqrt{g_{22}(r_0, 0)}.$$

We then have

$$r''(s) + \Gamma_{11}^1 (r'(s))^2 + \Gamma_{22}^1 (\theta'(s))^2 = 0,$$
$$\theta''(s) + \frac{2}{r} r'(s) \theta'(s) = 0.$$

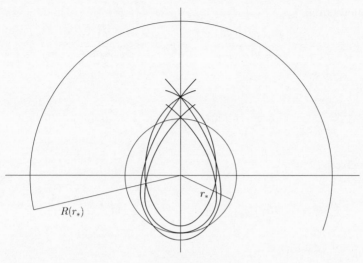

Figure 2.3.2

By the same method as in the previous example, we get for $r > r_0$

$$\frac{d\theta}{dr} = \frac{r_0}{r} \sqrt{\frac{k^2 r^2 (1 + k^2) + 1}{(r^2 - r_0^2)(k^2 r^2 + 1)}}.$$

Using the inequality

$$k^2 + 1 - \frac{k^2 r^2 (k^2 + 1) + 1}{k^2 r^2 + 1} > 0,$$

we get

$$\frac{d\theta}{dr} < \frac{r_0 \sqrt{k^2 + 1}}{r \sqrt{r^2 - r_0^2}}$$

and hence

$$\theta(r_1) - \theta(r_0) < \int_{r_0}^{r_1} r_0 \sqrt{k^2 + 1} \frac{dr}{r \sqrt{r^2 - r_0^2}} = \sqrt{k^2 + 1} \tan^{-1} \frac{\sqrt{r_1^2 - r_0^2}}{r_0}.$$

Thus we have

$$\lim_{r_1 \to \infty} \theta(r_1) \le \frac{\sqrt{k^2 + 1}}{2} \pi.$$

It follows from

$$\frac{k^2 r^2(1+k^2)+1}{k^2 r^2+1} \geq k^2+1-\frac{k^2}{k^2 r_0^2+1}$$

that

$$\frac{d\theta}{dr} \geq \sqrt{k^2+1-\frac{k^2}{k^2 r_0^2+1}} \, \frac{1}{r\sqrt{r^2-r_0^2}}.$$

This gives

$$\theta(r_1)-\theta(r_0) \geq \sqrt{k^2+1-\frac{k^2}{k^2 r_0^2+1}} \, \tan^{-1}\frac{\sqrt{r_1^2-r_0^2}}{r_0},$$

and therefore we have

$$\lim_{r_1\to\infty} \theta(r_1) \geq \frac{\pi}{2}\sqrt{k^2+1-\frac{k^2}{k^2 r_0^2+1}}.$$

Summing up, we see that the following are true.

(1) If $k \leq \sqrt{3}$ (or equivalently if $c(M) \leq \pi$) then γ has no self-intersection for $\lim_{r_1\to\infty} \theta(r_1) \leq \pi$.
(2) If $k > \sqrt{3}$ (or equivalently if $c(M) > \pi$) then γ has at most finitely many self-intersections. Here the number of self-intersections of γ is at most $(k^2+1)/2 = \pi/(2\pi - c(M))$ if r is sufficiently far from the origin.
(3) S_1 contains a neighborhood around the origin.

The behavior of the complete geodesics in Examples 2.3.1 and 2.3.2 holds for Riemannian planes of positive curvature. This is the simplest case. More general properties of complete geodesics on Riemannian planes are discussed later, in Chapter 8.

Theorem 2.3.2 (see Sätze 7, 8, [20]). *Assume that M has positive Gaussian curvature and $c(M) > \pi$. If a complete geodesic γ has a self-intersection then the following are true:*

(1) *$\gamma(\mathbf{R})$ contains a unique geodesic loop $\gamma[a,b]$;*
(2) *each of the subarcs $\gamma|_{[b,\infty)}$ and $\gamma|_{(-\infty,a]}$ of γ is divergent, has no self-intersection and does not intersect the loop $\gamma[a,b]$ except at the base point.*

Furthermore, if $\sigma : [0,\infty) \to M$ has a self-intersection then:

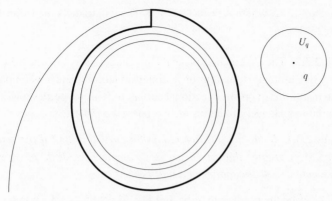

Figure 2.3.3

(3) $\sigma[0, \infty)$ *contains a unique geodesic loop, say* $\sigma[a, b]$, *such that* $\sigma[0, a) \cup \sigma(b, \infty)$ *does not meet the loop and* $\sigma[b, \infty)$ *has no self-intersection;*

(4) $\sigma[b, \infty)$ *is divergent;*

(5) $\sigma[0, \infty)$ *has at most finitely many self-intersections.*

Proof. We choose a subarc $\gamma|_{[a,b]}$ of γ that is a geodesic loop.

We will prove that $\gamma|_{[b,\infty)}$ has no self-intersection and does not meet the geodesic loop except at $\gamma(b)$. Suppose that $\gamma|_{[b,\infty)}$ has a self-intersection. We can then find a subarc $\gamma|_{[c,d]}$ with $b \leq c < d < \infty$ that is a geodesic loop also. Let D, D' be the disk domains bounded by $\gamma[a, b]$, $\gamma[c, d]$ respectively. Three cases occur.

First of all, if $D \cap D' = \emptyset$ then $c(M) > c(D) + c(D') > 2\pi$, a contradiction.

If $D \cap D' \neq \emptyset$ then we can find parameters $b' \in [a, b]$ and $d' \in [c, d]$ such that $\gamma(b') = \gamma(d') \in \partial D \cap \partial D'$. Thus $\gamma|_{[b',d']}$ is a geodesic loop whose inner angle at its base point is greater than π, a contradiction.

Suppose finally that one disk is contained entirely in the other. Let $D \subset D'$. Then the angle at $\gamma(c) = \gamma(d)$ of D' is greater than π, and hence $c(D') > 2\pi$, a contradiction. The uniqueness of a geodesic loop in $\gamma[0, \infty)$ is now clear. This reasoning also proves the injectivity of $\gamma|_{[b,\infty)}$ as well as of $\sigma|_{[b,\infty)}$.

For the rest of the proof of Theorem 2.3.2, we need only to verify that every geodesic $\gamma : [0, \infty) \to M$ is divergent. Suppose γ that is not divergent. Then we can find a monotone divergent sequence $\{t_j\}$ such that $\{\dot{\gamma}(t_j)\}$ converges to a unit vector $u \in T_pM$ for some point $p \in M$. We now find a disk domain bounded by a subarc of γ and a geodesic of small length meeting the subarc almost orthogonally (see Figure 2.3.3). Fixing a point $q \notin D$ and a neighborhood $U_q \subset M \setminus D$, we can choose a small geodesic segment bounding D such that

$c(D) > 2\pi - \varepsilon/2$, where $\varepsilon := c(U_q) > 0$. This contradicts the Cohn-Vossen theorem 2.2.1. □

Remark 2.3.2. The divergence property of geodesics is also proved by the convexity of Busemann functions on n-dimensional complete noncompact Riemannian manifolds of positive sectional curvature. This property shows that the exponential map at every point on M is proper (see [30]).

Corollary 2.3.1. *If M has positive Gaussian curvature and if the total curvature of M is greater than π then every complete geodesic of M without self-intersection is the boundary of a half-plane.*

Proof. Let $\mathcal{N}$ be the north pole of $\mathbf{S}^2$ and $H : M \to \mathbf{S}^2 \setminus \{\mathcal{N}\}$ a homeomorphism. From Theorem 2.3.2(2), the image under H of the two-point compactification of a complete geodesic $\gamma : \mathbf{R} \to M$ is a simple closed curve passing through $\mathcal{N}$, and hence it divides $\mathbf{S}^2$ into two disk domains. This concludes the proof. □

Theorem 2.3.3 (see Satz 10, [20]). *Let M have positive Gaussian curvature. If γ_1, γ_2 are complete geodesics on M then they have at least one point of intersection.*

Proof. The proof is discussed into three cases.

Assume first of all that each of the γ_i has a self-intersection. Then Theorem 2.3.2(1) implies that γ_i for $i = 1, 2$ contains a unique geodesic loop that bounds a disk domain D_i. Clearly $c(D_i) > \pi$ and $c(M) \leq 2\pi$. Thus we have $D_1 \cap D_2 \neq \emptyset$.

Assume secondly that γ_1, but not γ_2, has a self-intersection. Then Corollary 2.3.1 implies that $\gamma_1(\mathbf{R})$ is the boundary of a half-plane of M. Let $H_1 \cup H_2 = M \setminus \gamma_1(\mathbf{R})$ be half-planes bounded by $\gamma_1(\mathbf{R})$, thus $\lambda(\partial H_i) = 0$ for $i = 1, 2$. Suppose $\gamma_1(\mathbf{R}) \cap \gamma_2(\mathbf{R}) = \emptyset$. Then a geodesic loop lying on $\gamma_2(\mathbf{R})$ is contained entirely in one of the half-planes. If H_1 contains the geodesic loop then $c(H_1) > \pi$, contradicting Corollary 2.2.1.

Assume finally that neither γ_1 nor γ_2 has a self-intersection; further, suppose that they do not intersect. Then $\gamma_i(\mathbf{R})$ bounds an open half-plane H_i in M such that the intersection $H := H_1 \cap H_2$ forms an open strip homeomorphic to $\{(x, y) \in \mathbf{R}^2; \ 0 < x < 1\}$. We now apply (2.2.1) in Theorem 2.2.1 to H and obtain $c(H) \leq 0$, a contradiction. □

Theorem 2.3.4 (see Satz 11, [20]). *Assume that M has positive Gaussian curvature. Then, passing through every point of M there exists a complete geodesic without self-intersection.*

Note that if M has a positive Gaussian curvature and if $c(M) \le \pi$ then no complete geodesic on M has a self-intersection. Thus we consider only the case where $c(M) > \pi$. We see from Theorem 2.3.1 that M admits geodesic loops. If a complete geodesic γ has no self-intersection then the technique developed in the proof of Theorems 2.3.2 and 2.3.3 implies that $\gamma(\mathbf{R})$ passes through a point on every fixed geodesic loop. Therefore we see that every complete geodesic without self-intersection passes through a point of the disk domain bounded by a geodesic loop. This phenomenon is easily seen in Examples 2.3.1 and 2.3.2.

Proof of Theorem 2.3.4. We discuss only the case where $c(M) > \pi$. Let $p \in M$ be an arbitrary fixed point and choose a complete geodesic $\gamma_0 : \mathbf{R} \to M$ with $\gamma_0(0) = p$. Assume that γ_0 has a self-intersection. Let $\mathfrak{C}_0 := \gamma_0[a, b]$ be the unique geodesic loop in $\gamma_0(\mathbf{R})$ and D_0 the disk domain bounded by $\mathfrak{C}_0$. Choose an orthonormal-frame field $\{e_1, e_2\}$ along $\mathfrak{C}_0$ such that $e_1(\gamma_0(t)) = \dot{\gamma}_0(t)$ and $e_2(\gamma_0(t))$ is inward pointing with respect to D_0 for $t \in [a, b]$. The orientation of (e_1, e_2) is coherent with that of M. The unit circle $\mathbf{S}_p(1) \subset T_p M$ has the arc length parameterization $\theta : [0, 2\pi] \to \mathbf{S}_p(1)$ such that $\theta(0) := \dot{\gamma}_0(0)$. Thus the unit vectors at p are identified with values in $[0, 2\pi)$. Let $\gamma_u : \mathbf{R} \to M$ for $u \in [0, 2\pi)$ be a complete geodesic with $\gamma_u(0) = p$ and $\dot{\gamma}_u(0) = u$ for $u \in [0, 2\pi)$.

Suppose that there is no complete geodesic without self-intersection through p. Then $\gamma_u : \mathbf{R} \to M$ for every $u \in [0, 2\pi)$ contains a unique geodesic loop $\mathfrak{C}_u$ bounding a disk domain D_u. The continuity property of geodesics in their initial conditions implies that there is a small positive number h with the property that $\{e_1, e_2\}$ can be chosen continuously along $\mathfrak{C}_u$ for all $u \in [0, h)$ in such a way that $e_1(\gamma_u(t)) = \dot{\gamma}_u(t)$ and $e_2(\gamma_u(t))$ is inward pointing with respect to D_u, while the orientation of (e_1, e_2) is coherent with that of M. However, it is impossible to choose such an orthonormal frame field continuously along γ_π! Let $\mathbf{A}$ be the set of all parameters $\{u \in [0, \pi)\}$ with the property that there is an orthonormal-frame field along $\mathfrak{C}_u$ such that $e_1 = \dot{\gamma}_u$, e_2 is inward pointing with respect to D_u and the orientation of (e_1, e_2) is coherent with that of M. Clearly $0 \in \mathbf{A}$ and hence $\mathbf{A} \neq \emptyset$. Moreover, $\mathbf{A}$ contains an interval $I = [0, \mathbf{a})$. It follows from what we have supposed that $\gamma_{\mathbf{a}}$ has a self-intersection. A continuity argument implies that

$$\mathfrak{C}_{\mathbf{a}} = \lim_{u \to \mathbf{a}} \mathfrak{C}_u, \qquad D_{\mathbf{a}} = \lim_{u \to \mathbf{a}} D_u.$$

Thus we see that $\mathbf{a} \in \mathbf{A}$ and, in particular, I is open. Therefore we get $\pi \in \mathbf{A}$, a contradiction. $\square$

The following theorem 2.3.5, shows the existence of a maximal geodesic loop at a point on M.

Theorem 2.3.5 (see Satz 12, [20]). *Assume that M has positive Gaussian curvature and that $c(M) > \pi$. Let $p \in M$ be the base point of a geodesic loop. Then there exists a geodesic loop $\mathfrak{E}_p$ with base point at p having the following property: if D_p is the disk domain bounded by $\mathfrak{E}_p$ then the closure $\bar{D}_p$ of D_p is totally convex. Namely, every geodesic segment joining any points on $\bar{D}_p$ is contained entirely in $\bar{D}_p$.*

Remark 2.3.3. A set A is called *totally convex* iff every geodesic segment joining two points in A is contained entirely in A. The geodesic loop $\mathfrak{E}_p$ is *maximal* in the following sense. If $\mathfrak{E}'$ is a geodesic loop with base point at p and if D' is the disk domain bounded by $\mathfrak{E}'$ then $D' \subset D_p$.

Proof. Let $\mathfrak{E}$ be a fixed geodesic loop with base point at p and D the disk domain bounded by $\mathfrak{E}$. If α is the inner angle at p of D, we observe that $\alpha < \pi$.

Fix as anticlockwise the orientation of $\mathbf{S}_p(1)$ and let $[a, b] := \mathbf{S}_p(D)$ be the tangent unit vectors to D at p. Then

$$\alpha = b - a, \qquad a < b < a + \pi.$$

Let $\{\mathfrak{E}_\lambda\}_{\lambda \in \Lambda}$ be the set of geodesic loops at p, where Λ is an index set. Let D_λ and α_λ for every $\lambda \in \Lambda$ be respectively the disk domain and its inner angle at p. Setting

$$[a_\lambda, b_\lambda] := \mathbf{S}_p(D_\lambda), \qquad \lambda \in \Lambda,$$

we observe that

$$\alpha_\lambda = b_\lambda - a_\lambda, \qquad a_\lambda < b_\lambda < a_\lambda + \pi,$$

and also (see Figure 2.3.4)

$$b - \pi < a_\lambda < b < a + \pi, \qquad b - \pi < a < b_\lambda < a + \pi.$$

Note that

$$D_\lambda \cap D_\mu \neq \emptyset, \qquad D_\lambda \cap D \neq \emptyset, \qquad \text{for all } \lambda, \mu \in \Lambda.$$

The proofs of $a_\lambda < b$ and $a < b_\lambda$ can be checked by supposing otherwise; then $D \cap D_\lambda = \emptyset$ (see Figure 2.3.5).

In order to find the maximum geodesic loop at p we set

$$a_0 := \inf_{\lambda \in \Lambda} a_\lambda, \qquad b_0 := \sup_{\lambda \in \Lambda} b_\lambda.$$

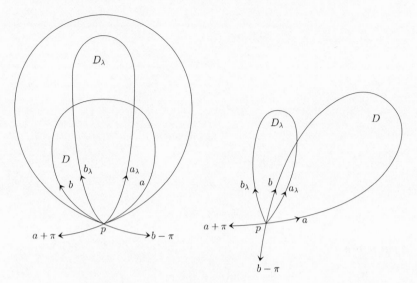

Figure 2.3.4

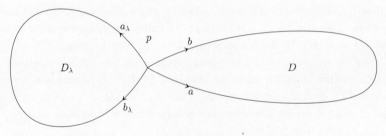

Figure 2.3.5

If a_0 is attained by some a_λ then $\mathfrak{C}_\lambda$ is the maximum geodesic loop at p. In fact, suppose that $b_\lambda < b_\mu$ for some $\mu \in \Lambda$. We then find, by using the short-cut principle, a geodesic loop $\mathfrak{C}'$ at p whose disk domain contains $D_\lambda \cup D_\mu$ (see Figure 2.3.4).

Thus we assume that

$$a_0 = \liminf_{\lambda \in \Lambda} a_\lambda.$$

We then have the following assertion.

Assertion 2.3.1. *If $a_0 = \liminf_{\lambda \in \Lambda} a_\lambda$ then there exists a geodesic loop $\mathfrak{C}_p$ at p that is tangent to a_0. Moreover, if b_0 is the other tangent vector to*

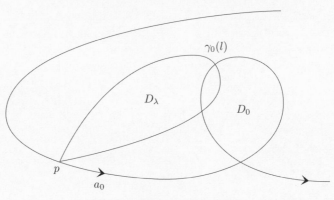

Figure 2.3.6

$\mathfrak{E}_p$ *at* p *then*

$$b_0 = \sup_{\lambda \in \Lambda} b_\lambda = \limsup_{\lambda \in \Lambda} b_\lambda.$$

Proof. Let $\gamma_0(t) := \exp_p t a_0$ be a complete geodesic. Suppose that γ_0 has no self-intersection. Then Corollary 2.3.1 implies that it divides M into half-planes. From $b_\lambda < a_0 + \pi$ for all $\lambda \in \Lambda$, there is a half-plane H containing D_λ such that $\partial H = \gamma_0(\mathbf{R})$. Then a contradiction is derived from Lemma 2.2.2.

Thus we see that γ_0 has a self-intersection. Theorem 2.3.2(1) implies that $\gamma_0(\mathbf{R})$ contains a unique geodesic loop, say $\mathfrak{E}_0$. Let D_0 be the disk domain bounded by $\mathfrak{E}_0$ and s its base point. Then $\mathfrak{E}_0 \cap \mathfrak{E}_\lambda \neq \emptyset$ for all $\lambda \in \Lambda$. In fact, if it is supposed otherwise then a contradiction is derived from $2\pi < c(D_0) + c(D_\lambda) < c(M)$.

We next assert that $p \in \mathfrak{E}_0$. Suppose $p \notin \mathfrak{E}_0$. Clearly $\mathfrak{E}_0 \cap \mathfrak{E}_\lambda \neq \emptyset$ for all $\lambda \in \Lambda$. Let $\ell > 0$ be such that $\gamma_0[0, \ell) \cap D_\lambda = \emptyset$ and $\gamma_0(\ell) \in \mathfrak{E}_\lambda$ for some $\lambda \in \Lambda$ (see Figure 2.3.6). We can then find a geodesic biangle consisting of $\gamma_0[0, \ell]$ and the subarc of $\mathfrak{E}_\lambda$ between p and $\gamma_0(\ell)$. Since the inner angle at $\gamma_0(\ell)$ of this biangle is greater than π, a length-decreasing deformation of this biangle with base point at p provides a geodesic loop $\mathfrak{E}_*$ with base point at p bounding the disk domain D_* that contains $D_\lambda \cup D_0$ in its interior. Let $[a_*, b_*] := \mathbf{S}_p(D_*)$. Then $a_* < a_0$, a contradiction to $a_0 = \inf_{\lambda \in \Lambda} a_\lambda$. If $\ell < 0$ then the disk domain D_* bounded by $\mathfrak{E}_*$ is obtained in the same manner. Then $b_* > a_0 + \pi$, a contradiction to the relation $b_\lambda \leq a_0 + \pi$ for all $\lambda \in \Lambda$. This proves $p \in \mathfrak{E}_0$. $\qquad \square$

We shall assert that $D_\lambda \cap D_0 \neq \emptyset$ for all $\lambda \in \Lambda$. Recall that

$$a_0 \leq a_\lambda, \qquad b_\lambda - \pi < a_0 \leq a_\lambda + \pi, \qquad \text{for all } \lambda \in \Lambda.$$

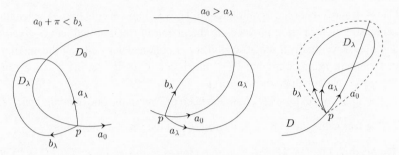

Figure 2.3.7

We then observe that positions of the two geodesic loops in which they have two points of intersection with $\mathfrak{E}_0$, as in Figure 2.3.7, do not occur. Also, there is no geodesic loop $\mathfrak{E}_\lambda$ having three points of intersection with $\mathfrak{E}_0$. In such a case a contradiction is derived from $a_\lambda + \pi < a_0$ or $a_0 < b_\lambda - \pi$.

Now suppose that D_λ for some $\lambda \in \Lambda$ is not contained entirely in D_0. We can then find a point $q \in \mathfrak{E}_0 \cap \mathfrak{E}_\lambda$ and subarcs of $\mathfrak{E}_0$ and $\mathfrak{E}_\lambda$ between p and q that form a geodesic biangle whose angle at q is greater than π. Thus the length-decreasing deformation of this geodesic biangle with base point at p provides a geodesic loop $\mathfrak{E}_*$ at p such that the disk domain D_* bounded by $\mathfrak{E}_*$ has the following property: if we set $[a_*, b_*] := \mathbf{S}_p(D_*)$ then $a_* < a_0$; this is a contradiction. Thus we have proved that all the geodesic loops at p are contained entirely in D_0, and so

$$D_\lambda \subset D_0 \qquad \text{for all } \lambda \in \Lambda.$$

Finally, let ℓ_λ be the length of $\mathfrak{E}_\lambda$. Suppose that there is a subsequence $\{\ell_i\}$ such that $\lim_{i \to \infty} \ell_i = \infty$. Choose a parameter t_0 such that $p = \gamma_0(t_0)$. Theorem 2.3.2(2) implies that $\gamma_0(t_0) \in M \setminus D_0$ and in particular that $\gamma_0(t_0 + 1) \in M \setminus D_0$. Since $\lim_{i \to \infty} a_i = a_0$, we have $\exp_p(t_0 + 1)a_i \in D_0$ for all i with $\ell_i > t_0 + 1$. Then a contradiction is derived from $\gamma_0(t_0 + 1) = \lim_{i \to \infty} \exp_p(t_0 + 1)a_i$.

The above argument shows that $\{\ell_i\}$ is bounded above. Therefore we can choose a convergent subsequence $\{\ell_k\}$. Letting $\ell := \lim_{k \to \infty} \ell_k$, we see that

$$p = \lim_{k \to \infty} \exp_p \ell_k a_k = \gamma_0(t_0).$$

Now the second statement of Assertion 2.3.1 follows from the maximal property of the disk domain D_0 bounded by $\mathfrak{E}_0$. Suppose that there exists a geodesic segment β joining two points $x, y \in \mathfrak{E}_0$ lying outside D_0. Suppose, further, that the geodesic biangle consisting of β and a subarc of $\mathfrak{E}_0$ bounds a disk domain E containing p. Then $c(E) > 2\pi$ follows from the

Gauss–Bonnet theorem, a contradiction. Thus E does not contain p. The length-decreasing deformation proceeds to the geodesic triangle with edges px, py and β, and eventually a geodesic loop at p is obtained that contains the maximal disk domain D_0, a contradiction to the choice of a_0. By setting $\mathfrak{E}_p := \mathfrak{E}_0$ and $D_p := D_0$ we conclude the assertion.

The total convexity of $\bar{D}_p$ has also been proved in the above paragraph. $\square$

The following theorem, 2.3.6, ensures that the set $\mathcal{S}_1$ is nonempty.

Theorem 2.3.6. *Let M have positive Gaussian curvature and $c(M) > \pi$. If $\sigma : \mathbf{R} \to M$ is a complete geodesic without self-intersection then there exists a point $q \in \sigma(\mathbf{R})$ that is not the base point of any geodesic loop.*

Proof. Suppose that every point on $\sigma(\mathbf{R})$ is the base point of some geodesic loop. Let $\mathfrak{E}_x$ for every $x \in \sigma(\mathbf{R})$ be the maximal geodesic loop with base point at x, as in Theorem 2.3.5.

We first assert that $\mathfrak{E}_x \cap \sigma(\mathbf{R})$ is a single point $\{x\}$. If it is supposed otherwise then there exists a geodesic loop at x that contains $\mathfrak{E}_x$ in its disk domain, a contradiction (see Figure 2.3.7).

Setting $x_1 := \mathfrak{E}_x \cap (\sigma(\mathbf{R}) \setminus \{x\})$ and, for $i \geq 1$, using induction on the maximal geodesic loop $\mathfrak{E}_i$ at x_i, which is the unique intersection $\mathfrak{E}_{i-1} \cap (\sigma(\mathbf{R}) \setminus \{x_{i-1}\})$, we see that the disk domain D_i bounded by $\mathfrak{E}_i$ is contained entirely in D_{i-1} and that x_i lies in the interior of the subarc of $\sigma(\mathbf{R})$ between x_{i-1} and x_{i-2}. If α_i is the inner angle at x_i of D_i then $c(D_i) = \pi + \alpha_i$ implies that $\{\alpha_i\}$ is strictly decreasing. A contradiction is derived from the positivity of the convexity radius over a compact set bounded by $\mathfrak{E}_x$. In fact, for any small positive number η there are large numbers $i > j$ such that $d(x_i, x_j) < \eta$. Clearly x_i is an interior point of $\mathfrak{E}_{i-1}$ and $\mathfrak{E}_j$ makes an acute angle at x_j that is less than α_1. Therefore $\mathfrak{E}_j$ intersects $\mathfrak{E}_{i-1}$ near x_i, a contradiction. This proves Theorem 2.3.6. $\square$

The global behavior of complete geodesics on M with positive curvature and $c(M) > \pi$ is now discussed. As is seen in Example 2.3.1 for a paraboloid, there are two distinct geodesic loops with the same base point. This can be seen by $\lim\inf_{r_0 \to 0} R(r_0) = \infty$. In fact this relation shows that if a point p on a parabola, as in Example 2.3.1, is sufficiently far from the origin then there are two geodesic loops with base point at p, one of which passes through a point close to the origin.

From Theorem 2.3.5 we see that if $p \in M$ is the base point of some geodesic loop then there exists *a maximal geodesic loop* $\mathfrak{E}_p$ at p. Let D_p be the open disk domain bounded by $\mathfrak{E}_p$. The inner angle of D_p at p is less than π. Thus the set of all unit vectors at p tangent to D_p forms an open interval $\mathbf{S}_p(D_p)$ of

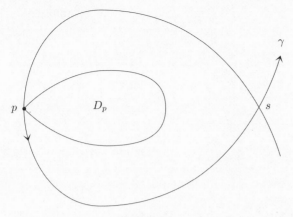

Figure 2.3.8

the unit circle $\mathbf{S}_p(1) \subset T_p M$. Set

$$\mathfrak{I}_p := \{\gamma : \mathbf{R} \to M; \ \gamma(0) = p, \ \pm\dot{\gamma}(0) \notin \mathbf{S}_p(D_p)\}.$$

Then Corollary 2.2.1 (and see also the proof of Theorem 2.3.5) implies that any $\gamma \in \mathfrak{I}_p$ cannot be the boundary of a half-plane and hence has at least one self-intersection. Let $\mathfrak{C}$ be the unique geodesic loop lying on $\gamma(\mathbf{R})$, D the corresponding disk domain and $s \in \mathfrak{C}$ the base point of $\mathfrak{C}$. The total convexity of $\bar{D}_p$ implies that $D_p \subset D$. Therefore every geodesic $\gamma \in \mathfrak{I}_p$ has a unique geodesic loop containing D_p in its disk domain (see Figure 2.3.8).

We next discuss complete geodesics emanating from p and passing through points in D_p. Set

$$\mathfrak{J}_p := \{\gamma : \mathbf{R} \to M; \ \dot{\gamma}(0) \in \mathbf{S}_p(D_p)\}.$$

We then observe that the following possibilities occur.

In the first case there exists a complete geodesic $\gamma \in \mathfrak{J}_p$ that has no self-intersection. This occurs if $\mathcal{P} \neq \emptyset$ and if γ passes through a point on $\mathcal{P}$ (Definition 2.3.1).

In the second case, $\gamma \in \mathfrak{J}_p$ has a self-intersection. Let $\mathfrak{C} \subset \gamma$ be the unique geodesic loop with base point at s and D the disk domain bounded by $\mathfrak{C}$ (see Figure 2.3.9). Three possibilities occur for the point s:

(1) $s \notin D_p$;
(2) $s = p$ and $D \subset D_p$;
(3) $s \in D_p$ and $D \subset D_p$.

Now the properties of the special sets $\mathcal{S}_0$, $\mathcal{S}_1$ and $\mathcal{P}$ will be discussed.

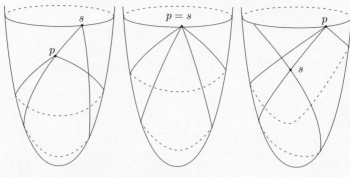

Figure 2.3.9

Theorem 2.3.7 (see Sätze 14, 15 and 16, [20]). *Assume that M has positive Gaussian curvature. Then the following statements are true.*

(1) *For every point $p \in \mathcal{P}$, the exponential map $\exp_p : T_p M \to M$ is a diffeomorphism.*

(2) *No complete geodesic emanating from a point on $\mathcal{S}_1$ has a self-intersection.*

(3) *Every geodesic joining any two points on $\mathcal{S}_1$ lies in $\mathcal{S}_0$.*

(4) *If $c(M) > \pi$ then every geodesic joining two points on $\mathcal{P}$ lies in $\mathcal{S}_1$.*

Proof. Clearly no geodesic $\gamma : [0, \infty) \to M$ with $\gamma(0) = p \in \mathcal{P}$ has a self-intersection; then γ has no conjugate pair on it. In fact, if it is supposed otherwise then a subarc $\gamma|_{[0,t]}$ for sufficiently large $t > 0$ is not minimizing, and hence $\gamma(t)$ is joined to p by a minimizing geodesic. Thus p is the corner of a geodesic biangle, a contradiction to the choice of p. This proves (1).

For (2), suppose that a complete geodesic γ with $\gamma(0) = p \in \mathcal{S}_1$ has a self-intersection. Let $\gamma[a, b]$ be the unique geodesic loop at $\gamma(a) = \gamma(b)$ and $D_{\gamma(a)}$ the corresponding disk domain. Since $p \in \mathcal{S}_1$, we may consider that $a > 0$. Clearly $\gamma[0, b]$ is the fine boundary of the compact set $\gamma[0, a] \cup \bar{D}_{\gamma(a)}$. The length-decreasing deformation applies to this fine boundary with base point at p and provides a geodesic loop at p. This contradicts the choice of $p \in \mathcal{S}_1$ and proves (2).

For (3), suppose that there is a geodesic segment $q_1 q_2$ joining $q_1, q_2 \in \mathcal{S}_1$ such that a point $r \in q_1 q_2$ is the base point of some geodesic loop. Theorem 2.3.5 implies that there is a maximal geodesic loop $\mathfrak{E}_r$ at r. As was seen in the proof of (2), the complete extension σ of $q_1 q_2$ has no self-intersection. We then observe that $\mathfrak{E}_r \cap (\sigma(\mathbf{R}) \setminus \{r\})$ is a single point, say r_1. We may consider that the points are either in the order q_1, r, r_1, q_2 or in the order q_1, r, q_2, r_1. In both cases we observe that $q_1 r \cup \mathfrak{E}_r$ forms the fine boundary of a compact set and hence obtain a geodesic loop at q_1, a contradiction to the choice of $q_1 \in \mathcal{S}_1$.

For (4), suppose that there exists a geodesic segment $p_1 p_2$ with $p_1, p_2 \in \mathcal{P}$ such that a point $q \in p_1 p_2$ lies on a geodesic loop bounding a disk domain D. Let γ be the complete extension of $p_1 p_2$ and let s be the base point of the geodesic loop. Then γ divides M into two half-planes, say H_1 and H_2. We may assume that $s \in H_2$ and hence that $H_1 \setminus D$ has all its corners on $\gamma(\mathbf{R})$, with all the inner angles less than π. We then observe from $q \in p_1 p_2$ that at least one of the two points p_1, p_2 is not contained in D. In fact, suppose $p_1, p_2 \in D$. Then $H_1 \cap D$ has at least two components D_1, D_2 such that $p_i \in \bar{D}_i$ and $p_1 \notin \bar{D}_2$. By the length-decreasing deformation procedure, we can find a geodesic joining p_1 to a point, away from D, on γ and lying in $H_1 \setminus D_2$, a contradiction to the choice of p_1. If $p_1 \notin D$, we can then find a geodesic joining p_1 to a point on γ by the same manner. Thus a contradiction is derived. □

Remark 2.3.4. In the proof of the statement (1) we have used only the length-decreasing deformation method, and hence (1) is valid for all Riemannian planes.

At the end of this section we introduce a result by Bangert (see [**10**]) that generalizes Theorem 2.3.4. The proof uses the Lusternik–Schnirelmann method for the length-decreasing deformation of homotopy curves.

Theorem 2.3.8 (see [**10**]). *Every complete Riemannian plane admits at least one complete and divergent geodesic without self-intersection.*

3

The ideal boundary

The concept of a (geometric) ideal boundary was originally invented by Gromov, see [7], and he defined Hadamard manifolds and nonnegatively curved complete noncompact 2-manifolds. In this chapter, we will present the ideal boundary $M(\infty)$ of a noncompact manifold M admitting a curvature at infinity $\lambda_\infty(M)$, this was first defined in [91] and is a very powerful tool to with which study the global geometric properties of such an M. Some new aspects and theorems will be included in this chapter, in particular a detailed study of the topological structure of the ideal boundary $M(\infty)$ and the compactification $\overline{M}^\infty = M \cup M(\infty)$, which were introduced briefly in [93, 97]. We will establish a triangle comparison theorem for triangular domains having small total absolute curvature. This will be applied to prove that the scaling limit of M is isometric to the Euclidean cone over the ideal boundary $M(\infty)$ with Tits metric, provided that the curvature at infinity of M is finite. A variant of the triangle comparison theorem was given in [98] with the aim of studying the limits of two-dimensional manifolds under an L^p curvature bound, $p \geq 1$. In the last section, we will study the asymptotic behavior and exhaustion property of Busemann functions, which have been treated in [78, 79, 91].

3.1 The curvature at infinity

Let $c : I \to M$ be a piecewise-smooth curve in a two-dimensional Riemannian manifold M, where $I \subset \mathbf{R}$ is an interval. We fix a side of c to determine the normal vector field $\mathbf{e}$ and the inner angles ω_i at the vertices x_i of c. Let $\kappa(s)$ be the geodesic curvature of c at $c(s)$, $s \in I$, with respect to the normal vector field **e**, and define

$$\lambda_+(c(A)) := \int_A (\kappa(s))_+ \, ds + \sum_{x_i \in c(A)} (\pi - \omega_i)_+,$$

$$\lambda_-(c(A)) := \int_A (\kappa(s))_- \, ds + \sum_{x_i \in c(A)} (\pi - \omega_i)_-,$$

for any Borel subset A of I, where $(x)_+ := \max\{x, 0\}$ and $(x)_- := -\min\{x, 0\}$ for $x \in \mathbf{R}$. It is easy to verify that $\lambda_\pm$ are positive Radon measures on $c(I)$. Therefore $\lambda := \lambda_+ - \lambda_-$ is a signed Radon measure on $c(I)$. If $c \subset \partial M$ then $\lambda(c) := \lambda(c(I))$ is compatible with the total geodesic curvature defined in Section 1.9 and in (2.1.7). Set $\lambda_{\text{abs}} := \lambda_+ + \lambda_-$.

Definition 3.1.1 (The curvature at infinity). Let M be a finitely connected complete two-dimensional Riemannian manifold, possibly with piecewise-smooth boundary. The *curvature at infinity of M* is defined to be

$$\lambda_\infty(M) := 2\pi \chi(M) - \pi \chi(\partial M) - c(M) - \lambda(\partial M),$$

which exists only when at least one of

$$\int_M G_+ \, dM + \lambda_+(\partial M) \qquad \text{and} \qquad \int_M G_- \, dM + \lambda_-(\partial M)$$

is finite.

The Cohn-Vossen theorem says that $\lambda_\infty(M) \geq 0$ if $\lambda_\infty(M)$ exists. The Gauss–Bonnet theorem says that $\lambda_\infty(M) = 0$ if M is compact. When $\lambda_\infty(M)$ exists, both $\int_M G_+ \, dM$ and $\lambda_+(\partial M)$ are finite, because of $\lambda_\infty(M) \geq 0$.

Example 3.1.1

(1) For the Euclidean plane $\mathbf{R}^2$ we have

$$\lambda_\infty(\mathbf{R}^2) = 2\pi.$$

(2) If M is a paraboloid then $\lambda_\infty(M) = 0$.
(3) For the hyperbolic plane $H^2(-1)$ we have $\lambda_\infty(H^2(-1)) = +\infty$.
(4) If M is the universal covering space of $\{(x, y) \in \mathbf{R}^2; x^2 + y^2 \geq 1\}$ then $\lambda_\infty(M) = +\infty$.

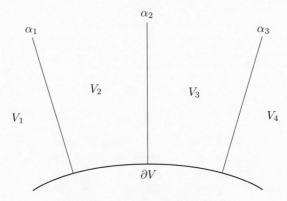

Figure 3.1.1 Riemannian half-plane

Lemma 3.1.1

(1) *If C is a core of a noncompact M for which $\lambda_\infty(M)$ exists then*

$$\lambda_\infty(M) = \sum_V \lambda_\infty(V),$$

where V runs over all connected components of $\overline{M \setminus C}$.

(2) *If a Riemannian half-plane V is split into Riemannian half-planes V_i, $i = 1, \ldots, k$, by disjoint proper curves $\alpha_i : [0, +\infty) \to V, i = 1, \ldots, k-1$ (see Figure 3.1.1), and if $\lambda_\infty(V)$ and the $\lambda_\infty(V_i)$ all exist, then*

$$\lambda_\infty(V) = \lambda_\infty(V_1) + \cdots + \lambda_\infty(V_k).$$

(3) *If a Riemannian half-cylinder V is split into Riemannian half-planes V_i, $i = 1, \ldots, k$, by disjoint proper curves $\alpha_i : [0, +\infty) \to V, i = 1, \ldots, k$ (see Figure 3.1.2), and if $\lambda_\infty(V)$ and the $\lambda_\infty(V_i)$ all exist, then*

$$\lambda_\infty(V) = \lambda_\infty(V_1) + \cdots + \lambda_\infty(V_k).$$

Proof. The proofs are straightforward (cf. the proof of Theorem 2.2.1). $\square$

Proposition 3.1.1. *If M, M' and M'' are such that their curvatures at infinity all exist, $M = M' \cup M''$ and $M' \cap M'' \subset \partial M' \cap \partial M''$ then*

$$\lambda_\infty(M) = \lambda_\infty(M') + \lambda_\infty(M'').$$

Proof. There exists a core C of M such that $C' := C \cap M'$ and $C'' := C \cap M''$ are cores of M' and M'' respectively. For any connected component V of $\overline{M \setminus C}$,

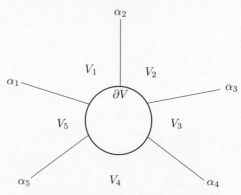

Figure 3.1.2 Riemannian half-cylinder

we can find all the components V_i, $i = 1, 2, \ldots$, of $\overline{M' \setminus C'}$ and $\overline{M'' \setminus C''}$ that are contained in V. Then, since V and the V_i together satisfy the assumption in either (2) or (3) of Lemma 3.1.1, we obtain

$$\lambda_\infty(M) = \sum_V \lambda_\infty(V) = \sum_V \sum_i \lambda_\infty(V_i) = \lambda_\infty(M') + \lambda_\infty(M'').$$ $\qquad \square$

Exercise 3.1.1. Prove the following.

(1) The curvature at infinity $\lambda_\infty(M)$ of M is an invariant of any deformation inside a compact set.
(2) Let M be such that $\lambda_\infty(M)$ exists, and let $\alpha : [0, +\infty) \to M$ be a curve such that $\lambda(\alpha)$ exists and is finite and $M \setminus \alpha$ is connected. Then the curvature at infinity of the completion of $M \setminus \alpha$ exists and is equal to $\lambda_\infty(M)$.
(3) If M and M' are such that their curvatures at infinity exist, and if M' is isometrically embedded into M, then

$$\lambda_\infty(M') \le \lambda_\infty(M).$$

3.2 Parallelism and pseudo-distance between curves

From now on, let M be a finitely connected complete noncompact two-dimensional Riemannian manifold, possibly with boundary, for which the curvature at infinity $\lambda_\infty(M)$ exists. For any curve $\alpha : I \to M$ from an interval I of

R to M, we put

$$E(\alpha) := \{t \in I;\ \alpha|_{I \cap [t-\epsilon, t+\epsilon]} \text{ for any } \epsilon > 0 \text{ is not a minimal segment}\}$$

and

$$\mathcal{C}_M := \{\alpha : [0, +\infty) \to M;\ \alpha \text{ is a simple proper piecewise-smooth curve}$$
$$\text{such that } \lambda_{\text{abs}}(\alpha|_{E(\alpha)}) < +\infty\}.$$

Obviously, if $\alpha \cap \partial M = \emptyset$ then $\lambda(\alpha|_{E(\alpha)}) = \lambda(\alpha)$.

Remark 3.2.1. Any piecewise-smooth curve $\alpha : [0, +\infty) \to M$ entirely contained in ∂M is an element of $\mathcal{C}_M$, because $\lambda(\alpha|_{E(\alpha)}) = \lambda_+(\alpha) \leq \lambda_+(\partial M) < +\infty$ with respect to a suitably chosen side of α.

We say that *curves $c_i : I_i \to M$, $i = 1, \ldots, k$, together bound a region D of M* if there exist lifts $\tilde{c}_i : I_i \to \partial D$ of c_i, $i = 1, \ldots, k$, into the fine boundary ∂D of D (see Section 1.9) such that each $\tilde{c}_i$ is an into-homeomorphism and $\bigcup_{i=1}^{k} \tilde{c}_i(I_i) = \partial D$.

Two curves α and β in $\mathcal{C}_M$ are said to *cross each other (or simply cross)* if both $\alpha(t)$ and $\beta(t)$ tend to a common end of M as $t \to +\infty$ and if there exists no curve joining $\alpha(0)$ and $\beta(0)$ that, together with α and β, bounds a region of M in the above sense. Two curves α and β in $\mathcal{C}_M$ are said to *cross each other near infinity (or simply cross near infinity)* if $\alpha|_{[a,+\infty)}$ and $\beta|_{[b,+\infty)}$ for any $a, b \geq 0$ cross each other.

Let α and β be two curves in $\mathcal{C}_M$ such that $\alpha(t)$ and $\beta(t)$ tend to a common end of M and do not cross near infinity, i.e., $\alpha|_{[a,+\infty)}$ and $\beta|_{[b,+\infty)}$ for some $a, b \geq 0$ do not cross each other. Then there exists a core C of M such that $\alpha|_{[a,+\infty)} \cap C = \{\alpha(a)\}$ and $\beta|_{[b,+\infty)} \cap C = \{\beta(b)\}$. Let V be the connected component of $\overline{M \setminus C}$ containing both $\alpha|_{[a,+\infty)}$ and $\beta|_{[b,+\infty)}$. When V is a Riemannian half-cylinder then $\alpha|_{[a,+\infty)}$ and $\beta|_{[b,+\infty)}$ together divide V into two closed regions, say $D(\alpha, \beta)$ and $D(\beta, \alpha)$, where the orientation of $D(\alpha, \beta)$ (resp. $D(\beta, \alpha)$) induced from V is compatible with the direction of the parameter of $\beta|_{[b,+\infty)}$ (resp. $\alpha|_{[a,+\infty)}$); see Figure 3.2.1.

When V is a Riemannian half-plane, then $\alpha|_{[a,+\infty)}$ and $\beta|_{[b,+\infty)}$ together divide V into three closed regions, one of which, say D, touches both α and β. We define only one of $D(\alpha, \beta)$ and $D(\beta, \alpha)$ by

$$D =: \begin{cases} D(\alpha, \beta) & \text{if the orientation of } D \text{ is compatible} \\ & \text{with the direction of the parameter of } \beta, \\ D(\beta, \alpha) & \text{otherwise;} \end{cases}$$

see Figure 3.2.2.

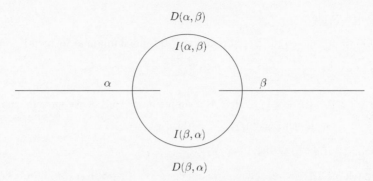

Figure 3.2.1 The case where V is a Riemannian half-cylinder.

Now set $I(\hat{\alpha}, \hat{\beta}) := D(\hat{\alpha}, \hat{\beta}) \cap \partial V$ for any $(\hat{\alpha}, \hat{\beta}) \in \{(\alpha, \beta), (\beta, \alpha)\}$. In the case where $\alpha|_{[a,+\infty)} = \beta|_{[b,+\infty)}$, we agree that

$$D(\alpha, \beta) := \alpha[a, +\infty) = \beta[b, +\infty) \quad \text{and} \quad I(\alpha, \beta) := \{\alpha(a)\} = \{\beta(b)\}.$$

Note that although for given α and β we have many choices of $D(\alpha, \beta)$, depending on $a, b \geq 0$ and the core C, we will pick up just one of them. Of course, this notation is defined when M is a Riemannian half-plane H or a Riemannian half-cylinder N.

For a Riemannian half-plane H, a relation $\leq$ on $\mathcal{C}_H$ is defined by the following: for two curves $\alpha, \beta \in \mathcal{C}_H$, $\alpha \leq \beta$ is true if there is a core C of H for which $D(\alpha, \beta)$ is defined. Note that both $\alpha \leq \beta$ and $\beta \leq \alpha$ hold iff α and β have a common subarc in $\mathcal{C}_H$. The relation $\leq$ is reflexive and transitive but not antisymmetric.

Definition 3.2.1. A function $d_\infty \colon \mathcal{C}_M \times \mathcal{C}_M \to \mathbf{R} \cup \{+\infty\}$ is defined as follows. Let $\alpha, \beta \in \mathcal{C}_M$ be two curves. If $\alpha(t)$ and $\beta(t)$ tend to different ends as $t \to +\infty$ then $d_\infty(\alpha, \beta) := +\infty$. Assume that they tend to a common end. If α and β cross near infinity then $d_\infty(\alpha, \beta) := 0$. If they do not, we can find a V as above corresponding to α and β. If V is a Riemannian half-plane then

$$d_\infty(\alpha, \beta) := \begin{cases} \lambda_\infty(D(\alpha, \beta)) & \text{if } \alpha \leq \beta, \\ \lambda_\infty(D(\beta, \alpha)) & \text{if } \beta \leq \alpha. \end{cases}$$

If $\alpha \leq \beta$ and $\beta \leq \alpha$ then $D(\alpha, \beta) = D(\beta, \alpha)$ is just the image of a common subarc of α and β, and we agree that $d_\infty(\alpha, \beta) = \lambda_\infty(D(\alpha, \beta)) = \lambda_\infty(D(\beta, \alpha)) = 0$. If V is a Riemannian half-cylinder then

$$d_\infty(\alpha, \beta) := \min\{\lambda_\infty(D(\alpha, \beta)), \lambda_\infty(D(\beta, \alpha))\}.$$

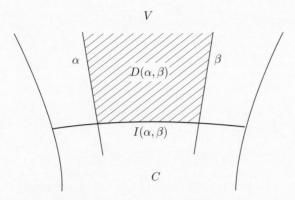

Figure 3.2.2 The case where V is a Riemannian half-plane.

The function d_∞ is obviously nonnegative and symmetric, and will later be proved to be a pseudo-distance function on $\mathcal{C}_M$ (see Theorem 3.3.1 below).

Lemma 3.2.1. *If a sequence $\{c_i\}_{i=1,2,...,}$ of piecewise-smooth curves (pointwise) converges to a piecewise-smooth curve c then*

(1) $\liminf_{i \to \infty} \lambda_{\mathrm{abs}}(c_i) \geq \lambda_{\mathrm{abs}}(c),$

(2) $\liminf_{i \to \infty} \lambda_{\mathrm{abs}}(c_i|_{E(c_i)}) \geq \lambda_{\mathrm{abs}}(c|_{E(c)}).$

Proof. (1): We first claim:

Sublemma 3.2.1. *The following, (i) and (ii), hold:*

(i) *There exists a smooth approximation $\tilde{c}_i$ of c_i tending to c as $i \to \infty$ such that for, any two numbers $a < b$,*

$$\liminf_{i \to \infty} \lambda_{\mathrm{abs}}(c_i|_{[a,b]}) = \liminf_{i \to \infty} \lambda_{\mathrm{abs}}(\tilde{c}_i|_{[a,b]}).$$

(ii) *In addition, for any two numbers $a < b$ there exist a_i and b_i, with $a \leq a_i < b_i \leq b$ for every i, such that $\lim_{i \to \infty} a_i = a$, $\lim_{i \to \infty} b_i = b$ and*

$$\lim_{i \to \infty} \lambda(\tilde{c}_i|_{[a_i,b_i]}) = \lambda(c|_{[a,b]}).$$

If the sublemma is assumed to be true then (i) and (ii) together imply that

$$\liminf_{i \to \infty} \lambda_{\mathrm{abs}}(c_i|_{[a,b]}) \geq |\lambda(c|_{[a,b]})| \qquad \text{for any two numbers } a < b,$$

which proves (1) of the lemma on noting the arbitrariness of a and b.

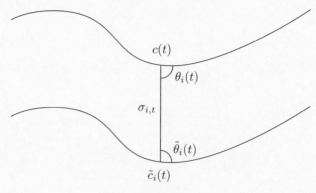

$$c(t)$$

$$\theta_i(t)$$

$$\sigma_{i,t}$$

$$\tilde{\theta}_i(t)$$

$$\tilde{c}_i(t)$$

Figure 3.2.3

Proof of Sublemma 3.2.1. (i): The proof of (i) is easy and will be left to the reader.

(ii): For any $t \in [a, b]$ and $i = 1, 2, \ldots$, denote by $\sigma_{i,t}$ a minimal segment from $c(t)$ to $\tilde{c}_i(t)$, and set

$$\delta_i := \sup_{t \in [a,b]} d(c(t), \tilde{c}_i(t)),$$

$$\theta_i(t) := \angle(\dot{c}(t), \dot{\sigma}_{i,t}(0)), \qquad \tilde{\theta}_i(t) := \angle(\dot{\tilde{c}}_i(t), -\dot{\sigma}_{i,t}(L(\sigma_{i,t})));$$

see Figure 3.2.3.

We can choose $\tilde{c}_i$ in such a way that $\tilde{c}_i(t) \neq c(t)$ for any t. Then $\sigma_{i,t}$ is nontrivial, so that $\theta_i(t)$ and $\tilde{\theta}_i(t)$ are defined for all $t \in [a, b]$. Assume that i is sufficiently large. We now prove

Sublemma 3.2.2. *For any $\epsilon > 0$ and $t \in [a, b]$ there exists $T = T(\epsilon, t) \in [a, b]$ such that $|t - T| \leq \delta_i/\epsilon$ and $|\cos \theta_i(T) + \cos \tilde{\theta}_i(T)| \leq \epsilon$.*

Proof of Sublemma 3.2.2. It suffices to prove that if three numbers $t_1 < t_2$ and $\epsilon > 0$ satisfy $|\cos \theta_i(t) + \cos \tilde{\theta}_i(t)| > \epsilon$ for any $t \in (t_1, t_2)$ then $t_2 - t_1 < \delta_i/\epsilon$. Since i has been assumed to be sufficiently large, $\sigma_{i,t}$ is uniquely determined, so that $\theta_i(t)$ and $\tilde{\theta}_i(t)$ are continuous in $t \in [a, b]$.

In the case where $\cos \theta_i(t) + \cos \tilde{\theta}_i(t) > \epsilon$ for any $t \in (t_1, t_2)$, the first variation formula implies that

$$d(c(t_2), \tilde{c}_i(t_2)) - d(c(t_1), \tilde{c}_i(t_1)) = \int_{t_1}^{t_2} (\cos \theta_i(t) + \cos \tilde{\theta}_i(t)) \, dt$$

$$> \epsilon (t_2 - t_1),$$

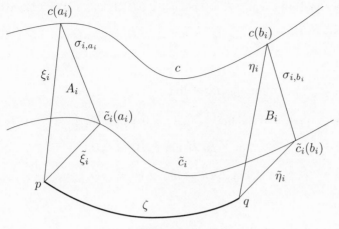

Figure 3.2.4

the left-hand side of which is less than or equal to δ_i, and thus

$$t_2 - t_1 < \delta_i/\epsilon.$$

In the case where $\cos \theta_i(t) + \cos \tilde{\theta}_i(t) < -\epsilon$ for any $t \in (t_1, t_2)$, a similar discussion also yields

$$t_2 - t_1 < \delta_i/\epsilon. \qquad \square$$

Let us continue the proof of part (ii) of Sublemma 3.2.1. We can find a sequence $\{\epsilon_i\}$ of positive numbers such that $\lim_{i \to \infty} \epsilon_i = \lim_{i \to \infty} \delta_i/\epsilon_i = 0$. Applying Sublemma 3.2.2 we set $a_i := T(\epsilon_i, a)$ and $b_i := T(\epsilon_i, b)$ and obtain

$$a \le a_i, \quad \lim_{i \to \infty} a_i = a, \quad b_i \le b, \quad \lim_{i \to \infty} b_i = b,$$
$$\lim_{i \to \infty}(\theta_i(a_i) + \tilde{\theta}_i(a_i)) = \lim_{i \to \infty}(\theta_i(b_i) + \tilde{\theta}_i(b_i)) = \pi. \tag{3.2.1}$$

We now find a point p (resp. q) so close to $c(a)$ (resp. $c(b)$) that a minimal segment joining p and $c(a)$ (resp. $c(b)$) is unique. Let ξ_i, $\tilde{\xi}_i$, η_i, $\tilde{\eta}_i$ be four minimal segments respectively from p, p, q, q to $c(a_i), \tilde{c}_i(a_i), c(b_i), \tilde{c}_i(b_i)$, and let A_i and B_i be the domains bounded respectively by the triangles $\sigma_{i,a_i} \cup \xi_i \cup \tilde{\xi}_i$ and $\sigma_{i,b_i} \cup \eta_i \cup \tilde{\eta}_i$; see Figure 3.2.4. We can find a curve ζ joining p and q in such a way that the two quadrangles $c \cup \xi_i \cup \zeta \cup \eta_i$ and $\tilde{c}_i \cup \tilde{\xi}_i \cup \zeta \cup \tilde{\eta}_i$ each bound a disk, say D_i and $\tilde{D}_i$ respectively. As i tends to infinity, D_i and $\tilde{D}_i$ tend to the same limit disk D, whose boundary is a quadrangle with vertices $c(a), p$, q and $c(b)$. It follows that $\lim_{i \to \infty}(\angle_{c(a_i)}A_i + \angle_{\tilde{c}_i(a_i)}A_i) = \pi$, where $\angle_x R$ denotes the inner angle of a region R at a point $x \in \partial R$. By this and (3.2.1) we have

$$\lim_{i \to \infty} \angle_{c(a_i)} D_i = \lim_{i \to \infty} \angle_{\tilde{c}_i(a_i)} \tilde{D}_i = \angle_{c(a)} D,$$

as well as

$$\lim_{i \to \infty} \angle_{c(b_i)} D_i = \lim_{i \to \infty} \angle_{\tilde{c}_i(b_i)} \tilde{D}_i = \angle_{c(b)} D.$$

By $\lim_{i \to \infty} \angle_p A_i = 0$ and $\lim_{i \to \infty} \angle_q B_i = 0$, we obtain

$$\lim_{i \to \infty} \angle_p D_i = \lim_{i \to \infty} \angle_p \tilde{D}_i = \angle_p D,$$

$$\lim_{i \to \infty} \angle_q D_i = \lim_{i \to \infty} \angle_q \tilde{D}_i = \angle_q D.$$

Moreover

$$\lim_{i \to \infty} c(D_i) = \lim_{i \to \infty} c(\tilde{D}_i) = c(D).$$

Thus, applying the Gauss–Bonnet theorem,

$$\lambda(c|_{[a,b]}) = \lim_{i \to \infty} \lambda(c|_{[a_i,b_i]}) = \lim_{i \to \infty} \lambda(\tilde{c}_i|_{[a_i,b_i]}),$$

which completes the proof of Sublemma 3.2.2.   □

Thus the proof of part (1) of Lemma 3.2.1 is complete.

Let us show part (2) of Lemma 3.2.1. Since $\lim_{i \to \infty} c_i \cap \partial M \subset c \cap \partial M$, we obtain $\lim_{i \to \infty} c_i|_{E(c_i)} \supset c|_{E(c)}$, which together with (1) proves (2). This completes the proof of Lemma 3.2.1.   □

Definition 3.2.2 (Proper convergence). Let $\alpha \colon [0, +\infty) \to M$ be a curve, and let $\{\alpha_i \colon [0, a_i) \to M\}_{i=1,2,\dots}$ be a sequence of curves such that each $\lambda(\alpha_i)$ exists and $\lim_{i \to \infty} a_i = +\infty$, where $0 \le a_i \le +\infty$. We say that $\{\alpha_i\}$ *properly converges to* α if $\{\alpha_i\}$ converges to α as $i \to \infty$ and the following, (P1) and (P2), hold:

(P1) $\lim_{t \to +\infty} \liminf_{i \to \infty} d(p, \alpha_i|_{[t,a_i)}) = +\infty$ for a fixed point $p \in M$,
(P2) $\lim_{t \to +\infty} \limsup_{i \to \infty} \lambda_{\mathrm{abs}}(\alpha_i|_{[t,a_i) \cap E(\alpha_i)}) = 0.$

This notion of convergence is called *proper convergence*.

Example 3.2.1

(1) Let $\alpha \in \mathcal{C}_M$ and $\alpha_i := \alpha|_{[0,a_i]}$, $a_i \to +\infty$. Then $\{\alpha_i\}$ properly converges to α.
(2) When a sequence of minimal segments converges to a ray, this is a proper convergence.

Proposition 3.2.1. *If a sequence $\{\alpha_i : [0, a_i) \to M\}$ of curves properly converges to a curve $\alpha : [0, +\infty) \to M$ then $\alpha \in \mathcal{C}_M$.*

Proof. (P1) implies that α is proper. (P2) and Lemma 3.2.1 together prove the existence and finiteness of $\lambda(\alpha|_{E(\alpha)})$. $\qquad\square$

Definition 3.2.3 (Parallelism). Two curves $\alpha, \beta \in \mathcal{C}_M$ are said to be *parallel* if there exist two sequences $\{\alpha_i : [0, a_i] \to M\}$ and $\{\beta_i : [0, b_i] \to M\}$ of piecewise-smooth curves properly converging respectively to α and β such that $\alpha_i(a_i) = \beta_i(b_i)$ for each i.

Obviously, if two curves $\alpha, \beta \in \mathcal{C}_M$ are parallel then both $\alpha(t)$ and $\beta(t)$ tend to a common end as $t \to +\infty$.

Let X be a (not necessarily two-dimensional) Riemannian manifold possibly with boundary.

Definition 3.2.4 (Asymptotic relation). A ray σ in X is said to be *asymptotic* to a ray τ in X if there exist a sequence $\{\sigma_i : [0, s_i] \to X\}$ of minimal segments tending to σ and a sequence $\{t_i\}$ of positive numbers tending to $+\infty$ such that $\sigma_i(s_i) = \tau(t_i)$ for each i. We call such a ray σ a *coray of* τ.

It follows that when a ray in M is asymptotic to another ray in M they are parallel. Notice that the asymptotic relation is not necessarily symmetric.

Proposition 3.2.2. *If a sequence of rays σ_i, $i = 1, 2, \ldots$, in X asymptotic to a ray τ in X converges to a ray σ in X then σ is asymptotic to τ.*

Proof. Assume that a sequence of rays σ_i in X asymptotic to a ray τ in X converges to a ray σ. Then, for each i, there is a sequence of minimal segments $\sigma_{i,j} : [0, s_{i,j}] \to X$, $j = 1, 2, \ldots$, such that $\lim_{j \to \infty} \sigma_{i,j} = \sigma_i$ and $\sigma_i(s_{i,j}) = \tau(t_{i,j})$ for any j, where $t_{i,j}$ is some sequence tending to ∞ as $j \to \infty$. If we take a sufficiently large $j(i)$ against each i then $\sigma_{i,j(i)}$ converges to σ and $t_{i,j(i)}$ to ∞ as $i \to \infty$. $\qquad\square$

Let H be a Riemannian half-plane for which the curvature at infinity exists.

Lemma 3.2.2. *Let c be a compact arc in H, and let $\{p_i\}_{i=1,2,\ldots}$, be a sequence of points in H tending to infinity (i.e., $d(p_i, c)$ tends to $+\infty$). Assume that given curves α_i and β_i, $i = 1, 2, \ldots$, all connect c to p_i and that a subarc of c, α_i and β_i, for each i, bounds a compact contractible region D_i (see Figure 3.2.5). If the sequences $\{\alpha_i\}$ and $\{\beta_i\}$ properly converge to two curves α and β in $\mathcal{C}_H$ respectively then the inner angle of D_i at p_i tends to zero as $i \to \infty$, and we have $d_\infty(\alpha, \beta) = 0$.*

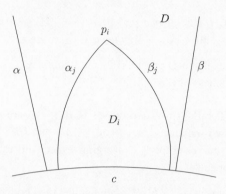

Figure 3.2.5

Proof. The limit region $D := \lim_{i \to \infty} D_i$ is bounded by α, the subarc of c from $\alpha(0)$ to $\beta(0)$, and β. Thus, it follows that $\lim_{i \to \infty} (\lambda(\partial D_i) - (\pi - \angle_{p_i} D_i)) = \lambda(\partial D)$, which, together with the Gauss–Bonnet theorem, shows that for any compact subset K of H,

$$
\begin{aligned}
c(D \cap K) &= \lim_{i \to \infty} c(D_i \cap K) = \lim_{i \to \infty} (c(D_i) - c(D_i \setminus K)) \\
&= \lim_{i \to \infty} (2\pi - \lambda(\partial D_i) - c(D_i \setminus K)) \\
&= \lim_{i \to \infty} (\pi - \lambda(\partial D) + \angle_{p_i} D_i - c(D_i \setminus K)).
\end{aligned}
$$

Since $c(D_i \setminus K) \le \int_{H \setminus K} G_+ \, dH$,

$$
\pi - c(D \cap K) - \lambda(\partial D) + \limsup_{i \to \infty} \angle_{p_i} D_i \le \int_{H \setminus K} G_+ \, dH.
$$

If K is increasing and tends to H then $c(D \cap K)$ converges to $c(D)$ and $\int_{H \setminus K} G_+ \, dH$ to zero. Thus

$$
d_\infty(\alpha, \beta) + \limsup_{i \to \infty} \angle_{p_i} D_i \le 0. \qquad \square
$$

Proposition 3.2.3. *If $\alpha, \beta \in \mathcal{C}_H$ are parallel then $d_\infty(\alpha, \beta) = 0$.*

Proof. If α and β cross near infinity, the conclusion is trivial. Assume that α and β do not cross at infinity and that $\alpha \le \beta$. We can find a compact subarc $c \colon [0, \ell] \to \partial H$ of ∂H such that the direction of the parameter of c is compatible with the orientation of ∂H and then assume $\alpha \cap c = \{\alpha(0)\}$ and $\beta \cap c = \{\beta(0)\}$ by extending or cutting α and β. Since α and β are parallel, there exist sequences $\{\alpha_i \colon [0, a_i] \to H\}$ and $\{\beta_i \colon [0, b_i] \to H\}$ of curves properly converging to α and β such that $\alpha_i(a_i) = \beta_i(b_i) =: p_i$, $\alpha_i \cap c = \{\alpha_i(0)\} = \{\alpha(0)\}$ and $\beta_i \cap c = \{\beta_i(0)\} = \{\beta(0)\}$ for every i (see Figure 3.2.6). Denote by H_i the

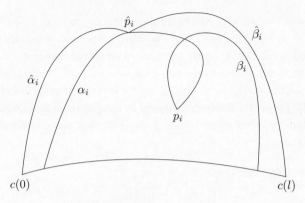

Figure 3.2.6 The case $p_i \notin H_i$.

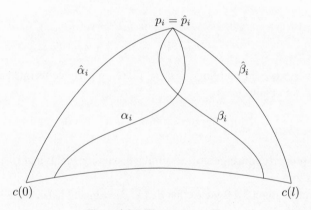

Figure 3.2.7 The case $p_i \in H_i$.

completion of $H \setminus \Delta(\alpha_i \cup \beta_i)$, where $\Delta(A)$ denotes the union of a subset A of H and of all bounded connected components of $H \setminus A$. Let $\hat{p}_i$ be a point in $\partial \Delta(\alpha_i \cup \beta_i)$ such that

$$d_{H_i}(\hat{p}_i, c) = \max_{x \in \partial \Delta(\alpha_i \cup \beta_i)} d_{H_i}(x, c) \quad \text{when } p_i \notin H_i,$$

where d_{H_i} denotes the interior distance function on H_i (i.e., $d_{H_i}(x, y)$ is the infimum of the lengths of all curves in H_i joining x and y) and

$$\hat{p}_i := p_i \quad \text{when } p_i \in H_i$$

(see Figure 3.2.7). Now find a minimal segment $\hat{\alpha}_i : [0, \hat{a}_i] \to H_i$ (resp. $\hat{\beta}_i : [0, \hat{b}_i] \to H_i$) in H_i joining $c(0)$ (resp. $c(\ell)$) to $\hat{p}_i$. Note that c, $\hat{\alpha}_i$ and $\hat{\beta}_i$ together bound a compact contractible region in H containing $\Delta(\alpha_i \cup \beta_i)$. Substituting a subsequence we assume that $\{\hat{\alpha}_i\}$ and $\{\hat{\beta}_i\}$ converge to two curves

$\hat\alpha, \hat\beta \colon [0, +\infty) \to H$ respectively. It follows that $\hat\alpha \le \alpha \le \beta \le \hat\beta$. Let us show that $\lim_{i \to \infty} \hat\alpha_i = \hat\alpha$ and $\lim_{i \to \infty} \hat\beta_i = \hat\beta$ are properly convergent. Condition (P1) in Definition 3.2.2 is easily checked, so let us verify only condition (P2). Let $\xi, \eta \colon [0, +\infty) \to \partial H$ be the two curves respectively emanating from $c(0)$ and $c(\ell)$ such that $\xi|_{(0,+\infty)}$ and $\eta|_{(0,+\infty)}$ are the two components of $\partial H \setminus c$. Since $\{\alpha_i\}$ and $\{\beta_i\}$ properly converge to α and β, there exists a sequence $\{s_t\}_{t \ge 0}$ of positive numbers tending to $+\infty$ as $t \to +\infty$ such that, for all sufficiently large i and for all $t \ge 0$,

$$\hat\alpha_i[t, \hat a_i] \cap (\alpha_i \cup \beta_i \cup \xi \cup \eta)$$
$$\subset \alpha_i[s_t, a_i] \cup \beta_i[s_t, b_i] \cup \xi[s_t, +\infty) \cup \eta[s_t, +\infty),$$

and hence

$$\lambda_{\mathrm{abs}}(\hat\alpha_i|_{[t,\hat a_i] \cap E(\hat\alpha_i)}) \le \lambda_{\mathrm{abs}}(\alpha_i|_{[s_t,a_i] \cap E(\alpha_i)}) + \lambda_{\mathrm{abs}}(\beta_i|_{[s_t,a_i] \cap E(\beta_i)})$$
$$+ \lambda_{\mathrm{abs}}(\xi|_{[s_t,+\infty) \cap E(\xi)}) + \lambda_{\mathrm{abs}}(\eta|_{[s_t,+\infty) \cap E(\eta)}).$$

Since $\lim_{i \to \infty} \alpha_i = \alpha$ and $\lim_{i \to \infty} \beta_i = \beta$ are properly convergent and since $\xi, \eta \in \mathcal{C}_H$, we obtain

$$\lim_{t \to +\infty} \limsup_{i \to \infty} \lambda_{\mathrm{abs}}(\hat\alpha_i|_{[t,\hat a_i] \cap E(\hat\alpha_i)}) = 0.$$

Thus $\{\hat\alpha_i\}$ properly converges to $\hat\alpha$. A similar discussion yields that $\{\hat\beta_i\}$ properly converges to $\hat\beta$.

Applying Lemma 3.2.2 yields that $d_\infty(\hat\alpha, \hat\beta) = 0$, and hence

$$d_\infty(\alpha, \beta) = \lambda_\infty(D(\alpha, \beta)) \le \lambda_\infty(D(\hat\alpha, \hat\beta)) = d_\infty(\hat\alpha, \hat\beta) = 0. \qquad \square$$

Lemma 3.2.3. *For any two curves $\alpha, \beta \in \mathcal{C}_H$ crossing each other near infinity, there exist two curves $\sigma_-, \sigma_+ \in \mathcal{C}_H$ such that*

$$\sigma_- \le \alpha, \beta \le \sigma_+ \qquad and \qquad d_\infty(\sigma_\pm, \alpha) = d_\infty(\sigma_\pm, \beta) = 0.$$

Proof. Extending α and β if necessary, we assume that $\alpha(0)$ and $\beta(0)$ both lie on ∂H. Let $\partial H \colon \mathbf{R} \to \partial H$ denote a unit-speed parameterization that is positive with respect to the orientation of H, and let H' be the closure of the component of $H \setminus (\alpha \cup \beta)$ containing $\partial H(-t)$ for all sufficiently large $t \ge 0$ (see Figure 3.2.8). Since both α and β are simple, there exist $p_i \in H' \cap \alpha \cap \beta, i = 1, 2, \ldots,$ tending to infinity. For each i, we can find a minimal segment σ_i in H' from a fixed point $p_0 \in \partial H'$ to p_i; we denote by D_i the compact region in H' bounded by σ_i. The closure of $\bigcup_i D_i$ is a region in H' bounded by the limit ray $\lim_{i \to \infty} \sigma_i =: \sigma_-$.

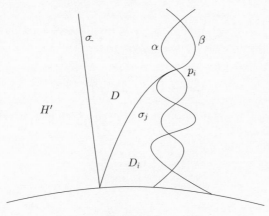

Figure 3.2.8

A discussion similar to that in Proposition 3.2.3 shows that σ_i properly converges to σ_-. Applying Proposition 3.2.3 yields $\sigma_- \in \mathcal{C}_H$ and

$$d_\infty(\sigma_-, \alpha) = d_\infty(\sigma_-, \beta) = 0.$$

The existence of σ_+ is proved in the same way. $\square$

Lemma 3.2.4

(1) *If three curves* $\alpha, \beta, \gamma \in \mathcal{C}_H$ *satisfy* $\alpha \leq \beta \leq \gamma$ *then*

$$d_\infty(\alpha, \gamma) = d_\infty(\alpha, \beta) + d_\infty(\beta, \gamma).$$

(2) *If four curves* $\alpha, \beta, \gamma_-, \gamma_+ \in \mathcal{C}_H$ *satisfy that* α *and* β *cross near infinity and that* $\gamma_- \leq \alpha, \beta \leq \gamma_+$ *then*

$$d_\infty(\alpha, \gamma_\pm) = d_\infty(\beta, \gamma_\pm)$$

(see Figure 3.2.10).

(3) *If three curves* $\alpha, \beta, \gamma \in \mathcal{C}_H$ *satisfy that* α *and* β *cross near infinity and that so do* β *and* γ, *and if* $\alpha \leq \gamma$, *then*

$$d_\infty(\alpha, \gamma) = 0$$

(see Figure 3.2.9).

Proof. (1): Obvious (cf. Lemma 3.1.1).

(2): Applying Lemma 3.2.3 for α and β in $D(\gamma_-, \gamma_+)$ for some core, we can find $\sigma_\pm$ such that $\gamma_- \leq \sigma_- \leq \alpha, \beta \leq \sigma_+ \leq \gamma_+$ and $d_\infty(\sigma_\pm, \alpha) =$

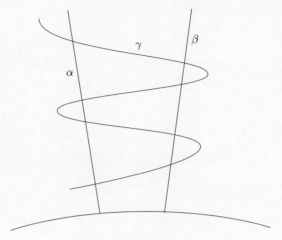

Figure 3.2.9

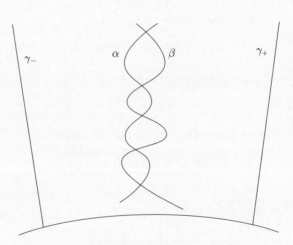

Figure 3.2.10

$d_\infty(\sigma_\pm, \beta) = 0$. Hence, by (1),

$$d_\infty(\alpha, \gamma_\pm) = d_\infty(\sigma_\pm, \gamma_\pm) = d_\infty(\beta, \gamma_\pm).$$

(3): By Lemma 3.2.3, there exist $\sigma_-, \sigma_+ \in \mathcal{C}_H$ such that $\sigma_- \leq \alpha, \beta, \gamma \leq \sigma_+$ and $d_\infty(\sigma_-, \alpha) = d_\infty(\sigma_-, \beta) = d_\infty(\beta, \sigma_+) = d_\infty(\gamma, \sigma_+) = 0$. Hence

$$d_\infty(\alpha, \gamma) = d_\infty(\sigma_-, \sigma_+) = d_\infty(\sigma_-, \beta) + d_\infty(\beta, \sigma_+) = 0. \qquad \square$$

Proposition 3.2.4. *The function* $d_\infty \colon \mathcal{C}_H \times \mathcal{C}_H \to \mathbf{R} \cup \{+\infty\}$ *is a pseudo-distance function.*

Proof. It suffices to prove the triangle inequality

$$d_\infty(\alpha, \gamma) \le d_\infty(\alpha, \beta) + d_\infty(\beta, \gamma) \qquad \text{for any } \alpha, \beta, \gamma \in \mathcal{C}_H.$$

If $d_\infty(\alpha, \gamma) = 0$ then the inequality is trivial. We can assume without loss of generality that $d_\infty(\alpha, \gamma) > 0$ and $\alpha \le \gamma$. If $\beta \le \alpha \le \gamma$, $\alpha \le \beta \le \gamma$ or $\alpha \le \gamma \le \beta$ holds then applying (1) of Lemma 3.2.4 yields the triangle inequality. If none of these hold, (2) and (3) of Lemma 3.2.4 prove the triangle inequality. $\square$

3.3 Riemannian half-cylinders and their universal coverings

Let N be a Riemannian half-cylinder for which the curvature at infinity exists.

Lemma 3.3.1. *If $\lambda_\infty(N) = 0$ then $d_\infty(\alpha, \beta) = 0$ for any $\alpha, \beta \in \mathcal{C}_N$.*

Proof. Let $\alpha, \beta \in \mathcal{C}_N$ be two curves. If α and β cross near infinity then the conclusion is trivial. Otherwise,

$$d_\infty(\alpha, \beta) \le \lambda_\infty(D(\alpha, \beta)) + \lambda_\infty(D(\beta, \alpha)) = \lambda_\infty(N) = 0. \qquad \square$$

Let $pr \colon \tilde{N} \to N$ be the universal covering. For any curve $\alpha \in \mathcal{C}_N$, denote by α^n for $n \in \mathbf{Z}$ all the different lifts in $\tilde{N}$ of α such that $\cdots \le \alpha^{-1} \le \alpha^0 \le \alpha^1 \le \cdots$. Let σ be a ray in N such that $\sigma \cap \partial N = \{\sigma(0)\}$. Then the σ^n for all $n \in \mathbf{Z}$ do not intersect each other and in particular do not cross each other. For $m < n$, let $I(\sigma^m, \sigma^n)$ be the subarc of $\partial \tilde{N}$ from $\sigma^m(0)$ to $\sigma^n(0)$, and let $D(\sigma^m, \sigma^n)$ be the Riemannian half-plane in $\tilde{N}$ bounded by σ^n, $I(\sigma^m, \sigma^n)$ and σ^m; see Figure 3.3.1. (The boundary $\partial \tilde{N}$ could be considered to be a core of $\tilde{N}$.)

Exercise 3.3.1. Prove that any two curves $\alpha, \beta \in \mathcal{C}_{D(\sigma^m, \sigma^n)}$, $m < n$, satisfy

$$d_\infty(pr \circ \alpha, pr \circ \beta) \le d_\infty(\alpha, \beta).$$

Lemma 3.3.2. *For any two integers $m < n$ and for any curve c in $D(\sigma^m, \sigma^n)$ connecting σ^m and σ^n, we have*

$$|\lambda(c)| \ge (n - m)\left(-\lambda(\partial N) - \int_N G_+ \, dN\right) - \pi.$$

Proof. Applying the Gauss–Bonnet theorem to the compact region, say D, in $D(\sigma^m, \sigma^n)$ that is surrounded by σ^m, $I(\sigma^m, \sigma^n)$, σ^n and c yields

$$(n - m)\int_N G_+ \, dN \ge c(D) \ge -\pi - \lambda(c) - (n - m)\lambda(\partial N)$$

with respect to a suitable side of c, which completes the proof. $\square$

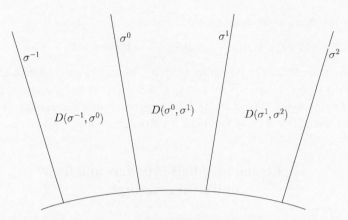

Figure 3.3.1

Lemma 3.3.3. *If $\lambda_\infty(N) > 0$ then there exists a Riemannian sub-half-cylinder N' of N such that*

$$-\lambda(\partial N') - \int_{N'} G_+\, dN > 0.$$

Proof. Let N' be any Riemannian sub-half-cylinder of N such that $\partial N'$ intersects σ perpendicularly at only one point. Then N' is a neighborhood of the end of N, and it suffices to prove that if N' is decreasing and $N \setminus N'$ tends to N then $-\lambda(\partial N') - \int_{N'} G_+\, dN$ tends to $\lambda_\infty(N)$. In fact, if $\lambda_\infty(N) < +\infty$ (or equivalently if $c(N)$ is finite) then

$$-\lambda(\partial N') - \int_{N'} G_+\, dN = \lambda_\infty(N) - \int_{N'} G_-\, dN,$$

and the right-hand side of this tends to $\lambda_\infty(N)$. If $\lambda_\infty(N) = +\infty$ (or equivalently if $c(N) = -\infty$) then

$$-\lambda(\partial N') = -\lambda(\partial N) - c(N \setminus N'),$$

which tends to $+\infty$ and, besides, $\int_{N'} G_+\, dN$ tends to zero. $\qquad\square$

Lemma 3.3.4. *If $\lambda_\infty(N) > 0$ then any curve $\tilde{\alpha} : [0, +\infty) \to \tilde{N}$ with $pr \circ \tilde{\alpha} \in \mathcal{C}_N$ is entirely contained in $D(\sigma^m, \sigma^n)$ for integers $m < n$.*

Proof. Let N' be as in Lemma 3.3.3, and let $\tilde{\alpha}: [0, +\infty) \to \tilde{N}$ be a curve such that $pr \circ \tilde{\alpha} \in \mathcal{C}_N$. Then there exists a number $t \geq 0$ such that $pr \circ \tilde{\alpha}|_{[t,+\infty)} \subset N'$. It follows that

$$|\lambda(\tilde{\alpha}|_{[t,+\infty)})| < k \left(-\lambda(\partial N') - \int_{N'} G_+ dN \right) - \pi$$

for sufficiently large $k \in \mathbf{N}$, so that Lemma 3.3.2 completes the proof. $\qquad\square$

Proposition 3.3.1. *The function* $d_\infty \colon \mathcal{C}_N \times \mathcal{C}_N \to \mathbf{R} \cup \{+\infty\}$ *is a pseudo-distance function.*

Proof. By Lemma 3.3.1, it suffices to consider the case where $\lambda_\infty(N) > 0$. To prove the triangle inequality

$$d_\infty(\alpha, \gamma) \le d_\infty(\alpha, \beta) + d_\infty(\beta, \gamma) \qquad \text{for any } \alpha, \beta, \gamma \in \mathcal{C}_N$$

we find a ray σ from ∂N. By Lemma 3.3.4, there exist two integers $m < n$ and lifts $\tilde{\alpha}, \tilde{\beta}, \tilde{\gamma}$ in $D(\sigma^m, \sigma^n)$ of α, β, γ such that

$$d_\infty(\alpha, \beta) = d_\infty(\tilde{\alpha}, \tilde{\beta}) \qquad \text{and} \qquad d_\infty(\beta, \gamma) = d_\infty(\tilde{\beta}, \tilde{\gamma}).$$

Applying Proposition 3.2.4 to $H := D(\sigma^m, \sigma^n)$ yields

$$d_\infty(\alpha, \gamma) \le d_\infty(\tilde{\alpha}, \tilde{\gamma}) \le d_\infty(\tilde{\alpha}, \tilde{\beta}) + d_\infty(\tilde{\beta}, \tilde{\gamma}) = d_\infty(\alpha, \beta) + d_\infty(\beta, \gamma). \qquad \square$$

For a finitely connected 2-manifold M admitting a curvature at infinity, we obtain the following:

Theorem 3.3.1. *The function* $d_\infty \colon \mathcal{C}_M \times \mathcal{C}_M \to \mathbf{R} \cup \{+\infty\}$ *is a pseudo-distance function.*

Proof. Since $\mathcal{C}_M$ splits into components $\mathcal{C}_V$ for all connected components V of $\overline{M \setminus C}$, where C is a core of M, the theorem follows from Propositions 3.2.4 and 3.3.1. $\qquad \square$

Lemma 3.3.5. *Assume that* $\lambda_\infty(N) > 0$, *and let* $\{\tilde{\alpha}_i \colon [0, a_i) \to \tilde{N}\}$ *be a sequence of piecewise-smooth curves converging to a curve* $\tilde{\alpha} \in \mathcal{C}_{\tilde{N}}$. *If* $\{pr \circ \tilde{\alpha}_i\}$ *properly converges the to* $pr \circ \tilde{\alpha}$ *then there exist two integers* $m < n$, *independent of* i, *such that the* $\tilde{\alpha}_i$ *for all* i *and* $\tilde{\alpha}$ *are entirely contained in* $D(\sigma^m, \sigma^n)$.

Proof. By Lemma 3.3.3, substituting a sub-half-cylinder of N we may assume that

$$\epsilon := -\lambda(\partial N) - \int_N G_+ \, dN > 0.$$

From Lemma 3.3.4, there exist two integers $m' < n'$ such that $\tilde{\alpha}$ is contained in the interior of $D(\sigma^{m'}, \sigma^{n'})$. Define two integers m and n to satisfy

$$m' - m = n - n' = \min\{k \in \mathbf{N}; k > \pi/\epsilon\}.$$

Suppose that there exists a subsequence $\{\tilde{\alpha}_{j(i)}\}$ of $\{\tilde{\alpha}_i\}$ such that no $\alpha_{j(i)}$ is contained in $D(\sigma^m, \sigma^n)$. Assume without loss of generality that every $\tilde{\alpha}_{j(i)}$ intersects both σ^m and $\sigma^{m'}$ (see Figure 3.3.2). Since $\tilde{\alpha}_{j(i)}$ tends to $\tilde{\alpha}$ as $i \to \infty$, each $\tilde{\alpha}_{j(i)}$ has a subarc $\tilde{\alpha}_{j(i)}|_{[t_i, t'_i]}$ connecting σ^m and $\sigma^{m'}$ such that t_i and t'_i both

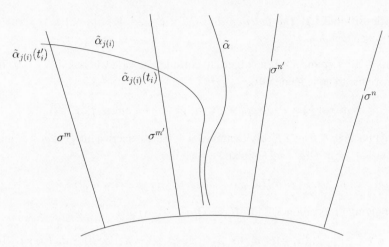

Figure 3.3.2

tend to infinity. Lemma 3.3.2 implies that

$$\lambda_{\mathrm{abs}}(\tilde{\alpha}_{j(i)}|_{[t_i, a_{j(i)})}) \geq |\lambda(\tilde{\alpha}_{j(i)}|_{[t_i, t_i']})| \geq (m' - m)\epsilon - \pi > 0,$$

which contradicts (P2), Definition 3.2.2, for $pr \circ \tilde{\alpha}_{j(i)}$. □

Theorem 3.3.2. *If α, $\beta \in \mathcal{C}_M$ are parallel then $d_\infty(\alpha, \beta) = 0$.*

Proof. When $\lambda_\infty(N) = 0$, the theorem follows from Lemma 3.3.1, so assume that $\lambda_\infty(N) > 0$, and let α, $\beta \in \mathcal{C}_N$ be parallel. Find two sequences $\{\alpha_i : [0, a_i] \to N\}$ and $\{\beta_i : [0, b_i] \to N\}$ of piecewise-smooth curves properly converging to α and β such that $\alpha_i(a_i) = \beta_i(b_i)$. There exist lifts $\tilde{\alpha}_i$, $\tilde{\alpha}$, $\tilde{\beta}_i$, $\tilde{\beta}$ of α_i, α, β_i, β for every i, such that $\{\tilde{\alpha}_i\}$ and $\{\tilde{\beta}_i\}$ properly converge to $\tilde{\alpha}$ and $\tilde{\beta}$ respectively and $\tilde{\alpha}_i(a_i) = \tilde{\beta}_i(b_i)$. By Lemma 3.3.5, there exist two integers $m < n$ independent of i such that $\tilde{\alpha}_i$, $\tilde{\alpha}$, $\tilde{\beta}_i$, $\tilde{\beta}$ are all contained in $D(\sigma^m, \sigma^n)$. Since $\{\tilde{\alpha}_i\}$ and $\{\tilde{\beta}_i\}$ properly converge to $\tilde{\alpha}$ and $\tilde{\beta}$, applying Proposition 3.2.3 yields $d_\infty(\tilde{\alpha}, \tilde{\beta}) = 0$. □

3.4 The ideal boundary and its topological structure

Let us now give the definition of the ideal boundary of a finitely connected 2-manifold M for which the curvature at infinity exists.

Definition 3.4.1 (Ideal boundary). The *ideal boundary $M(\infty)$ of M* is defined to be the quotient space of $\mathcal{C}_M$ modulo $d_\infty(\cdot, \cdot) = 0$. The pseudo-distance function d_∞ on $\mathcal{C}_M$ induces a distance function on $M(\infty)$, which we call the *Tits* *distance* and also denote as d_∞. Denote by $\alpha(\infty)$ the class in $M(\infty)$ represented by $\alpha \in \mathcal{C}_M$.

The ideal boundary $M(\infty)$ has the decomposition

$$M(\infty) = \bigcup_V V(\infty),$$

where V is any connected component of $\overline{M \setminus C}$ for a fixed core C of M.

Lemma 3.4.1. *For any* $x \in M(\infty)$ *and any compact subset* K *of* M, *there exists a ray* σ *from* K *such that* $\sigma(\infty) = x$.

Proof. Let $K \subset M$ be a compact subset, and $x \in M(\infty)$. We can find a curve $\alpha \in \mathcal{C}_M$ such that $\alpha(\infty) = x$. For a sequence $\{t_i \to +\infty\}$ of positive numbers, let σ_i for $i = 1, 2, \ldots$, be minimal segments from K to $\alpha(t_i)$. Then there exists a subsequence $\{\sigma_{j(i)}\}$ of $\{\sigma_i\}$ converging to a ray σ from K. Since α and σ are parallel, we obtain $\sigma(\infty) = x$. $\square$

Set $\overline{M}^\infty := M \cup M(\infty)$ (a disjoint union) and call it the *compactification of* M. We will define a topology of $\overline{M}^\infty$. Let D be any finitely connected closed noncompact region in M such that $\partial D \setminus C_D$ for a core C_D of D consists of the images of finitely many curves in $\mathcal{C}_M$. Then the curvature at infinity of D exists, and $D(\infty)$ is defined. There is a natural embedding $D(\infty) \subset M(\infty)$, so that $\overline{D}^\infty := D \cup D(\infty)$ is embedded into $\overline{M}^\infty$. In the same way, the curvature at infinity of $D' := \overline{M \setminus D}$ exists and we have $\overline{D'}^\infty = D' \cup D'(\infty) \subset \overline{M}^\infty$ and $\overline{D}^\infty \cup \overline{D'}^\infty = \overline{M}^\infty$.

Let $\{\mathcal{N}_x\}_{x \in M}$ be a canonical fundamental neighborhood system of M and, for $x \in M(\infty)$, let $\mathcal{N}_x$ be the family of all $\overline{D}^\infty$, where D is as above and satisfies $x \notin D'(\infty)$.

Definition 3.4.2 (Topology of $\overline{M}^\infty$). A topology of $\overline{M}^\infty$ is defined to have $\{\mathcal{N}_x\}_{x \in \overline{M}^\infty}$ as a fundamental neighborhood system. We usually employ the restricted topology of $\overline{M}^\infty$ on $M(\infty)$.

Trivially, the inclusion map $M \hookrightarrow \overline{M}^\infty$ is an into-homeomorphism. For a Riemannian half-plane H admitting a curvature at infinity, the relation $\leq$ on $\mathcal{C}_H$ induces a total order relation on $H(\infty)$, say $\leq$ also, which is compatible with the restricted topology on $H(\infty)$. Note that the Tits metric d_∞ is not necessarily compatible with the topology of $M(\infty)$ defined here (see Example 3.5.1 below).

Once the canonical topology of $\overline{M}^\infty$ has been defined, any curve $\alpha \in \mathcal{C}_M$ naturally extends to a curve defined on $[0, +\infty]$ continuously, i.e., $\alpha(t)$ tends to $\alpha(\infty) =: \alpha(+\infty)$ as $t \to +\infty$. Note however that, since we do not know at present whether $\overline{M}^\infty$ is a Hausdorff space, the limit of $\alpha(t)$ as $t \to +\infty$

is not necessarily unique. We will prove later that $\overline{M}^\infty$ is a two-dimensional topological manifold with boundary.

A curve $\alpha: [0, +\infty] \to \overline{M}^\infty$ is called a *piecewise-smooth curve* if $\alpha|_{[0,+\infty)} \in \mathcal{C}_M$ and $\alpha(+\infty) = \alpha(\infty)$.

Lemma 3.4.2. *Let* $\{\alpha_i: [0, a_i] \to \overline{M}^\infty\}_{i=1,2,...}$ *be a sequence of piecewise-smooth curves, where* $0 \leq a_i \leq +\infty$. *If* $\{\alpha_i|_{[0,a_i)}\}$ *converges to a curve* $\alpha \in \mathcal{C}_M$ *then*

$$\lim_{i\to\infty} \alpha_i(a_i) = \alpha(\infty).$$

Proof. Find any fixed neighborhood $\overline{D}^\infty \in \mathcal{N}_{\alpha(\infty)}$. It will suffice to show that $\alpha_i(a_i) \in \overline{D}^\infty$ for all sufficiently large i. Suppose the contrary. Since $\alpha[t_0, +\infty)$ $\subset \text{int } D$ for some large $t_0 \geq 0$, there is a sequence $b_i \to \infty$ such that $\alpha_i(b_i) \in \text{int } D$ for all sufficiently large i. Since $\alpha_i(a_i) \notin \overline{D}^\infty$, we have $\alpha_i[b_i, a_i] \cap \partial D \neq \emptyset$ for all sufficiently large i, so that α is parallel to a curve contained in $\partial D = \partial D'$ that belongs to $\mathcal{C}_{D'}$. This proves that $\alpha(\infty) \in D'(\infty)$, which contradicts $\overline{D}^\infty \in \mathcal{N}_{\alpha(\infty)}$. $\square$

Let C be a core of M. The set $\overline{\partial M \setminus C}$ consists of finitely many disjoint curves in $\mathcal{C}_M$. Denote these boundary curves by $\partial_1, \ldots, \partial_k : [0, +\infty) \to \partial M$. We set

$$\mathcal{R}_C := \{\text{all rays in } M \text{ from } C\} \cup \{\partial_1, \ldots, \partial_k\}.$$

For any $p \in M \setminus C$ denote by $\mathcal{R}'_{C,p}$ the set of minimal segments from C to p, and put

$$\mathcal{R}_{C,p} := \begin{cases} \mathcal{R}'_{C,p} & \text{if } p \text{ lies on none of the } \partial_i, \\ \mathcal{R}'_{C,p} \cup \{\partial_{i(p)}|_{[0,\ell(p)]}\} & \text{if } p \text{ lies on } \partial_{i(p)} \text{ for some } i(p), \end{cases}$$

where $\ell(p) \geq 0$ is a number such that $p = \partial_{i(p)}(\ell(p))$. Note that $\mathcal{R}_C$ and $\mathcal{R}_{C,p}$ are sequentially compact with respect to the pointwise-convergence topology.

The following lemma is important.

Lemma 3.4.3. *Let* $\sigma, \tau \in \mathcal{R}_C$ *be two different curves for which* $D(\sigma, \tau)$ *is defined with respect to the core* C *(see Section 3.2) and contains no curves in* $\mathcal{R}_C$ *except* σ *and* τ. *Then there exist a sequence* $\{p_i\}$ *in* $D(\sigma, \tau)$ *with* $\lim_{i\to\infty} d(p_i, C) = +\infty$ *and two sequences of curves* $\sigma_i, \tau_i \in \mathcal{R}_{C,p_i}$ *that converge to* σ *and* τ *respectively.*

Proof. Find a continuous arc $c : [0, \ell] \to D(\sigma, \tau)$ connecting σ to τ that does not intersect σ and τ except at $c(0)$ and $c(\ell)$. Let $\epsilon > 0$ be a small number. We first claim that there exists a connected neighborhood $U_\epsilon \subset D(\sigma, \tau)$ of the end of $D(\sigma, \tau)$ such that any curve in $\mathcal{R}_{C,p}$ with $p \in U_\epsilon$ does not intersect $c[\epsilon, \ell - \epsilon]$. In fact, if not there are sequences $p_i \in D(\sigma, \tau)$ and $\gamma_i \in \mathcal{R}_{C,p_i}$ such that $\lim_i d(p_i, C) = +\infty$ and each γ_i intersects $c[\epsilon, \ell - \epsilon]$. A limit of γ_i intersects $c[\epsilon, \ell - \epsilon]$ and is an element of $\mathcal{R}_C$, which contradicts the assumption of the lemma. Thus the above claim has been proved.

We now set

$$E_\epsilon \ (\text{resp. } F_\epsilon) := \{p \in U_\epsilon; \text{ some curve in } \mathcal{R}_{C,p}$$
$$\text{intersects } c[0, \epsilon] \ (\text{resp. } c[\ell - \epsilon, \ell])\}.$$

It then follows from the claim that $E_\epsilon \cup F_\epsilon = U_\epsilon$. Since any limit of minimal curves intersecting a compact set A is a minimal curve intersecting A, both E_ϵ and F_ϵ are closed in U_ϵ. Therefore, from the connectivity of U_ϵ, we have $E_\epsilon \cap F_\epsilon \neq \emptyset$.

Let $\epsilon_i \to 0$ be a sequence of small positive numbers. Without loss of generality we may assume that $d(C, U_{\epsilon_i}) \to \infty$. Take a point p_i in the nonempty intersection $E_{\epsilon_i} \cap F_{\epsilon_i}$ for each i. Then for every i there exist two arcs $\sigma_i, \tau_i \in \mathcal{R}_{C,p_i}$ such that σ_i intersects E_{ϵ_i} and τ_i intersects F_{ϵ_i}. This completes the proof. $\square$

Let $\sigma, \tau \in \mathcal{R}_C$, $p_i \in D(\sigma, \tau)$ and $\sigma_i, \tau_i \in \mathcal{R}_{C,p_i}$ be as in Lemma 3.4.3. Then the union $\sigma_i \cup \tau_i$ divides $D(\sigma, \tau)$ into two closed domains, one of which is a compact subset, say D_i, with $\bigcup_i D_i = D(\sigma, \tau)$. Lemma 3.2.2 (cf. Proposition 3.2.3) implies:

Corollary 3.4.1. *The inner angle of D_i at p_i tends to zero. The curves σ and τ are parallel in $D(\sigma, \tau)$, and $\lambda_\infty(D(\sigma, \tau)) = 0$.*

Let σ_x for $x \in M(\infty)$ denote a ray from C to x, i.e., $d(\sigma_x(t), C) = t$ and $\sigma_x(\infty) = x$. The existence of σ_x follows from Lemma 3.4.1. A closed region D_x in M for $x \in M(\infty)$ is defined as follows. For each $x \in M(\infty)$, if the component V of $\overline{M \setminus C}$ with $V(\infty) \ni x$ satisfies $\lambda_\infty(V) = 0$ then $D_x := V$. Otherwise,

$$D_x := \bigcup \{D(\sigma, \tau); \sigma, \tau \in \mathcal{R}_C \text{ are such that } \sigma(\infty) = \tau(\infty) = x,$$
$$D(\sigma, \tau) \text{ is defined and } \lambda_\infty(D(\sigma, \tau)) = 0\}.$$

By Lemma 3.4.2, there exist $\sigma_x^-, \sigma_x^+ \in \mathcal{R}_C$ for any $x \in M(\infty)$ such that $D_x = D(\sigma_x^-, \sigma_x^+)$ provided that the component V corresponding to x satisfies $\lambda_\infty(V) > 0$.

Lemma 3.4.4. *We have*

$$\bigcup_{x \in M(\infty)} D_x = \overline{M \setminus C}.$$

Proof. Let $p \in \overline{M \setminus C}$ be a point. If some curve in $\mathcal{R}_C$ passes through p then obviously we have $p \in D_x$ for some $x \in M(\infty)$. Assume that no curve in $\mathcal{R}_C$ passes through p. Then there exist $\sigma, \tau \in \mathcal{R}_C$ such that $p \in D(\sigma, \tau) \setminus (\sigma \cup \tau)$. The sequential compactness of $\mathcal{R}_C$ shows that we can choose σ and τ such that $D(\sigma, \tau)$ is minimal with respect to the inclusion relation. This $D(\sigma, \tau)$ satisfies the assumption of Lemma 3.4.3, so that $\sigma(\infty) = \tau(\infty) =: x$. Thus we obtain $p \in D(\sigma, \tau) \subset D_x$. $\qquad\square$

Define a map $\Pi_C \colon \overline{M \setminus C} \to M(\infty)$ by the following: for any $p \in \overline{M \setminus C}$, we can find a point $x \in M(\infty)$ with $p \in D_x$ and set $\Pi_C(p) := x$. Note that, for a given point $p \in M$, an $x \in M(\infty)$ with $p \in D_x$ is not necessarily unique, because a ray may branch at a boundary point of M. For the definition of Π_C we choose one such x. In the following lemma, we are concerned with the uniqueness of x. Setting $\partial M(\infty) := \{\partial_1(\infty), \ldots, \partial_k(\infty)\}$, where $\partial_1, \ldots, \partial_k$ are defined as in the proof of Lemma 3.4.2, we have:

Lemma 3.4.5. *For any neighborhood U in $M(\infty)$ of $\partial M(\infty)$, there exists a compact subset K_U of M with the following property: for each $p \in M \setminus K_U$, if a point $x \in M(\infty)$ with $p \in D_x$ belongs to U then such a point x is unique.*

Proof. If the property does not hold then there exists a sequence $\{\sigma_i\}$ of rays from C such that $\sigma_i \cap \partial M$ is nonempty and tends to infinity as $i \to \infty$ and such that $\sigma_i(\infty) \notin U$. Replacing the sequence by a subsequence, we assume that σ_i tends to a ray σ. Then σ is parallel to at least one of $\partial_1, \ldots, \partial_k$. Applying Theorem 3.3.2 yields $\sigma(\infty) \in \partial M(\infty)$. On the contrary, however, the fact $\sigma_i(\infty) \notin U$ and Lemma 3.4.2 together imply that $\sigma(\infty) \notin \operatorname{int} U$. $\qquad\square$

Theorem 3.4.1

(1) *If $\lambda_\infty(H) = 0$ then $H(\infty)$ consists of a single point.*

(2) *If $\lambda_\infty(H) > 0$ then $H(\infty)$ is homeomorphic to a nontrivial line segment.*

Proof. (1): Any curves $\alpha, \beta \in \mathcal{C}_H$ with $\alpha \le \beta$ satisfy

$$d_\infty(\alpha, \beta) = \lambda_\infty(D(\alpha, \beta)) \le \lambda_\infty(H) = 0.$$

(2): Let $z, w \in H(\infty)$ be any fixed points such that $\partial H(-\infty) < z < w < \partial H(+\infty)$, where $\partial H(-\infty)$ and $\partial H(+\infty)$ are defined to be the points in $\partial H(\infty)$ such that $\partial H(-\infty) \le \partial H(+\infty)$. It suffices to prove that $[z, w]$, the set of all $x \in H(\infty)$ with $z \le x \le w$, is homeomorphic to $[0, 1]$. For a fixed core C_H of H, set $D_{z,w} := D(\sigma_z^-, \sigma_w^+)$ and let K_U be as in Lemma 3.4.5 for $U := H(\infty) \setminus [z, w]$. We can find a curve $c \colon [0, 1] \to D_{z,w} \setminus K_U$ connecting σ_z^+ to σ_w^- in such a way that $\Pi \circ c$ is monotone nondecreasing in the sense of the order relation $\leq$ on $H(\infty)$, where $\Pi := \Pi_{C_H} : \overline{H \setminus C_H} \to M(\infty)$. We will construct a homeomorphism $f : [0, 1] \to [z, w]$. Let us first define $f(j/2^k)$ for all $k = 0, 1, \ldots$, and $j = 0, 1, 2, 3, \ldots, 2^k$ inductively. Set $f(0) := z$ and $f(1) := w$. For an integer $k \geq 0$ we assume $f(j/2^k)$ for all $j = 0, 1, 2, 3, \ldots, 2^k$ to be defined. For all $j = 0, 1, 2, 3, \ldots, 2^k - 1$, define

$$f\left(\frac{2j+1}{2^{k+1}}\right)$$

$$:= \Pi \circ c \left(\frac{1}{2} \inf (\Pi \circ c)^{-1} f\left(\frac{j+1}{2^k}\right) - \frac{1}{2} \sup (\Pi \circ c)^{-1} f\left(\frac{j}{2^k}\right) \right)$$

Since $\{f(j/2^k); k \geq 0, 0 \leq j \leq 2^k\} \subset [z, w]$ is dense and f preserves the order relations, f is extended to a homeomorphism from $[0, 1]$ to $[z, w]$ by taking completion. $\qquad \square$

Let $pr : \tilde{N} \to N$ be the universal covering of a Riemannian half-cylinder N admitting a curvature at infinity, and let σ be a ray such that $\sigma \cap \partial N = \{\sigma(0)\}$. Set $H_n := D(\sigma^{-n}, \sigma^n)$ for $n \in \mathbf{N}$ (see Section 3.3). For any two natural numbers $m < n$ we have $H_m \subset H_n$, which induces a natural embedding $H_m(\infty) \hookrightarrow H_n(\infty)$ preserving the topology and the Tits distance d_∞. Therefore $\{H_n(\infty)\}$ is an inductive system. Denote by $\tilde{N}(\infty)$ the inductive limit of $\{H_n(\infty)\}$, and define the map $pr_\infty : \tilde{N}(\infty) \to N(\infty)$ as follows. For $x \in \tilde{N}(\infty)$, we find an $n \in \mathbf{N}$ with $x \in H_n(\infty)$ and take an $\alpha \in C_{H_n}$ such that $\alpha(\infty) = x$, so that $pr_\infty(x) := (pr \circ \alpha)(\infty)$. It can be easily checked that pr_∞ is open, continuous and d_∞-nonincreasing. Setting $\overline{\tilde{N}}^\infty := \tilde{N} \cup \tilde{N}(\infty)$ we have the natural projection $\overline{pr}^\infty : \overline{\tilde{N}}^\infty \to \overline{N}^\infty$, defined by

$$\overline{pr}^\infty(x) := \begin{cases} pr(x) & \text{for } x \in \tilde{N}, \\ pr_\infty(x) & \text{for } x \in \tilde{N}(\infty). \end{cases}$$

Let $G := \pi_1(N)$ ($\approx \mathbf{Z}$). There exists a $g_0 \in G$ such that $g_0 \circ \sigma^n = \sigma^{n+1}$ for any $n \in \mathbf{Z}$. It follows that G is generated by g_0 only. An action of G to $\tilde{N}(\infty)$ is defined by

$$g \cdot \alpha(\infty) := (g \circ \alpha)(\infty)$$

for any $g \in G$ and any $\alpha(\infty) \in \tilde{N}(\infty)$, $\alpha \in C_{\tilde{N}}$. This action is continuous and d_∞-isometric.

Proposition 3.4.1. *If $\tilde{N}(\infty)$ contains different two points then $pr_\infty : \tilde{N}(\infty) \to N(\infty)$ is a covering map with transformation group G and is a d_∞-local isometry.*

Proof. The map $\varphi: N(\infty) \to \tilde{N}(\infty)/G$, $\alpha(\infty) \mapsto \{\alpha^n(\infty)\}_{n\in\mathbf{Z}}$ is a homeomorphism and a d_∞-isometry, and $\varphi \circ pr_\infty$ is just the projection $\tilde{N}(\infty) \to \tilde{N}(\infty)/G$. Moreover, G is fixed-point free unless $\tilde{N}(\infty)$ consists of a single point. This completes the proof. $\square$

Theorem 3.4.2

(1) *If* $\lambda_\infty(N) = 0$ *then* $N(\infty)$ *and* $\tilde{N}(\infty)$ *each consist of a single point.*
(2) *If* $\lambda_\infty(N) > 0$ *then* $\tilde{N}(\infty)$ *is homeomorphic to* $\mathbf{R}$ *and* $N(\infty)$ *to* S^1.

Proof. (1): Assume that $\lambda_\infty(N) = 0$. Then Lemma 3.3.1 implies that $N(\infty)$ consists of a single point. Since the Riemannian half-planes $H_n \subset \tilde{N}$, $n \in \mathbf{N}$ satisfy $\lambda_\infty(H_n) = 2n\lambda_\infty(N) = 0$, Theorem 3.4.1 implies that each $H_n(\infty)$ consists of a single point and that so does $\tilde{N}(\infty)$.

(2): Assume that $\lambda_\infty(N) > 0$. Since, by (2) of Theorem 3.4.1, each $H_n(\infty)$ is homeomorphic to a nontrivial line segment, so is $\tilde{N}(\infty)$ to $\mathbf{R}$. Proposition 3.4.1 completes the proof. $\square$

Since each connected component of $\overline{M \setminus C}$ is either a Riemannian half-plane or a Riemannian half-cylinder, Theorems 3.4.1 and 3.4.2 together determine the topological structure of $M(\infty)$ completely and lead to the following theorem.

Theorem 3.4.3. *The compactification* $\overline{M}^\infty$ *of* M *is a two-dimensional compact topological manifold with boundary.*

3.5 The structure of the Tits metric d_∞

Let us consider a Riemannian half-plane H admitting a curvature at infinity.

Lemma 3.5.1. *Let* C_H *be a core of* H. *Assume that a sequence* $\{\sigma_i\}$ *of rays from* C_H *converges to a ray* σ *and that* $\sup_{i,j} d_\infty(\sigma_i, \sigma_j) < +\infty$. *Then we have*

$$\lim_{i\to\infty} d_\infty(\sigma_i, \sigma) = 0.$$

Proof. Without loss of generality, it may be assumed that $H' := D(\sigma_1, \sigma)$ is defined and contains every σ_i entirely. It follows that for each i and any compact subset K of H',

$$d_\infty(\sigma_1, \sigma_i) = \pi - \lambda(\partial D(\sigma_1, \sigma_i) \cap K) - c(D(\sigma_1, \sigma_i) \cap K)$$
$$- \lambda(\partial D(\sigma_1, \sigma_i) \setminus K) - c(D(\sigma_1, \sigma_i) \setminus K).$$

Here, we have

$$-\lambda_-(\partial H' \setminus K) \le \lambda(\partial D(\sigma_1, \sigma_i) - K) \le \lambda_+(\partial H' \setminus K),$$

$$-\int_{H' \setminus K} G_- \, dH \le c(D(\sigma_1, \sigma_i) \setminus K) \le \int_{H' \setminus K} G_+ \, dH,$$

$$\lim_{i \to \infty} \lambda(\partial D(\sigma_1, \sigma_i) \cap K) = \lambda(\partial H' \cap K),$$

$$\lim_{i \to \infty} c(D(\sigma_1, \sigma_i) \cap K) = c(H' \cap K).$$

Therefore

$$\pi - \lambda(\partial D(\sigma_1, \sigma) \cap K) - c(D(\sigma_1, \sigma) \cap K)$$

$$-\lambda_+(\partial H' \setminus K) - \int_{H' \setminus K} G_+ \, dH$$

$$\le \liminf_{i \to \infty} d_\infty(\sigma_1, \sigma_i) \le \limsup_{i \to \infty} d_\infty(\sigma_1, \sigma_i)$$

$$\le \pi - \lambda(\partial D(\sigma_1, \sigma) \cap K) - c(D(\sigma_1, \sigma) \cap K)$$

$$+\lambda_-(\partial H' \setminus K) + \int_{H' \setminus K} G_- \, dH. \tag{3.5.1}$$

As K tends to H', the first formula in (3.5.1) converges to $d_\infty(\sigma_1, \sigma)$, so that

$$d_\infty(\sigma_1, \sigma) \le \liminf_{i \to \infty} d_\infty(\sigma_1, \sigma_i),$$

which is finite by the assumption above. This relationship implies that $\lambda_-(\partial H') + \int_{H'} G_- \, dH$ is finite, and hence that the last formula of (3.5.1) also converges to $d_\infty(\sigma_1, \sigma)$ as K tends to H'. This completes the proof. $\quad\square$

Proposition 3.5.1. *The metric space $(H(\infty), d_\infty)$ is complete.*

Proof. Let $\{x_i\}$ be a Cauchy sequence in $H(\infty)$ with respect to d_∞, and find rays σ_{x_i} from C_H to x_i. Then there exists a subsequence $\{\sigma_{x_{j(i)}}\}$ of $\{\sigma_{x_i}\}$ converging to a ray σ from C_H. Applying Lemma 3.5.1 shows that $d_\infty(x_{j(i)}, \sigma(\infty))$ tends to zero as $i \to \infty$, which implies that $d_\infty(x_i, \sigma(\infty))$ tends to zero also. $\quad\square$

Theorem 3.5.1

(1) *If $\lambda_\infty(H) < +\infty$ then $(H(\infty), d_\infty)$ is isometric to the compact line segment of length $\lambda_\infty(H)$.*

(2) *If $\lambda_\infty(H) = +\infty$ then each connected component of $(H(\infty), d_\infty)$ is isometric to a closed interval of $\mathbf{R}$ and moreover the one-dimensional Hausdorff measure of $(H(\infty), d_\infty)$ is*

$$\mathcal{H}^1(H(\infty), d_\infty) = +\infty.$$

Proof. For any $x \in H(\infty)$, define a map

$$\iota_x \colon I_x := \{y \in H(\infty); d_\infty(x, y) < +\infty\} \to \mathbf{R}$$

by

$$\iota_x(y) := \begin{cases} d_\infty(x, y) & \text{if } x \leq y, \\ -d_\infty(x, y) & \text{if } x \geq y, \end{cases}$$

for any $y \in H(\infty)$. Obviously, this is isometric and preserves the order relation. Since d_∞ is complete, the image $\iota_x(I_x)$ for any $x \in H(\infty)$ is a closed subset of $\mathbf{R}$.

Let us prove that (I_x, d_∞) for any $x \in H(\infty)$ is connected. In fact, if (I_x, d_∞) for an $x \in H(\infty)$ is disconnected then there exists a nonempty interval (a, b) of $\mathbf{R}$ such that $\iota_x(I_x) \cap (a, b) = \emptyset$ and $a, b \in \iota_x(I_x)$. Therefore $\iota_x^{-1}(a) < \iota_x^{-1}(b)$ and $(\iota_x^{-1}(a), \iota_x^{-1}(b)) = \emptyset$, which contradicts the fact that $H(\infty)$ is homeomorphic to a compact interval of $\mathbf{R}$ (see Theorem 3.4.1).

Thus, $\iota_x(I_x)$ for any $x \in H(\infty)$ is a closed interval of $\mathbf{R}$. This proves (1) and the first assertion of (2).

We will prove that $\mathcal{H}^1(H(\infty), d_\infty) = +\infty$ if $\lambda_\infty(H) = +\infty$ as follows. Suppose that $\lambda_\infty(H) = +\infty$ and $\mathcal{H}^1(H(\infty), d_\infty) < +\infty$. Note that the family of connected components of $(H(\infty), d_\infty)$ is just equal to $\{I_x; x \in H(\infty)\}$. By $\mathcal{H}^1(H(\infty), d_\infty) < +\infty$, each connected component of $(H(\infty), d_\infty)$ is isometric to a compact line segment. The fact that $H(\infty)$ with its canonical topology is homeomorphic to $[0, 1]$ shows that, for any two components $[x_1, y_1]$ and $[x_2, y_2]$ of $(H(\infty), d_\infty)$ for which $y_1 < x_2$, there exists a point $z \in H(\infty)$ such that $y_1 < z < x_2$. Therefore the cardinal number of the set of components of $(H(\infty), d_\infty)$ is not less than $2^{\aleph_0}$, which contradicts $\mathcal{H}^1(H(\infty), d_\infty) < +\infty$. $\square$

For a Riemannian half-cylinder N admitting a curvature at infinity, we have:

Theorem 3.5.2

(1) If $0 < \lambda_\infty(N) < +\infty$ then $(\tilde{N}(\infty), d_\infty)$ is isometric to $\mathbf{R}$ and $(N(\infty), d_\infty)$ to the circle of length $\lambda_\infty(N)$.

(2) If $\lambda_\infty(N) = +\infty$ then each connected component of $(N(\infty), d_\infty)$ is a closed line segment, and moreover

$$\mathcal{H}^1(N(\infty), d_\infty) = +\infty.$$

Proof. (1): Assume that $0 < \lambda_\infty(N) < +\infty$. Recalling the Riemannian half-plane $H_n = D(\sigma^{-n}, \sigma^n) \subset \tilde{N}$, $n \in \mathbf{N}$, defined in Section 3.4, we have $\lambda_\infty(H_n) = 2n\lambda_\infty(N) < +\infty$. Theorem 3.5.1 then implies that $(H_n(\infty), d_\infty)$ for any $n \in \mathbf{N}$ is isometric to the compact line segment of length $2n\lambda_\infty(H)$.

Hence $(\tilde{H}(\infty), d_\infty)$ is isometric to $\mathbf{R}$ and $(N(\infty), d_\infty)$ to the circle of length $\lambda_\infty(N)$ (see Proposition 3.4.1).

(2): Since $\lambda_\infty(N) = \lambda_\infty(D(\sigma^0, \sigma^1))$ and $D(\sigma^0, \sigma^1)(\infty) \setminus \{\sigma^1(\infty)\}$ is a fundamental domain of the covering $pr_\infty \colon \tilde{N}(\infty) \to N(\infty)$, applying Theorem 3.5.1 to $H := D(\sigma^0, \sigma^1)$ proves (2).   □

Combining Theorems 3.5.1 and 3.5.2 implies:

Theorem 3.5.3. *Each connected component of* $(M(\infty), d_\infty)$ *is a complete one-dimensional Riemannian manifold, possibly with boundary unless it consists of a single point. Moreover*

$$\mathcal{H}^1(M(\infty), d_\infty) = \lambda_\infty(M).$$

In general, the topology of $M(\infty)$ is not compatible with the Tits distance d_∞, as is seen in the following example.

Example 3.5.1

(1) Let M be the hyperbolic plane $H^2(-1)$. Then on the one hand we have $d_\infty(x, y) = +\infty$ for any different $x, y \in M(\infty)$. On the other hand, $M(\infty)$ is homeomorphic to S^1.

(2) Let M be a nonpositively curved Riemannian cylinder that contains a simple closed geodesic dividing M into two Riemannian half-cylinders N_1 and N_2 such that N_1 is of curvature ≤ -1 outside some compact subset and N_2 is isometric to $S^1 \times [0, +\infty)$. Then the universal covering space $\tilde{M}$ of M is a Riemannian plane with nonpositive curvature, and its ideal boundary $\tilde{M}(\infty)$ is isometric to the disjoint union of the interval $[0, \pi]$ and a discrete continuum.

Remark 3.5.1. If M is a two-dimensional Hadamard manifold, i.e., a Riemannian plane with nonpositive curvature, then the topology of the ideal boundary $M(\infty)$ of M as defined here is equivalent to the sphere topology defined in [7], and the Tits metric defined here is equivalent to that in [7].

Lemma 3.5.2. *The canonical topology of* $H(\infty)$ *is stronger than the topology induced from the Tits distance function* d_∞.

Proof. Suppose that there exist a point $x \in H(\infty)$ and a sequence $\{x_i\}$ of points in $H(\infty)$ such that $d_\infty(x_i, x)$ tends to zero as $i \to \infty$ but x_i does not tend to x. We can find rays σ_{x_i} from C_H to x_i and we assume that σ_{x_i} tends to a ray σ as $i \to \infty$ by taking a subsequence. Then Lemma 3.5.1 proves that $d_\infty(x_i, \sigma(\infty))$

tends to zero, which implies that

$$d_\infty(x, \sigma(\infty)) \leq \limsup_{i \to \infty} (d_\infty(x, x_i) + d_\infty(x_i, \sigma(\infty))) = 0,$$

i.e., $x = \sigma(\infty)$. On the contrary, however, Lemma 3.4.2 shows that x_i tends to $\sigma(\infty)$, a contradiction. $\square$

Lemma 3.5.2 extends easily to the following theorem, the proof of which is omitted.

Theorem 3.5.4. *The canonical topology of $M(\infty)$ is stronger than the topology induced from the Tits distance function d_∞.*

We shall next establish the Gauss–Bonnet theorem for $\overline{M}^\infty$. Although the Gaussian curvature on the (nonsmooth) boundary of a convex body in $\mathbf{R}^3$ is not defined in general, the notion of the curvature measure is defined instead, as the integral of the area of the image of the Gauss map, where the Gauss map is defined here as a multivalued map to $S^2(1)$. Similarly, the set functions $\lambda_\pm(\cdot)$ can be extended to be measures on the piecewise-smooth boundary ∂M. So, on the one hand it is natural to call $\lambda(\cdot) = \lambda_+(\cdot) - \lambda_-(\cdot)$ the (geodesic) curvature measure on ∂M. On the other hand, we can define the notion of the curvature measure on $M(\infty)$ naturally as follows. The *(geodesic) curvature measure λ on $M(\infty) \setminus \partial M(\infty)$* is defined to be the one-dimensional Hausdorff measure of $(M(\infty), d_\infty)$. If V is any component of $\overline{M \setminus C}$ for a fixed core C of M then we have $\lambda(V(\infty) \setminus \partial M(\infty)) = \lambda_\infty(V)$. Let $x \in \partial M(\infty) \cap V(\infty)$ be any point. When $\lambda_\infty(V) > 0$, we consider x to have inner angle $\pi/2$ and define $\lambda(\{x\}) := \pi/2$. When $\lambda_\infty(V) = 0$ and V is a Riemannian half-plane, then we consider x to have inner angle zero and define $\lambda(\{x\}) := 2\pi$. This convention defines the (geodesic) curvature measure over $\partial \overline{M}^\infty$, which satisfies

$$\lambda(\partial \overline{M}^\infty) = \lambda(\partial M) + \lambda_\infty(M) + \sum_i (\pi - \theta_i),$$

where the θ_i are the inner angles of $\overline{M}^\infty$ at $\partial M(\infty)$.

Consider any interior point of $\overline{M}^\infty$ in $M(\infty)$ to have total (Gaussian) curvature 2π, so that

$$c(\overline{M}^\infty) = c(M) + 2\pi \cdot \#\{x \in M(\infty); x \text{ is an interior point of } \overline{M}^\infty\}.$$

Then follows

Theorem 3.5.5 (The Gauss–Bonnet theorem for $\overline{M}^\infty$). *We have*

$$c(\overline{M}^\infty) + \lambda(\overline{M}^\infty) = 2\pi \chi(\overline{M}^\infty).$$

Exercise 3.5.1. Prove Theorem 3.5.5.

3.6 Triangle comparison

In this section, we study the relation between the angles and the lengths of the edges of almost-flat triangles in M; this relation will be almost equivalent to that in Euclidean space (see Theorem 3.6.1 below). A more refined and advanced version of Theorem 3.6.1 is seen in [98]. Results from the present section will be needed in the final section of this chapter to prove that if M is contracted by the scaling of its metric then it converges to the cone over the ideal boundary with Tits metric (see Theorem 3.7.2 below).

Notation 3.6.1. Let $\omega \colon \mathbf{R}^k \to \mathbf{R}$ denote some function such that

$$\lim_{\epsilon_1 \to 0} \cdots \lim_{\epsilon_k \to 0} \omega(\epsilon_1, \ldots, \epsilon_k) = 0,$$

and let ω be used like Landau's symbol; for example, we have

(1) $\quad \lim_{\epsilon_k \to 0} \omega(\epsilon_1, \ldots, \epsilon_k) = \omega(\epsilon_1, \ldots, \epsilon_{k-1})$,

(2) $\quad\quad\quad \omega(\epsilon) f = \omega(\epsilon) \quad$ if f is a bounded function,

(3) $\quad\quad \sin(\theta + \omega(\epsilon)) = \sin \theta + \omega(\epsilon)$,

(4) $\quad\quad \cos(\theta + \omega(\epsilon)) = \cos \theta + \omega(\epsilon)$,

(5) $\quad\quad \omega(\epsilon) f + \omega(\epsilon) g = \omega(\epsilon)(|f| + |g|)$,

where the left-hand side reduces to the right-hand side in each of (1)–(5).

Let $\triangle ABC$ denote a triangle in M with vertices $A, B, C \in M$ and edges α, β, γ that are minimal segments in M respectively joining B to C, C to A, A to B. Note that, since M may have a nonempty boundary, the edges α, β and γ are not necessarily geodesics. Assume that α, β and γ together bound a contractible region in M, denoted also by $\triangle ABC$. Set $a := d(B, C) = L(\alpha)$, $b := d(C, A) = L(\beta)$ and $c := d(A, B) = L(\gamma)$ and denote by $\angle A$ (resp. $\angle B$, $\angle C$) the inner angle of $\triangle ABC$ at A (resp. B, C). Such a triangle $\triangle ABC$ is said to be ϵ-*almost-flat*, $\epsilon \geq 0$, if the following condition holds:

$$\int_{\triangle ABC} |G| \, dM + \lambda_{\text{abs}}(\alpha) + \lambda_{\text{abs}}(\beta) + \lambda_{\text{abs}}(\gamma) \leq \epsilon. \tag{AF}$$

Note that a 0-almost-flat triangle is Euclidean. Under (AF), the Gauss–Bonnet theorem for $\triangle ABC$ implies that

$$|\angle A + \angle B + \angle C - \pi| \leq \epsilon,$$

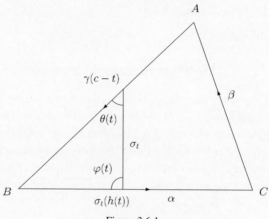

Figure 3.6.1

and, although α, β, γ may have break points at the boundary, we have

$$\angle(\dot{\alpha}(t-0), \dot{\alpha}(t+0)) \le \epsilon \qquad \text{for any } t \in (0, a);$$

the same inequalities hold for β and γ.

Assume now that an ϵ-almost-flat triangle $\triangle ABC$ satisfies $\angle A \le \pi$ and $\angle B \le \pi/2$. Let σ_t for $t \in [0, c]$ be a minimal segment in $\triangle ABC$ from $\gamma(c-t)$ to α, and let $h(t) := L(\sigma_t)$; see Figure 3.6.1. Denote by $\theta(t)$ the inner angle at $\gamma(c-t)$ of the region, say D_t, bounded by $\gamma|_{[c-t,c]}$, σ_t and a subarc of α from B to $\sigma_t(h(t))$, and denote by $\varphi(t)$ the inner angle at $\sigma_t(h(t))$ of D_t. The domain D_t is monotone increasing in t. Note that the first variation formula implies that $\varphi(t) = \pi/2$ provided that $\sigma_t(h(t))$ is an interior and nonbreak point of α. There exists a $t_0 \in [0, c]$ such that

$$\sigma_t(h(t)) \begin{cases} \ne C & \text{for any } t \in [0, t_0), \\ = C & \text{for any } t \in (t_0, c], \end{cases}$$

and in particular,

$$\varphi(t) \begin{cases} = \pi/2 + \omega(\epsilon) & \text{for any } t \in [0, t_0), \\ \ge \pi/2 & \text{for any } t \in (t_0, c]. \end{cases}$$

By setting $\sigma_t^- := \lim_{s \nearrow t} \sigma_s$, $\sigma_t^+ := \lim_{s \searrow t} \sigma_s$ it follows that

$$\angle(\dot{\sigma}_{t_0}^-(h(t_0)), \dot{\alpha}) = \pi/2 + \omega(\epsilon) \qquad \text{and} \qquad \sigma_{t_0}^+(h(t_0)) = C.$$

Lemma 3.6.1. *We have*

$$h(t) = t \sin(\angle B + \omega(\epsilon)) \qquad \text{for any } t \in [0, t_0].$$

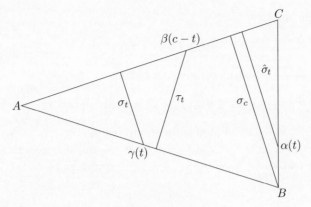

Figure 3.6.2

Moreover, if $\angle C \le \pi/2 + \omega(\epsilon)$ then

$$h(c) = c \sin(\angle B + \omega(\epsilon)).$$

Proof. The Gauss–Bonnet theorem and the condition (AF) show that

$$|\angle B + \theta(t) + \varphi(t) - \pi| \le \omega(\epsilon) \qquad \text{for any } t \in [0, c].$$

For any $t \in [0, t_0)$, we have $\varphi(t) = \pi/2 + \omega(\epsilon)$ and hence

$$\theta(t) = \pi/2 - \angle B + \omega(\epsilon).$$

From the first variation formula,

$$h(t) = \int_0^t \cos\theta(t)\, \mathrm{d}t = t \sin(\angle B + \omega(\epsilon)) \qquad \text{for any } t \in [0, t_0].$$

If $\angle C \le \pi/2 + \omega(\epsilon)$ then $\varphi(t) = \pi/2 + \omega(\epsilon)$ holds for any $t \in [0, c]$, and consequently

$$h(c) = c \sin(\angle B + \omega(\epsilon)). \qquad \square$$

Lemma 3.6.2. *If an ϵ-almost-flat triangle $\triangle ABC$ satisfies*

$$\angle B, \angle C \ge \pi/2$$

then

$$a \le (b + c)\,\omega(\epsilon).$$

Proof. Let σ_t (resp. τ_t) be a minimal segment in $\triangle ABC$ from $\gamma(t)$ to β (resp. $\beta(c - t)$ to γ), and let t_0 (resp. t_1) be as defined above in connection with σ_t (resp. τ_t); see Figure 3.6.2. If $t_0 < c$ and $t_1 < b$ then, since $\sigma_{t_0}^+$ intersects $\tau_{t_1}^+$, the triangle inequality implies

$$a \leq L(\sigma_{t_0}) + L(\tau_{t_0}).$$

Since $\angle A = \omega(\epsilon)$, applying Lemma 3.6.1 yields

$$L(\sigma_{t_0}) = t_0\,\omega(\epsilon) \leq b\,\omega(\epsilon) \qquad \text{as well as} \qquad L(\tau_{t_1}) \leq c\,\omega(\epsilon).$$

Thus, we obtain the conclusion of the lemma in this case.

If at least one of $t_0 = c$ and $t_1 = b$ holds, we can assume without loss of generality that $t_0 = c$. Then, since $\angle A = \omega(\epsilon)$, applying Lemma 3.6.1 yields

$$d_{\triangle ABC}(B, \beta) \leq c\,\omega(\epsilon). \tag{3.6.1}$$

If $d_{\triangle ABC}(B, \beta) = a$, this and (3.6.1) together imply the lemma. Assume now that $d_{\triangle ABC}(B, \beta) < a$. Since $\angle(\dot{\sigma}_c(L(\sigma_c)), \dot{\beta}) = \pi/2 + \omega(\epsilon)$, the condition (AF) shows that $\angle(\dot{\sigma}_c(0), \dot{\alpha}(0)) \leq \omega(\epsilon)$. Let $\hat{\sigma}_t$ for $t \in [0, a]$ be a minimal segment in $\triangle ABC$ from $\alpha(t)$ to β. For any $t \in [0, a - d_{\triangle ABC}(B, \beta))$, the triangle inequality implies that $\hat{\sigma}_t(L(\hat{\sigma}_t)) \neq C$ and hence that $\angle(\dot{\hat{\sigma}}_t(L(\hat{\sigma}_t)), \dot{\beta}) = \pi/2 + \omega(\epsilon)$, which together with $\angle C \geq \pi/2$ and (AF) gives

$$\angle(-\dot{\alpha}(t), \dot{\hat{\sigma}}_t(0)) = \pi + \omega(\epsilon).$$

The same discussion as in the proof of Lemma 3.6.1 yields

$$0 \leq L(\hat{\sigma}_t) \leq d_{\triangle ABC}(B, \beta) - (1 - \omega(\epsilon))t$$

for any $t \in [0, a - d_{\triangle ABC}(B, \beta)]$, which implies that

$$t \leq (1 + \omega(\epsilon))d_{\triangle ABC}(B, \beta).$$

In particular, when $t := a - d_{\triangle ABC}(B, \beta)$ we have

$$a \leq (2 + \omega(\epsilon))d_{\triangle ABC}(B, \beta),$$

which together with (3.6.1) completes the proof. $\qquad\square$

Lemma 3.6.3. *If an ϵ-almost-flat triangle $\triangle ABC$ satisfies $\angle C = \pi/2 + \omega(\epsilon)$ then*

(1)
$$\frac{b}{c} = \cos \angle A + \omega(\epsilon) = \sin \angle B + \omega(\epsilon),$$

(2)
$$\frac{a}{c} = \cos \angle B + \omega(\epsilon) = \sin \angle A + \omega(\epsilon),$$

(3)
$$\frac{a^2 + b^2}{c^2} = 1 + \omega(\epsilon).$$

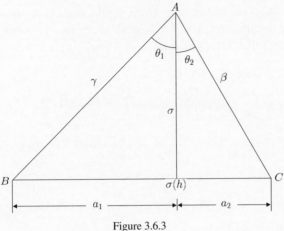

Figure 3.6.3

Proof. Find a minimal segment σ in $\triangle ABC$ from A to α, and set $h := L(\sigma) = d_{\triangle ABC}(A, \alpha)$. Lemma 3.6.1 implies that

$$\frac{h}{c} = \sin \angle B + \omega(\epsilon). \tag{3.6.2}$$

Since the inner angle at $\sigma(h)$ of the region bounded by σ, β and the subarc of α from $\sigma(h)$ to C is equal to $\pi/2 + \omega(\epsilon)$, applying Lemma 3.6.2 yields

$$d(\sigma(h), C) \leq (h + b)\omega(\epsilon);$$

this and the triangle inequality imply that

$$\frac{b}{h} = 1 + \omega(\epsilon).$$

Combining this and (3.6.2) yields

$$\frac{b}{c} = (1 + \omega(\epsilon))(\sin \angle B + \omega(\epsilon)) = \sin \angle B + \omega(\epsilon) = \cos \angle A + \omega(\epsilon),$$

where the last equality follows from $\angle A + \angle B = \pi/2 + \omega(\epsilon)$. Thus, (1) has been proved.

(2) is proved in the same way, and (3) is implied by (1) and (2). $\qquad\square$

Theorem 3.6.1 (The cosine law). *Any ϵ-almost-flat triangle $\triangle ABC$ satisfies*

$$a^2 = b^2 + c^2 - 2bc \cos \angle A + (a + b + c)^2 \omega(\epsilon).$$

Proof. First consider the case where $\angle A \geq \pi/2$. Then $\angle B$, $\angle C \leq \pi/2 + \epsilon$. We divide $\triangle ABC$ into two triangles by taking a minimal segment σ in $\triangle ABC$ from A to α; see Figure 3.6.3. Set $h := L(\sigma) = d_{\triangle ABC}(A, \alpha)$, $a_1 := d(B, \sigma(h))$,

$a_2 := d(C, \sigma(h))$ and denote by θ_1 (resp. θ_2) the angle at A between σ and γ (resp. σ and β). Since $\angle A = \theta_1 + \theta_2$, applying Lemma 3.6.3 to each of the two triangles shows that on the one hand

$$\cos \angle A = \cos \theta_1 \cos \theta_2 - \sin \theta_1 \sin \theta_2$$

$$= \left(\frac{h}{b} + \omega(\epsilon)\right)\left(\frac{h}{c} + \omega(\epsilon)\right) - \left(\frac{a_1}{b} + \omega(\epsilon)\right)\left(\frac{a_2}{c} + \omega(\epsilon)\right)$$

$$= \frac{h^2 - a_1 a_2}{bc} + \frac{bc + bh + ch + a_1 c + a_2 b}{bc} \omega(\epsilon). \qquad (3.6.3)$$

On the other hand, by (3) of Lemma 3.6.3,

$$b^2 = (1 + \omega(\epsilon))\left(a_1^2 + h^2\right), \qquad c^2 = (1 + \omega(\epsilon))\left(a_2^2 + h^2\right),$$

which together with $a = a_1 + a_2$ implies that

$$b^2 + c^2 - a^2 = 2(h^2 - a_1 a_2) + \left(a_1^2 + a_2^2 + h^2\right)\omega(\epsilon).$$

Combining this and (3.6.3) yields

$$\cos \angle A = \frac{b^2 + c^2 - a^2}{2bc} + \frac{bc + bh + ch + a_1 c + a_2 b + a_1^2 + a_2^2 + h^2}{bc}\omega(\epsilon).$$

Here, since $h \le a + b$ and $a_1, a_2 \le a$, the factor

$$\left(bc + bh + ch + a_1 c + a_2 b + a_1^2 + a_2^2 + h^2\right)\omega(\epsilon)$$

reduces to $(a + b + c)^2 \omega(\epsilon)$. The proof in this case is completed.

Next consider the case where $\angle A < \pi/2$. Then $\min\{\angle B, \angle C\} < \pi/2 + \epsilon$. We can assume without loss of generality that $\angle B < \pi/2 + \epsilon$. In the same way as in the first case, we find a minimal segment σ in $\triangle ABC$ from C to γ with length $h := d_{\triangle ABC}(C, \gamma)$, which divides $\triangle ABC$ into two triangles. Applying Lemma 3.6.3 to each of these triangles yields

$$h = b \sin \angle A + b\omega(\epsilon) = a \sin \angle B + a\omega(\epsilon)$$

and hence

$$a^2 \sin^2 \angle B = b^2 \sin^2 \angle A + (a + b)^2 \omega(\epsilon). \qquad (3.6.4)$$

However, by Lemma 3.6.3,

$$d(A, \sigma(h)) = b\left(\cos \angle A + \omega(\epsilon)\right), \qquad d(B, \sigma(h)) = a\left(\cos \angle B + \omega(\epsilon)\right);$$

these, together with $c = d(A, \sigma(h)) + d(B, \sigma(h))$, imply that

$$a^2 \cos^2 \angle B = c^2 + b^2 \cos^2 \angle A - 2bc \cos \angle A + (a + b + c)^2 \omega(\epsilon).$$

Combining this and (3.6.4) completes the proof. □

3.7 Convergence to the limit cone

Definition 3.7.1 (Gromov–Hausdorff convergence). Let X and Y be two metric spaces. A (not necessarily continuous) map $f \colon X \to Y$ is called an ϵ-*approximation* if the following, (1) and (2), hold:

(1) $|d(p, q) - d(f(p), f(q))| < \epsilon$ for any $p, q \in X$,
(2) $B(f(X), \epsilon) = Y$.

A sequence $\{X_i\}_{i=1,2,\dots}$ of compact metric spaces is said to *converge to a metric space* X_∞ *in the Gromov–Hausdorff sense* if there exists a sequence of ϵ_i-approximations $f_i \colon X_i \to X_\infty$, $i = 1, 2, \dots$, with $\epsilon_i \to 0$. This concept of convergence, called *Gromov–Hausdorff convergence*, induces a topology of the set of compact metric spaces called the *Gromov–Hausdorff topology*. (To define the topology we need a net for a directed set rather than a sequence.)

We shall also define a version of the Gromov–Hausdorff convergence for pointed noncompact spaces. Let (X_∞, x_∞) be a pointed metric space and $\{(X_i, x_i)\}_{i=1,2,\dots}$, a sequence of pointed metric spaces such that all closed bounded subsets of X_i and X_∞ are compact. The sequence $\{(X_i, x_i)\}_{i=1,2,\dots}$ is said to *converge to* (X_∞, x_∞) if for any $r > 0$ there exist a sequence $r_i \searrow r$ and ϵ_i-approximations $f_{r,i} \colon \overline{B(x_i, r_i)} \to \overline{B(x_\infty, r)}$, $i = 1, 2, \dots$, with $\epsilon_i \to 0$ such that $f_{r,i}(x_i) = x_\infty$ for every i. This concept of convergence is called *pointed Gromov–Hausdorff convergence*.

Definition 3.7.2 (Euclidean cone). The *Euclidean cone* "cone X" *over a metric space* X is defined to be the quotient space $X \times [0, +\infty)/X \times \{0\}$ equipped with a metric ρ defined by

$$\rho((p, a), (q, b)) := \sqrt{a^2 + b^2 - 2ab \cos \min\{d(p, q), \pi\}}$$

for any $(p, a), (q, b) \in \text{cone } X$ $(p, q \in X, a, b \geq 0)$. Let o denote the point in cone X corresponding to $X \times \{0\}$ and call it the *vertex of* cone X.

The purpose of this section is to prove that if $\lambda_\infty(M) < +\infty$ then the pointed space $((1/r)M, p)$ tends to (cone $M(\infty), o)$ as $r \to +\infty$ in the sense of the

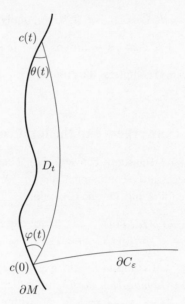

Figure 3.7.1

pointed Gromov–Hausdorff convergence (see Theorem 3.7.2 below), where $p \in M$ is any fixed point, and $(1/r)M$ denotes the space M equipped with metric multiplied by $1/r$.

Assume that $\lambda_\infty(M) < +\infty$ and a number $0 < \epsilon \ll 1$ is fixed. Then there exists a core C_ϵ of M such that each curve that is a component of $\overline{\partial M \setminus C_\epsilon}$ touches the topological boundary of C_ϵ orthogonally and that

$$\int_{M \setminus C_\epsilon} |G| \, dM + \lambda_{\text{abs}}(\partial M \setminus C_\epsilon) \le \epsilon.$$

Note that any triangle $\triangle ABC$ contained in $M \setminus C_\epsilon$ is ϵ-almost-flat.

Lemma 3.7.1. *If $c : [0, +\infty) \to \partial M$ is a component of $\overline{\partial M \setminus C_\epsilon}$ then*

$$d(c(0), c(t)) \ge t \cos \epsilon.$$

Proof. Let D_t be a region bounded by $c|_{[0,t]}$ and a minimal segment, say σ_t, in $M \setminus C_\epsilon$ from $c(0)$ to $c(t)$, and set $\theta(t) := \angle_{c(t)} D_t$, $\varphi(t) := \angle_{c(0)} D_t$; see Figure 3.7.1. Then the Gauss–Bonnet theorem implies that

$$c(D_t) = \theta(t) + \varphi(t) - \lambda(c|_{[0,t]}) - \lambda(\sigma_t) \qquad \text{for any } t \ge 0$$

with respect to suitable sides of σ_t and c. Here, by the assumption for C_ϵ,

$$c(D_t) + \lambda(c|_{[0,t]} \setminus \sigma_t) \le \epsilon$$

and besides, by the minimal property of σ_t,

$$\lambda(\sigma_t) = \lambda(\sigma_t \cap \partial C_\epsilon) \leq 0.$$

Thus we obtain $\theta(t) \leq \epsilon$ for any $t \geq 0$, which together with the first variation formula completes the proof. $\qquad\square$

Lemma 3.7.2. *There exists $T = T(\epsilon, L(\partial C_\epsilon)) > 0$ such that if $\sigma, \tau \in \mathcal{R}_{C_\epsilon}$ are two curves for which $D(\sigma, \tau)$ is defined and $\lambda_\infty(D(\sigma, \tau)) < \pi - 5\epsilon$ then a minimal segment γ_τ in $D(\sigma, \tau)$ joining $\sigma(T)$ to $\tau(T)$ does not intersect ∂C_ϵ.*

Proof. Set

$$T := \frac{L(\partial C_\epsilon)}{\cos \epsilon - \cos 2\epsilon},$$

and let $\sigma, \tau \in \mathcal{R}_{C_\epsilon}$ be such that $D(\sigma, \tau)$ is defined and $\lambda_\infty(D(\sigma, \tau)) < \pi - 5\epsilon$. For $t \geq 0$, denote by γ_t a minimal segment in $D(\sigma, \tau)$ from $\sigma(t)$ to $\tau(t)$ and by D_t the noncompact closed region in $D(\sigma, \tau)$ bounded by $\sigma|_{[t,+\infty)}$, γ_t and $\tau|_{[t,+\infty)}$. Since $\lambda_{\mathrm{abs}}(\sigma), \lambda_{\mathrm{abs}}(\tau) \leq \epsilon$ and γ_t is locally concave with respect to D_t, the Gauss–Bonnet theorem and the assumption for C_ϵ prove that

$$5\epsilon < \pi - \lambda_\infty(D(\sigma, \tau)) = c(D_t) + \lambda(\partial D_t) \leq \theta(t) + \varphi(t) + \epsilon$$

for any $t \geq 0$, where

$$\theta(t) := \angle(-\dot\sigma(t), \dot\gamma_t(0)) \qquad \text{and} \qquad \varphi(t) := \angle(-\dot\tau(t), -\dot\gamma_t(L(\gamma_t))).$$

Hence, by the first variation formula, for almost all $t \geq 0$,

$$\frac{\mathrm{d}}{\mathrm{d}t} d_{D(\sigma,\tau)}(\sigma(t), \tau(t)) = \cos\theta(t) + \cos\varphi(t)$$

$$\leq 2\cos\frac{\theta(t) + \varphi(t)}{2} < 2\cos 2\epsilon,$$

which together with $d_{D(\sigma,\tau)}(\sigma(0), \tau(0)) \leq L(\partial C_\epsilon)$ implies that

$$d_{D(\sigma,\tau)}(\sigma(t), \tau(t)) - L(\partial C_\epsilon) < 2t \cos 2\epsilon \qquad \text{for any } t > 0. \quad (3.7.1)$$

Suppose now that γ_{t_0} for a $t_0 \geq 0$ intersects $I(\sigma, \tau)$ and that p is a point in $\gamma_{t_0} \cap I(\sigma, \tau)$. We remark that σ and τ may be boundary curves and so we have, by applying Lemma 3.7.1,

$$d_{D(\sigma,\tau)}(\sigma(t_0), \tau(t_0)) \geq d_{D(\sigma,\tau)}(\sigma(0), \sigma(t_0)) + d_{D(\sigma,\tau)}(\tau(0), \tau(t_0))$$

$$- d_{D(\sigma,\tau)}(\sigma(0), p) - d_{D(\sigma,\tau)}(\tau(0), p)$$

$$\geq 2t_0 \cos\epsilon - L(\partial C_\epsilon).$$

Combining this and (3.7.1) yields $t_0 < T$. This completes the proof. $\qquad\square$

Set, for simplicity, $\Pi := \Pi_{C_\epsilon}$, and for $r > 0$ define a map $f_{\epsilon,r} \colon M \to$ cone $M(\infty)$ by

$$f_{\epsilon,r}(p) := \begin{cases} o & \text{for } p \in C_\epsilon, \\ (\Pi_{C_\epsilon}(p), d(p, C_\epsilon)/r) & \text{for } p \in M \setminus C_\epsilon. \end{cases}$$

We shall eventually prove that $f_{\epsilon,r}$ is an approximation. For the proof, the following lemma is essential.

Lemma 3.7.3. *Let p and q be two points contained in a common connected component of $M \setminus C_\epsilon$. Assume that $H := D(\sigma^-_{\Pi(p)}, \sigma^+_{\Pi(q)})$ contains p and q (see Section 3.4 for the definition of $\sigma^\pm_x$). Denote by d_H the interior distance function on H and by ρ_H the distance function on the Euclidean cone over $(H(\infty), \hat{d}_\infty)$, where $\hat{d}_\infty$ is the intrinsic Tits metric on $H(\infty)$. Then we have*

$$\frac{d_H(p, q)}{r} = \rho_H(f_{\epsilon,r}(p), f_{\epsilon,r}(q)) + \omega\left(\epsilon, \frac{1}{r}\right),$$

where $\omega(\cdot, \cdot)$ is independent of p and q.

Proof. We first prove the lemma under the assumption $\lambda_\infty(H) < \pi - 5\epsilon$. Applying Lemma 3.7.2, we obtain a minimal segment γ_T from $\sigma^-_{\Pi(p)}(T)$ to $\sigma^+_{\Pi(q)}(T)$ not intersecting ∂C_ϵ. Let us prove

Sublemma 3.7.1. *There exists a constant $c = c(T, L(\partial C_\epsilon)) > 0$ such that the diameter of the compact region, say D, bounded by the four curves $\sigma^-_{\Pi(p)}|_{[0,T]}$, γ_T, $\sigma^+_{\Pi(q)}|_{[0,T]}$ and $I(\sigma^-_{\Pi(p)}, \sigma^+_{\Pi(q)})$ is less than or equal to c.*

Proof of Sublemma 3.7.1. Fix a point $x_0 \in \partial D$. Find a continuous variation $\{c_t : [0, 1] \to D\}_{0 \le t \le 1}$ of curves constructed by a length-decreasing process in D such that $c_t(0) = c_t(1) = x_0$ for any $t \in [0, 1]$ and c_0 is the closed curve (or loop) bounding D. Here c_1 is either a locally minimal simple closed curve or a constant map with image $\{x_0\}$.

If $c_1[0, 1] \ne \{x_0\}$ then the compact region, say D', bounded by c_1 is locally concave except at x_0, and hence the Gauss–Bonnet theorem implies that $c(D') \ge \pi$, which contradicts $\int_{M \setminus C_\epsilon} |G| \, dM \le \epsilon \ll 1$.

Thus, it follows that $c_1[0, 1] = \{x_0\}$. Let $x \in D$ be any point. Since a number $t(x) \in [0, 1]$ with $x \in c_{t(x)}$ exists, we have

$$d(x_0, x) \le \frac{L(c_{t(x)})}{2} \le \frac{L(\partial D)}{2}$$

and hence diam $D \le L(\partial D) \le 2(2T + L(\partial C_\epsilon))$, which proves the sublemma.

$\square$

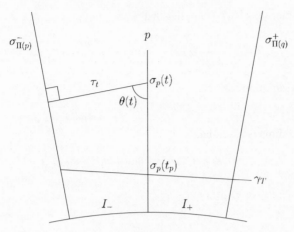

Figure 3.7.2

Let σ_p (resp. σ_q) be a minimal segment from C_ϵ to p (resp. q). Note that $\sigma_p, \sigma_q \subset H$. We denote by I_- (resp. I_+) the subarc of $I(\sigma_{\bar{\Pi}(p)}^-, \sigma_{\Pi(p)}^+)$ from $\sigma_{\bar{\Pi}(p)}^-(0)$ to $\sigma_p(0)$ (resp. from $\sigma_p(0)$ to $\sigma_{\Pi(p)}^+(0)$); see Figure 3.7.2.

Sublemma 3.7.2. *If $d(p, C_\epsilon)/r \geq \epsilon$ then*

$$\lambda(I_\pm) = \omega\left(\epsilon, \frac{1}{r}\right).$$

Proof of Sublemma 3.7.2. Since the assumption for C_ϵ implies that $0 = d_\infty(\sigma_{\bar{\Pi}(p)}^-, \sigma_{\Pi(p)}^+) = -\lambda(I_-) - \lambda(I_+) + \omega(\epsilon)$, we obtain

$$|\lambda(I_-) + \lambda(I_+)| \leq \omega(\epsilon). \tag{3.7.2}$$

First consider the case where $\lambda(I_-) \geq \lambda(I_+)$. Let τ_t be a minimal segment in H joining $\sigma_p(t)$ to $\sigma_{\bar{\Pi}(p)}^-$. The Gauss–Bonnet theorem for the region, say D_t, in H bounded by $\sigma_p|_{[0,t]}$ and τ_t yields

$$\theta(t) - \frac{\pi}{2} - \lambda(I_-) - \lambda(\tau_t) = \omega(\epsilon), \tag{3.7.3}$$

where $\theta(t) := \angle_{\sigma_p(t)} D_t$. For the proof of Sublemma 3.7.2, we may assume that $r\epsilon$ is larger than the constant c of Sublemma 3.7.1. Then $d(p, C_\epsilon) \geq r\epsilon > c$ and so σ_p intersects γ_T. Let t_p be the number such that $\sigma_p \cap \gamma_T = \{\sigma_p(t_p)\}$. It follows that $L(\tau_t) \leq c$ for any $t \in [0, t_p]$. The minimal properties of γ_T, σ_p and τ_t show that τ_t for any $t > t_p$ does not intersect I_+ but possibly does intersect I_-. Thus

$$\lambda(\tau_t) \geq 0 \qquad \text{for any } t \geq t_p,$$

which together with (3.7.3) implies that

$$\theta(t) \geq \frac{\pi}{2} + \lambda(I_-) + \omega(\epsilon) \qquad \text{for any } t \geq t_p.$$

By the first variation formula,

$$-L(\tau_{t_p}) \leq L(\tau_{d(p,C_\epsilon)}) - L(\tau_{t_p}) = \int_{t_p}^{d(p,C_\epsilon)} \cos\theta(t)\, dt$$

$$\leq (d(p, C_\epsilon) - t_p) \sin(-\lambda(I_-) + \omega(\epsilon)).$$

Since $d(p, C_\epsilon) \geq \epsilon r$ and $L(\tau_{t_p}) \leq c$,

$$\lambda(I_+) \leq \lambda(I_-) \leq \arcsin \min\left\{ \frac{c}{\epsilon r - t_p}, 1 \right\} + \omega(\epsilon) = \omega\left(\epsilon, \frac{1}{r}\right).$$

Combining this with (3.7.2) implies the conclusion of the sublemma. If $\lambda(I_+) \geq \lambda(I_-)$, we can prove the sublemma in the same way. $\qquad\square$

If $d(p, C_\epsilon)/r < \epsilon$ then the triangle inequality implies that

$$\frac{d_H(p, q)}{r} = \frac{d(q, C_\epsilon)}{r} + \omega\left(\epsilon, \frac{1}{r}\right),$$

which finally proves Lemma 3.7.3. Here we note that $d(\cdot, C_\epsilon) = d_H(\cdot, C_\epsilon)$ on H. If $d(q, C_\epsilon)/r < \epsilon$, the same method leads to the lemma.

Assume thus that $d(p, C_\epsilon)/r, d(q, C_\epsilon)/r \geq \epsilon$. Then, by Sublemma 3.7.1 and supposing that $r \gg 1/\epsilon$, the segment σ_p intersects γ_τ at a point that we denote by P; we can find s_p and t_p such that $\gamma_\tau(s_p) = \sigma_p(t_p) = P$. A point Q and numbers s_q, t_q are defined in a similar manner; see Figure 3.7.3.

We set

$$\theta := \angle(\dot\gamma_\tau(0), \dot\sigma_{\overline\Pi(p)}^-(T)), \qquad \theta' := \angle(\dot\gamma_\tau(s_p), \dot\sigma_p(t_p)),$$

$$\varphi := \angle(-\dot\gamma_\tau(L(\gamma_\tau)), \dot\sigma_{\overline\Pi(q)}^+(T)), \qquad \varphi' := \angle(-\dot\gamma_\tau(s_q), \dot\sigma_q(t_q)).$$

By Sublemma 3.7.2 and by applying the Gauss–Bonnet theorem for the region in H surrounded by $\sigma_{\overline\Pi(p)}^-|_{[0,T]}$, $\gamma_\tau|_{[0,s_p]}$, $\sigma_p|_{[0,t_p]}$ and I_-, we see that

$$\theta = \theta' + \omega\left(\epsilon, \frac{1}{r}\right) \qquad \text{as well as} \qquad \varphi = \varphi' + \omega\left(\epsilon, \frac{1}{r}\right).$$

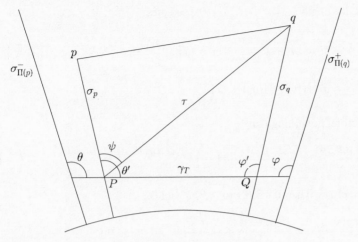

Figure 3.7.3

Finding a minimal segment τ from P to q, we now consider the triangle $\triangle q \, P \, Q$ with inner angles $\angle q$, $\angle P$ and $\angle Q$. Since

$$\frac{d_H(P, Q)}{r} = \omega\left(\frac{1}{r}\right), \qquad \frac{d_H(q, P)}{r} = \frac{d(q, C_\epsilon)}{r} + \omega\left(\frac{1}{r}\right),$$

and

$$\frac{d_H(q, Q)}{r} = \frac{d(q, C_\epsilon)}{r} + \omega\left(\frac{1}{r}\right),$$

Theorem 3.6.1 shows that

$$\cos \angle q = \frac{1}{2 \, d_H(q, P) \, d_H(q, Q)}$$

$$\times \left\{ d_H(q, P)^2 + d_H(q, Q)^2 - d_H(P, Q)^2 \right.$$

$$\left. + (d_H(q, P) + d_H(q, Q) + d_H(P, Q))^2 \, \omega\left(\epsilon, \frac{1}{r}\right) \right\}$$

$$\geq 1 - \omega\left(\epsilon, \frac{1}{r}\right)$$

and hence

$$\angle P + \varphi' = \pi - \angle q + \omega(\epsilon) = \pi - \omega\left(\epsilon, \frac{1}{r}\right).$$

Thus, the angle $\psi := \angle(\dot{\sigma}_p(t_p), \dot{\tau}(0))$ satisfies

$$\psi = \theta' - \angle P = \theta + \varphi' - \pi + \omega\left(\epsilon, \frac{1}{r}\right) = \theta + \varphi - \pi + \omega\left(\epsilon, \frac{1}{r}\right)$$

$$= \hat{d}_\infty(\Pi(p), \Pi(q)) + \omega\left(\epsilon, \frac{1}{r}\right).$$

Therefore, remarking that

$$\frac{d_H(P, p)}{r} = \frac{d(p, C_\epsilon)}{r} + \omega\left(\frac{1}{r}\right), \qquad \frac{d_H(P, q)}{r} = \frac{d(q, C_\epsilon)}{r} + \omega\left(\frac{1}{r}\right)$$

and applying Theorem 3.6.1 to $\triangle Ppq$ yields

$$\frac{d_H(p, q)^2}{r^2} = \frac{d(p, C_\epsilon)^2}{r^2} + \frac{d(q, C_\epsilon)^2}{r^2}$$
$$- 2\frac{d(p, C_\epsilon)}{r}\frac{d(q, C_\epsilon)}{r}\cos\hat{d}_\infty(\Pi(p), \Pi(q)) + \omega\left(\epsilon, \frac{1}{r}\right),$$

which proves the lemma.

We next prove the lemma under the condition $\lambda_\infty(H) \geq \pi - 5\epsilon$. Find a minimal segment γ joining p and q in H. If γ intersects C_ϵ then the triangle inequality shows that

$$\frac{d_H(p, q)}{r} = \frac{d(p, C_\epsilon)}{r} + \frac{d(q, C_\epsilon)}{r} + \omega\left(\frac{1}{r}\right),$$

which, together with $\lambda_\infty(H) \geq \pi - 5\epsilon$, leads to the conclusion of the lemma. Assume that γ does not intersect C_ϵ. Then the continuity of the map $t \mapsto \Pi(\gamma(t))$ implies that there exists an $x_0 \in \gamma$ such that $\lambda_\infty(H') = \pi - 6\epsilon$, where $H' := D(\sigma_{\Pi(p)}^-, \sigma_{\Pi(x_0)}^+)$. Applying the first case of the lemma to $p, x_0 \in H'$, we have

$$\frac{d_H(p, x_0)}{r} = \frac{d_{H'}(p, x_0)}{r} = \rho_{H'}(f_{\epsilon, r}(p), f_{\epsilon, r}(x_0)) + \omega\left(\epsilon, \frac{1}{r}\right)$$
$$= \frac{d_H(p, C_\epsilon)}{r} + \frac{d_H(x_0, C_\epsilon)}{r} + \omega\left(\epsilon, \frac{1}{r}\right),$$

and hence

$$\frac{d_H(p, q)}{r} = \frac{d_H(p, x_0)}{r} + \frac{d_H(x_0, q)}{r}$$
$$= \frac{d_H(p, C_\epsilon)}{r} + \frac{d_H(x_0, C_\epsilon)}{r} + \frac{d_H(x_0, q)}{r} + \omega\left(\epsilon, \frac{1}{r}\right)$$

$$\geq \frac{d_H(p, C_\epsilon)}{r} + \frac{d_H(q, C_\epsilon)}{r} + \omega\left(\epsilon, \frac{1}{r}\right)$$

$$= \rho_H(f_{\epsilon,r}(p), f_{\epsilon,r}(q)) + \omega\left(\epsilon, \frac{1}{r}\right).$$

Moreover, the reverse of the above inequality follows from the triangle inequality. This completes the proof of Lemma 3.7.3. □

Theorem 3.7.1. *The map $f_{\epsilon,r}: (1/r)B(C_\epsilon, r) \to B(o, 1; \text{cone } M(\infty))$ is an $\omega(\epsilon, 1/r)$-approximation, where $B(o, 1; \text{cone } M(\infty))$ is the metric ball centered at the vertex o and of radius 1 in cone $M(\infty)$.*

Proof. The surjectivity of $f_{\epsilon,r}$ is easily verified. Let $p, q \in M$. It suffices to show that

$$\frac{d(p, q)}{r} = \rho(f_{\epsilon,r}(p), f_{\epsilon,r}(q)) + \omega\left(\epsilon, \frac{1}{r}\right), \qquad (3.7.4)$$

where $\omega(\cdot, \cdot)$ is independent of p and q. If either p or q belongs to C_ϵ then (3.7.4) is obvious. Assume thus that $p, q \in M \setminus C_\epsilon$. Find a minimal segment γ joining p and q. If γ intersects C_ϵ then

$$\frac{d(p, q)}{r} = \frac{d(p, C_\epsilon)}{r} + \frac{d(q, C_\epsilon)}{r} + \omega\left(\frac{1}{r}\right)$$

$$\geq \rho(\Pi(p), \Pi(q)) + \omega\left(\epsilon, \frac{1}{r}\right). \qquad (3.7.5)$$

If p and q are contained in two different components of $M \setminus C_\epsilon$ then γ intersects C_ϵ, so that from (3.7.5) and $d_\infty(\Pi(p), \Pi(q)) = +\infty$ we obtain (3.7.4). Therefore, we assume that p and q are contained in a common connected component, say V, of $\overline{M \setminus C_\epsilon}$. Replacing p and q if necessary, we assume that $p, q \in H := D(\sigma_{\Pi(p)}^-, \sigma_{\Pi(q)}^+)$.

Let us now consider the case where V is a Riemannian half-plane. Then, by Lemma 3.7.3,

$$\frac{d(p, q)}{r} \leq \frac{d_H(p, q)}{r} = \rho_H(\Pi(p), \Pi(q)) + \omega\left(\epsilon, \frac{1}{r}\right)$$

$$= \rho(\Pi(p), \Pi(q)) + \omega\left(\epsilon, \frac{1}{r}\right). \qquad (3.7.6)$$

If γ intersects C_ϵ then combining (3.7.5) and (3.7.6) yields (3.7.4). If γ does not intersects C_ϵ then γ is contained in H and $d(p, q) = d_H(p, q)$, which together with (3.7.6) implies (3.7.4).

Consider now the case where V is a Riemannian half-cylinder. Set $H' := \overline{V \setminus H}$. It follows from Lemma 3.7.3 that

$$\frac{d(p, q)}{r} \leq \frac{d_H(p, q)}{r} = \rho_H(\Pi(p), \Pi(q)) + \omega\left(\epsilon, \frac{1}{r}\right),$$

$$\frac{d(p, q)}{r} \leq \frac{d_{H'}(p, q)}{r} = \rho_{H'}(\Pi(p), \Pi(q)) + \omega\left(\epsilon, \frac{1}{r}\right)$$

and also

$$\rho(\Pi(p), \Pi(q)) = \min\{\rho_H(\Pi(p), \Pi(q)), \rho_{H'}(\Pi(p), \Pi(q))\}.$$

If γ intersects C_ϵ then combining the above formulae and (3.7.5) proves (3.7.4). If γ does not intersects C_ϵ then

$$d(p, q) = \min\{d_H(p, q), d_{H'}(p, q)\}.$$

This completes the proof of the theorem.   □

Theorem 3.7.1 implies

Theorem 3.7.2. *Assume that $\lambda_\infty(M) < +\infty$ and $p \in M$ is any fixed point. Then the pointed space $((1/r)M, p)$ tends to $(\text{cone } M(\infty), o)$ as $r \to +\infty$ in the sense of the pointed Gromov–Hausdorff convergence.*

Next we consider the relation between straight lines and the Tits metric on the ideal boundary.

For any curve $\gamma : \mathbf{R} \to M$ with $\gamma|_{(-\infty,0]} \in \mathcal{C}_M$, let $\gamma(-\infty)$ denote the class of $\gamma|_{(-\infty,0]}$.

Theorem 3.7.3. *If γ is a straight line in M then*

$$d_\infty(\gamma(-\infty), \gamma(\infty)) \geq \pi.$$

Proof. The theorem directly follows from Corollary 2.2.1.   □

Remark 3.7.1. We can construct another proof using Theorem 3.7.1. Assume that γ is a straight line and $d_\infty(\gamma(-\infty), \gamma(\infty)) < \pi$. Then, by reversing the parameter of γ if necessary, there is a Riemannian half-plane $H = D(\gamma|_{(-\infty,-a]},$

$\gamma|_{[a,+\infty)})$ for an $a \geq 0$ such that $\lambda_\infty(H) = d_\infty(\gamma(-\infty), \gamma(\infty)) < \pi$. Applying Theorem 3.7.1 to H yields a contradiction, because any Gromov–Hausdorff limit of straight lines is also a straight line.

Theorem 3.7.4. *For any* $x, y \in M(\infty)$ *with* $d_\infty(x, y) > \pi$, *there exists a straight line* γ *in* M *such that* $\gamma(-\infty) = x$ *and* $\gamma(\infty) = y$.

Proof. Let $x, y \in M(\infty)$ and fix a core C of M. For each $t \geq 0$, find a minimal segment γ_t in M joining $\sigma_x(t)$ to $\sigma_y(t)$. Assume that

$$\liminf_{t \to +\infty} d(\gamma_t, C) = +\infty.$$

Then there exists a connected component V of $\overline{M \setminus C}$ such that $V(\infty) \ni x, y$. If V is a Riemannian half-cylinder then γ_t for t sufficiently large is contained in either $D(\sigma_x, \sigma_y)$ or $D(\sigma_y, \sigma_x)$. If V is a Riemannian half-plane then γ_t for t sufficiently large is contained in $D(\sigma_x, \sigma_y)$, provided that $D(\sigma_x, \sigma_y)$ is defined. Thus, without loss of generality it may be assumed that for some monotone increasing sequence $\{t_i\}_{i=1,2,\ldots}$ of positive numbers tending to $+\infty$, the segments γ_{t_i} are all contained in $D(\sigma_x, \sigma_y)$. Now, for simplicity, set $D := D(\sigma_x, \sigma_y)$. For each i, let D_i be the compact domain in D surrounded by $\sigma_x|_{[0,t_i]}$, $I(\sigma_x, \sigma_y)$, $\sigma_y|_{[0,t_i]}$ and γ_{t_i}. Then $\{D_i\}$ is a monotone increasing sequence covering D, so that

$$\lim_{i \to \infty} c(D_i) = c(D),$$

$$\lim_{i \to \infty} (\lambda(\partial D_i) - (\pi - \theta_i) - (\pi - \varphi_i)) = \lambda(\partial D),$$

where $\theta_i := \angle_{\sigma_x(t_i)} D_i$ and $\varphi_i := \angle_{\sigma_y(t_i)} D_i$. The Gauss–Bonnet theorem implies that

$$c(D_i) + \lambda(\partial D_i) = 2\pi.$$

Therefore

$$d_\infty(x, y) \leq \lambda_\infty(D) = \pi - \lambda(\partial D) - c(D)$$
$$= \pi - \lim_{i \to \infty} (\theta_i + \varphi_i) \leq \pi.$$

Now, supposing that $d_\infty(x, y) > \pi$ for given two points $x, y \in M(\infty)$, the above discussion implies the existence of a sequence $\{t_i\}$ of positive numbers tending to $+\infty$ such that $\{d(\gamma_{t_i}, C)\}$ is uniformly bounded, so that there exists a subsequence of $\{\gamma_{t_i}\}$ converging to a curve $\gamma : \mathbf{R} \to M$ that is a straight line. Since $\gamma|_{(-\infty,0]}$ and $\gamma|_{[0,+\infty)}$ are respectively parallel to σ_x and σ_y, we obtain $\gamma(-\infty) = x$ and $\gamma(\infty) = y$. $\square$

Remark 3.7.2. For two given points $x, y \in M(\infty)$ we cannot in general say whether there exists a straight line γ such that $\gamma(-\infty) = x$ and $\gamma(\infty) = y$. In fact, if $M = \mathbf{R}^2$ with a flat metric then there always exists such a straight line. However, we have a Riemannian plane with total curvature 0 (i.e., $M(\infty)$ is isometric to the unit circle) that contains no straight lines, as is seen in the following example.

Example 3.7.1 (cf. [64]). For two fixed positive numbers y_0, y_1 with $y_0 + \pi/2 < y_1$, let $f : (0, y_1) \to (0, +\infty)$ be a C^∞-function such that

$$f(0+) = +\infty,$$

$$f > 1, \quad f' < 0, \quad f'' > 0 \qquad \text{on } (0, y_0),$$

$$f = 1 \qquad \text{on } [y_0, y_0 + \pi/2],$$

$$f(y_1-) = 0, \quad f'(y_1-) = -\infty, \quad f^{(n)}(y_1-) = 0 \quad \text{for any } n \geq 2,$$

where $a+$ (resp. $a-$) means $y < a$ (resp. $> a$) and tending to a. For coordinates (x, y, z) of $\mathbf{R}^3$, the subset

$$\{(f(y), y, 0); y \in (0, y_1)\} \cup \{(0, y_1, 0)\}$$

is the image of a smooth xy-plane curve, which generates a surface of revolution, say M, with rotation axis y; see Figure 3.7.4. The surface M satisfies $c(M) = 0$, or equivalently $\lambda_\infty(M) = 2\pi$.

Let us now prove the nonexistence of a straight line in M. We divide M into the following three regions:

- $M_1 := M \cap \{(x, y, z) \in \mathbf{R}^3; y_0 + \pi/2 < y \leq y_1\}$, which is an open disk domain of $G > 0$;
- $M_2 := M \cap \{(x, y, z) \in \mathbf{R}^3; y_0 \leq y \leq y_0 + \pi/2\}$, which is a flat cylinder;
- $M_3 := M \cap \{(x, y, z) \in \mathbf{R}^3; 0 < y < y_0\}$, which is an open cylinder of $G < 0$.

Suppose that there is a straight line γ in M. On the one hand, if γ passes through a point in $M_1 \cup M_2$ then it intersects both M_1 and M_3, so that there are numbers $t_1 < t_2 < t_3$ with $\gamma(t_1), \gamma(t_3) \in \partial M_3$ and $\gamma(t_2) \in M_1$. Hence $L(\gamma|_{[t_1, t_3]}) > 2d(M_1, M_3) = \pi$. On the other hand, $d(\gamma(t_1), \gamma(t_3)) \leq \text{diam}\, \partial M_3 = \pi$. This is a contradiction of the minimal property of γ. Therefore γ must be contained in M_3. Since $d_\infty(\gamma(-\infty), \gamma(\infty)) = \pi$, each of the two half-planes bounded by γ has total curvature 0, which contradicts the fact that one of them is a subset of M_3. Thus M contains no straight lines.

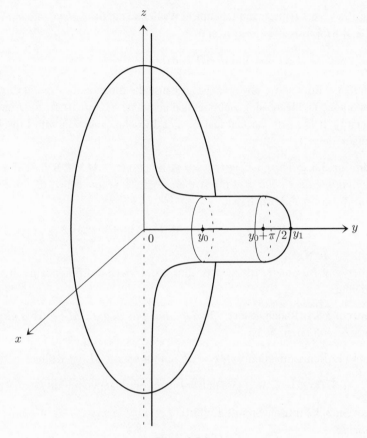

Figure 3.7.4

3.8 The behavior of Busemann functions

In this section, we will consider the asymptotic behavior and the exhaustion property of Busemann functions in connection with the Tits metric on the ideal boundary; these have been studied in [**78, 79, 42, 91**]. We here extend known theorems to the case where the boundary of the manifold may be nonempty.

First of all, we will give the definition of a Busemann function and prove its elementary properties. Assume for a while that M is a complete noncompact Riemannian manifold possibly with boundary.

Definition 3.8.1 (Busemann function). We define the *Busemann function F_σ* :
$M \to \mathbf{R}$ *with respect to a ray σ in M* by

$$F_\sigma(x) := \lim_{t \to \infty} \{t - d(x, \sigma(t))\} \qquad \text{for } x \in M.$$

Here, the limit above always exists because the function $t \mapsto t - d(x, \sigma(t))$
is monotone nondecreasing and bounded above by $d(x, \gamma(0))$, by the triangle
inequality. It is easily verified that F_σ is a Lipschitz function with Lipschitz
constant unity 1.

Definition 3.8.2 (The gradient vector). A function $f : M \to \mathbf{R}$ is said to be
differentiable at a point $x \in M$ if there exists a vector called the *gradient
vector* $\nabla f(x) \in T_x M$ such that

$$f(\exp_x v) = f(x) - \langle \nabla f(x), v \rangle + o(|v|)$$

for all $v \in T_x M$.

Then any Busemann function is almost everywhere differentiable, by the
following:

Theorem 3.8.1 (Rademacher). *Any Lipschitz function $f : M \to \mathbf{R}$ is almost
everywhere differentiable.*

For a Lipschitz function $f : M \to \mathbf{R}$ and for an $x \in M$, we define

$$\bar{\nabla} f(x) := T_x M \cap \overline{\{\nabla f(y); \ y \text{ is a differentiable point of } f\}}.$$

Theorem 3.8.2 (cf. Theorem 1.1, [**80**]). *Let σ be a ray in M. We have the
following.*

(1) $B(\sigma(t), t - a) \uparrow F_\sigma^{-1}(a, \infty)$ *as $t \to \infty$.*
(2) *For any $x \in M$ and $a \in F_\sigma(M)$ with $F_\sigma(x) > a$,*

$$F_\sigma(x) = a - d(x, F_\sigma^{-1}(a)).$$

(3) *For any ray γ asymptotic to σ,*

$$F_\sigma \circ \gamma(t) = t + F_\sigma \circ \gamma(0) \qquad \text{for any } t \geq 0.$$

(4) *For any ray γ asymptotic to σ and for any $a > 0$ with
$\gamma(a) \in \operatorname{int} M := M \setminus \partial M$, the Busemann function F_σ is differentiable at
$\gamma(a)$ and $\nabla F_\sigma(\gamma(a)) = \dot{\gamma}(a)$.*
(5) *For any $x \in \operatorname{int} M$, $\bar{\nabla} F_\sigma(x)$
coincides with the set of initial vectors of rays from x asymptotic to σ.*
(6) *For any differentiable point $x \in \operatorname{int} M$ of F_σ, we have $\bar{\nabla} F_\sigma(x) =
\{\nabla F_\sigma(x)\}$.*

Proof. (1): The monotonicity of $B(\sigma(t), t - a)$ follows from the triangle inequality. To prove $\bigcup_{t>a} B(\sigma(t), t - a) \subset F_\sigma^{-1}(a, \infty)$, we take any point $x \in B(\sigma(t), t - a)$ and deduce that $F_\sigma(x) \geq t - d(x, \sigma(t)) > a$, which means $x \in F_\sigma^{-1}(a, \infty)$. To prove $\bigcup_{t>a} B(\sigma(t), t - a) \supset F_\sigma^{-1}(a, \infty)$, we take any point $x \in F_\sigma^{-1}(a, \infty)$. Then there is a $t > a$ such that $t - d(x, \sigma(t)) > a$, which implies that $x \in B(\sigma(t), t - a)$. This completes the proof of (1).

(2): Let $x \in M$ and $a \in \mathbf{R}$ be such that $F_\sigma(x) > a$. Then it follows that

$$t - d(x, \sigma(t)) = a - d(x, B(\sigma(t), t - a)),$$

which together with (1) implies that

$$F_\sigma(x) = a - d\big(x, F_\sigma^{-1}(a, \infty)\big),$$

where $\partial F_\sigma^{-1}(a, \infty) = F_\sigma^{-1}(a)$. Thus we have (2).

(3): There is a sequence $\{\sigma_i : [0, \ell_i] \to M\}$ of minimal segments tending to γ such that $\sigma(t_i) = \sigma_i(\ell_i)$ for every i and $t_i, \ell_i \to +\infty$. It follows that, for an arbitrarily fixed $t \geq 0$,

$$|d(\gamma(t), \sigma(t_i)) - d(\sigma_i(t), \sigma(t_i))| \leq d(\gamma(t), \sigma_i(t)) \to 0 \qquad \text{as } i \to \infty,$$

and therefore

$$\begin{aligned}
F_\sigma \circ \gamma(t) &= \lim_{i \to \infty} \{t_i - d(\sigma_i(t), \sigma(t_i))\} \\
&= t + \lim_{i \to \infty} \{t_i - d(\sigma_i(0), \sigma(t_i))\} = t + F_\sigma \circ \gamma(0).
\end{aligned}$$

(4): Let γ be a ray asymptotic to σ and $a > 0$ be a number such that $\gamma(a) \in \text{int } M$. Set $\delta := (\text{conv } \gamma(a))/3$ and take a number $b \in (a, a + \delta)$. Denote by $S(p, r)$ the metric sphere centered at p and of radius $r > 0$. Since $\Sigma_\pm := S(\gamma(b \pm \delta), \delta)$ is contained in $B(\gamma(a), 3\delta)$, both $\Sigma_\pm$ are smooth hypersurfaces. Since a minimal segment from $\gamma(t)$ for $t \in (b - \delta, a]$ to $\Sigma_\pm$ is unique and is $\gamma|_{[t,b]}$, the function $d(\cdot, \Sigma_\pm)$ is differentiable at $\gamma(a)$ and its gradient is $-\dot\gamma(a)$, i.e.,

$$d(\exp_{\gamma(a)} v, \Sigma_\pm) = b - a - \langle \dot\gamma(a), v \rangle + o(|v|) \qquad \text{for any } v \in T_{\gamma(a)}M. \tag{3.8.1}$$

Setting $\alpha := F_\sigma(\gamma(b))$, we have, by (1),

$$M \setminus B(\gamma(b - \delta), \delta) \supset F_\sigma^{-1}(\alpha, \infty) \supset B(\gamma(b + \delta), \delta),$$

and hence

$$d(\cdot, \Sigma_-) \leq d(\cdot, F_\sigma^{-1}(\alpha)) \leq d(\cdot, \Sigma_+),$$

which together with (3.8.1) implies that

$$d(\exp_{\gamma(a)} v, F_\sigma^{-1}(\alpha)) = b - a - \langle \dot{\gamma}(a), v \rangle + o(|v|) \qquad \text{for any } v \in T_{\gamma(a)}M.$$

Since $d(\cdot, F_\sigma^{-1}(\alpha)) = \alpha - F_\sigma$, this completes the proof of (4).

(5): Let γ be any ray from an $x \in \text{int } M$ asymptotic to σ. Then, by (4), we have $\dot{\gamma}(a) = \nabla F_\sigma(\gamma(a))$ for any sufficiently small $a > 0$, which implies that $\dot{\gamma}(0) \in \bar{\nabla} F_\sigma(x)$.

Conversely, for any $v \in \bar{\nabla} F_\sigma(x)$ and $x \in \text{int } M$, there is a sequence $x_i \to x$ such that $\nabla F_\sigma(x_i) \to v$. If γ_i is a ray from x_i asymptotic to σ, it follows from (3) that $(F_\sigma \circ \gamma_i)'(0) = 1$ and hence $\dot{\gamma}_i(0) = \nabla F_\sigma(x_i) \to v$. A limit of γ_i is a ray from x with initial vector v and asymptotic to σ. This completes the proof of (5).

(6): Assume that F_σ is differentiable at a point $x \in \text{int } M$. By (5), for any $v \in \bar{\nabla} F_\sigma(x)$ we can find a ray γ from x whose initial vector is v. It follows from (3) that $(F_\sigma \circ \gamma)'(0) = 1$ and hence $\nabla F_\sigma(x) = v$. $\square$

Remark 3.8.1. If M has no boundary then the converse of (3) of Theorem 3.8.2 holds; precisely, if a unit-speed curve $\gamma : [0, +\infty) \to M$ satisfies

$$F_\sigma \circ \gamma(t) = t + F_\sigma \circ \gamma(0) \qquad \text{for any } t \geq 0$$

then γ is a ray asymptotic to σ. See Theorem 1.1(4) of [**80**] for a proof.

Now, let H be a Riemannian half-plane admitting a curvature at infinity and assume that $\sigma := \partial H|_{(-\infty, -a]}$ and $\tau := \partial H|_{[b, +\infty)}$ for two numbers $a, b \geq 0$ are (not necessarily geodesic) rays in H, where $\mathbf{R} \ni t \mapsto \partial H(t)$ is a positive arc length parameterization of ∂H. For each $s, t \geq 0$, we can find a compact contractible region $D_{s,t}$ in H bounded by a minimal segment, say $\gamma_{s,t}$, from $\sigma(s)$ to $\tau(t)$. It may be assumed that $D_{s,t}$ is monotone nondecreasing in s and t. The region $D_t := \bigcup_{s \geq 0} D_{s,t}$ is bounded by the ray $\gamma_t := \lim_{s \to \infty} \gamma_{s,t}$ asymptotic to σ. The angle $\theta_{s,t} := \angle_{\tau(t)}(H \setminus D_{s,t})$ is monotone nonincreasing in s and tends to $\theta_t := \angle_{\tau(t)}(H \setminus D_t)$ as $s \to \infty$. Since D_t is monotone nondecreasing in t, the ray γ_t is monotone nondecreasing in t with respect to the order relation on $\mathcal{C}_H$. Hence, $\{\gamma_t\}_{t \geq 0}$ satisfies only one of the two following conditions:

(1) for any compact subset K of H there is a $t_0 \geq 0$ such that γ_t for any $t \geq t_0$ does not intersect K;

(2) the ray γ_t converges to a straight line in H as $t \to \infty$.

We will now prove:

Lemma 3.8.1. *If $\{\gamma_t\}$ diverges then*

$$\lim_{t \to \infty} \theta_t = \lambda_\infty(H) \leq \pi. \qquad \bullet$$

Proof. Since each γ_t is parallel to σ, we have

$$0 = \lambda_\infty(D_t) = \pi - \lambda(\partial D_t) - c(D_t).$$

Since $\{\gamma_t\}$ diverges, it follows that

$$\lim_{t \to \infty}(\lambda(\partial D_t) - (\pi - \angle_{\sigma(t)}D_t)) = \lambda(\partial H) \qquad \text{and} \qquad \lim_{t \to \infty}c(D_t) = c(H).$$

Therefore

$$\lim_{t \to \infty}(\pi - \angle_{\sigma(t)}D_t) = \pi - \lambda(\partial H) - c(H) = \lambda_\infty(H),$$

so that, in particular, $\lambda(\partial H)$ is finite, which implies that the left-hand side of the above is equal to $\lim \theta_t$ and is not greater than π. $\qquad\square$

Lemma 3.8.2. *We have*

$$\frac{\mathrm{d}}{\mathrm{d}t}F_\sigma(\tau(t)) = \cos\theta_t \qquad \text{for almost all } t \geq 0.$$

Proof. Let $t_1 > t_0 \geq 0$ be any fixed numbers. Since F_σ is differentiable almost everywhere, Fubini's theorem shows that there is a sequence of piecewise-mooth curves $\tau_i : [t_0, t_1] \to H$, $i = 1, 2, \ldots$, tending to $\tau|_{[t_0,t_1]}$ such that

(1) F_σ is differentiable at almost all points on τ_i;
(2) the right derivative of τ_i uniformly converges to the right derivative of $\tau|_{[t_0,t_1]}$.

Then it follows that

$$F_\sigma(\tau_i(t_1)) - F_\sigma(\tau_i(t_0)) = \int_{t_0}^{t_1} \langle \nabla F_\sigma(\tau_i(t)), \dot\tau_i(t)\rangle \, \mathrm{d}t.$$

Denote by $V(t)$ (resp. $V_i(t)$) the set of initial vectors of rays in H from $\tau(t)$ (resp. $\tau_i(t)$) asymptotic to σ. Notice that, since $\{t \geq 0;\ \#V(t) \geq 2\}$ is of measure zero, we have $V(t) = \{\dot\gamma_t(0)\}$ for almost all t and that $\lim_{i\to\infty}V_i(t) \subset V(t)$ for any $t \in [t_0, t_1]$ by Proposition 3.2.2. Now, if F_σ is differentiable at $\tau_i(t)$ for a $t \in [t_0, t_1]$ then Theorem 3.8.2 (5), (6) implies that $V_i(t) = \{\nabla F_\sigma(\tau_i(t))\}$. Hence $\lim_{i\to\infty}\nabla F_\sigma(\tau_i(t)) = \dot\gamma_t(0)$ for almost all $t \in [t_0, t_1]$, so that

$$\lim_{i\to\infty}\langle \nabla F_\sigma(\tau_i(t)), \dot\tau_i(t)\rangle = \cos\theta_t \qquad \text{for almost all } t \in [t_0, t_1].$$

By the Lebesgue dominated convergence theorem,

$$F_\sigma(\tau(t_1)) - F_\sigma(\tau(t_0)) = \int_{t_0}^{t_1} \cos\theta_t \, \mathrm{d}t.$$

Since θ_t is almost everywhere continuous in t, this completes the proof. $\qquad\square$

Remark 3.8.2. If F_σ is differentiable at $\tau(t)$ for almost all $t \geq 0$ then Lemma 3.8.2 is obvious. However, the former will not be true in general.

Lemma 3.8.3

(1) *If* $\{\gamma_t\}$ *diverges then*

$$\lim_{t \to \infty} \frac{F_\sigma(\tau(t))}{t} = \cos \lambda_\infty(H) \qquad and \qquad \lambda_\infty(H) \leq \pi.$$

(2) *If* $\{\gamma_t\}$ *converges then*

$$\lim_{t \to \infty} \frac{F_\sigma(\tau(t))}{t} = -1 \qquad and \qquad \lambda_\infty(H) \geq \pi.$$

Proof. (1) follows from Lemmas 3.8.1 and 3.8.2.

(2): Assume that γ_t converges to a straight line γ as $t \to \infty$. Then $\lambda_\infty(H) \geq \pi$ follows from Corollary 2.2.1 (cf. Theorem 3.7.3). There is a sequence $s_i \to \infty$ such that $\gamma_{s_i,t}$ tends to γ_t as $i \to \infty$. If $i(t)$ is taken to be large enough, it follows that the distance between the image of $\gamma_{s_i,t}$ and a fixed point on M is uniformly bounded for all $t \geq 0$ and $i \geq i(t)$. Then the triangle inequality shows that $|d(\sigma(s_i), \tau(t)) - s_i - t|$ is uniformly bounded for all $t \geq 0$ and $i \geq i(t)$, so that

$$\lim_{t \to \infty} \frac{F_\sigma(\tau(t))}{t} = \lim_{t \to \infty} \left[\frac{1}{t} \lim_{i \to \infty} \{s_i - d(\sigma(s_i), \tau(t))\} \right] = -1. \qquad \square$$

Let M be a finitely connected, complete and noncompact 2-manifold admitting a curvature at infinity.

Theorem 3.8.3. *For any rays σ and τ in M, we have*

$$\lim_{t \to \infty} \frac{F_\sigma(\tau(t))}{t} = \cos \min\{d_\infty(\sigma(\infty), \tau(\infty)), \pi\}.$$

Proof. If $\sigma \cap \tau$ is unbounded then there are a number $a \in \mathbf{R}$ and a sequence $t_i \to \infty$ such that $\sigma(s_i) = \tau(s_i + a)$ for all i, which implies that

$$F_\sigma(\tau(t)) = \lim_{i \to \infty} \{s_i - d(\tau(s_i + a), \tau(t))\} = t - a.$$

Since $\sigma(\infty) = \tau(\infty)$, this proves the theorem.

Assume that $\sigma \cap \tau$ is bounded. Then there is a core C of M such that $\sigma[a, +\infty) \cap \tau[b, +\infty) = \emptyset$, $\sigma|_{[a,+\infty)} \cap C = \{\sigma(a)\}$ and $\tau|_{[b,+\infty)} \cap C = \{\tau(b)\}$. We can find an arbitrary divergent sequence $\{t_i\}$ of positive numbers and a ray γ_i from $\tau(t_i)$ asymptotic to σ for each i.

If a subsequence $\{\gamma_{j(i)}\}$ of $\{\gamma_i\}$ tends to some straight line γ then the same discussion as in the proof of Lemma 3.8.3(2) yields

$$\lim_{i \to \infty} \frac{F_\sigma(\tau(t_{j(i)}))}{t_{j(i)}} = -1 \quad \text{and} \quad d_\infty(\sigma(\infty), \tau(\infty)) \geq \pi.$$

Assume that a subsequence $\{\gamma_{j(i)}\}$ of $\{\gamma_i\}$ has no accumulation lines. Then the rays $\sigma(t)$ and $\tau(t)$ both tend to a common end as $t \to \infty$. If V denotes the connected component of $\overline{M \setminus C}$ corresponding to that end then the ray $\gamma_{j(i)}$ for each sufficiently large i is contained in V. Now, we consider a minimal segment $\gamma : [0, \ell] \to M$ joining a point $p \in \sigma$ to a point $q \in \tau$ that is contained in V. Since σ and τ may have break points at ∂M, the minimal segment γ is not necessarily contained in only one of $D(\sigma, \tau)$ and $D(\tau, \sigma)$. However, we can find a different minimal segment $\gamma' : [0, \ell] \to M$ from p to q that is contained in one of $D(\sigma, \tau)$ and $D(\tau, \sigma)$. In fact, there are $a, b \in [0, \ell]$ such that $\gamma(a) \in \sigma$, $\gamma(b) \in \tau$ and $\gamma(a, b) \cap H = \emptyset$, where H is one of $D(\sigma, \tau)$ and $D(\tau, \sigma)$. Then, the join of the subarc of σ from p to $\gamma(a)$ and $\gamma|_{[a,b]}$ and the subarc of τ from $\gamma(b)$ to q is the desired minimal segment γ'. Therefore, we can replace $\gamma_{j(i)}$ in such a way that $\gamma_{j(i)}$ is contained in one of $D(\sigma, \tau)$ and $D(\tau, \sigma)$ and still has no accumulation lines. We can assume without loss of generality that all the $\gamma_{j(i)}$ are contained in $D(\sigma, \tau) =: H$. Then it follows that $F_\sigma(\tau(t_{j(i)})) = F_\sigma^H(\tau(t_{j(i)}))$ for all i, where F_σ^H is the intrinsic Busemann function of H, i.e., the Busemann function defined by the intrinsic metric d_H of H. Hence, by Lemma 3.8.3,

$$\lim_{i \to \infty} \frac{F_\sigma(\tau(t_{j(i)}))}{t_{j(i)}} = \lim_{i \to \infty} \frac{F_\sigma^H(\tau(t_{j(i)}))}{t_{j(i)}} = \cos \lambda_\infty(H)$$
$$\text{and} \quad \lambda_\infty(H) \leq \pi. \tag{3.8.2}$$

If V is a Riemannian half-plane then we have $d_\infty(\sigma(\infty), \tau(\infty)) = \lambda_\infty(H)$ and thus the proof is complete.

Now assume that V is a Riemannian half-cylinder. Then $\gamma_{j(i)} \subset H$ implies that there is a sequence $s_k \to \infty$ such that $d_{V \setminus H}(\sigma(s_k), \tau(t_{j(i)})) \geq d(\sigma(s_k), \tau(t_{j(i)})) = d_H(\sigma(s_k), \tau(t_{j(i)}))$ for all k large enough compared with i and hence $F_\sigma^{V \setminus H}(\tau(t_{j(i)})) \leq F_\sigma(\tau(t_{j(i)}))$. Therefore

$$\lim_{i \to \infty} \frac{F_\sigma(\tau(t_{j(i)}))}{t_{j(i)}} \geq \lim_{i \to \infty} \frac{F_\sigma^{V \setminus H}(\tau(t_{j(i)}))}{t_{j(i)}} = \cos \min\{\lambda_\infty(V \setminus H), \pi\}. \tag{3.8.3}$$

The formulae (3.8.2) and (3.8.3) together imply that $d_\infty(\sigma(\infty), \tau(\infty)) =$

$\lambda_\infty(H) \le \min\{\lambda_\infty(V \setminus H), \pi\}$. Thus, we obtain

$$\lim_{i \to \infty} \frac{F_\sigma(\tau(t_{j(i)}))}{t_{j(i)}} = \cos d_\infty(\sigma(\infty), \tau(\infty))$$

$$\text{and} \qquad d_\infty(\sigma(\infty), \tau(\infty)) \le \pi.$$

By the arbitrariness of the sequence $\{t_i\}$, this completes the proof. $\qquad \square$

Theorem 3.8.4. *Let σ be a ray in M and let a sequence $p_i \in M$, $i = 0, 1, \ldots$, tend to a point $p_\infty \in M(\infty)$. If p_∞ has a neighborhood in $M(\infty)$ with finite d_∞-diameter, we have*

$$\lim_{i \to \infty} \frac{F_\sigma(p_i)}{d(p_0, p_i)} = \cos \min\{d(p_\infty, \sigma(\infty)), \pi\}.$$

Proof. Let C be a core of M. The assumption yields that there is a Riemannian half-plane $H = D(\alpha, \beta)$, with $\alpha, \beta \in \mathcal{R}_C$, such that $\lambda_\infty(H) < +\infty$ and $H(\infty)$ is a neighborhood of p_∞. Let $\tau \in \mathcal{R}_C$ be such that $\tau(\infty) = p_\infty$ and τ is contained in H. Setting $t_i := d(p_i, C)$ we have

$$\left| \frac{F_\sigma(p_i)}{t_i} - \frac{F_\sigma(\tau(t_i))}{t_i} \right| \le \frac{d(p_i, \tau(t_i))}{t_i},$$

where applying Theorem 3.7.1 to H yields that the right-hand side of the above relationship tends to zero as $i \to 0$. Thus, by Theorem 3.8.3 the proof is complete. $\qquad \square$

Definition 3.8.3 (Radius of metric space). Let (X, d) be a metric space and $x \in X$ any point. The *radius* $\mathrm{rad}(X, x, d)$ *of the pointed space* (X, x, d) is defined by

$$\mathrm{rad}(X, x, d) := \sup_{y \in X} d(x, y).$$

Now the *radius* $\mathrm{rad}(X, d)$ *of the metric space* (X, d) is defined by

$$\mathrm{rad}(X, d) := \inf_{x \in X} \mathrm{rad}(X, x, d).$$

It is easily verified that $\frac{1}{2} \mathrm{diam}(X, d) \le \mathrm{rad}(X, x, d) \le \mathrm{diam}(X, d)$ provided that (X, d) is an intrinsic metric space.

Remark 3.8.3. Notice that

(1) if M has more than one end then

$$\mathrm{diam}(M(\infty), d_\infty) = \mathrm{rad}(M(\infty), d_\infty) = +\infty;$$

(2) if $\partial M(\infty) = \emptyset$ then $\mathrm{diam}(M(\infty), d_\infty) = \mathrm{rad}(M(\infty), d_\infty)$;
(3) if $\partial M(\infty) \ne \emptyset$ then $\mathrm{diam}(M(\infty), d_\infty) = 2 \, \mathrm{rad}(M(\infty), d_\infty)$.

Theorem 3.8.4 can be used to prove:

Theorem 3.8.5. *For any ray σ in M we have the following:*

(1) *if* $\mathrm{rad}(M(\infty), \sigma(\infty), d_\infty) < \pi/2$ *then* F_σ *is an exhaustion function;*
(2) *if* $\mathrm{rad}(M(\infty), \sigma(\infty), d_\infty) > \pi/2$ *then* F_σ *is a nonexhaustion function.*

Exercise 3.8.1. Prove Theorem 3.8.5.

Lemma 3.8.4. *Let H, σ, τ be as defined after Remark 3.8.1. Assume that $\lambda_\infty(H) = \pi/2$, that σ, τ are geodesics and that there exists a compact subset $K \subset H$ such that $G \geq 0$ on $H \setminus K$. Then there exists a large number $T \geq 0$ such that $F_\sigma(\tau(t))$ is monotone nonincreasing in $t \in [T, +\infty)$.*

Proof. It follows from $\lambda_\infty(H) = \pi/2$ that $\{\gamma_t\}$ diverges. There exists a large $T \geq 0$ such that $K \subset D_T$. Then $c(D_t)$ is monotone nondecreasing in $t \geq T$ and tends to $c(H)$ as $t \to +\infty$. We have $0 = \lambda_\infty(D_t) = \pi - \lambda(\partial D_t) - c(D_t)$ and $\lambda(\partial D_t) = \lambda(\partial H) + \theta_t$, so that θ_t is monotone nonincreasing in $t \geq T$ and tends to $\lambda_\infty(H) = \pi/2$ as $t \to +\infty$. In particular we obtain $\theta_t \geq \pi/2$ for any $t \geq T$. Thus by Lemma 3.8.2 this completes the proof. $\qquad\square$

Theorem 3.8.6. *Assume that for a core C of M, $G \geq 0$ on $M \setminus C$, $\lambda(\partial M \setminus C) \geq 0$, and let σ and τ be two rays from C such that $d_\infty(\sigma(\infty), \tau(\infty)) = \pi/2$. Then there exists a number $T \geq 0$ such that $F_\sigma(\tau(t))$ is monotone nonincreasing in $t \geq T$. In particular, if $\mathrm{rad}(M(\infty), \sigma(\infty), d_\infty) = \pi/2$ for a ray σ in M then F_σ is a nonexhaustion function.*

Proof. Notice that the assumption implies that both $\sigma|_{[a,+\infty)}$ and $\tau|_{[a,+\infty)}$ for some large $a \geq 0$ are geodesics and that $\sigma(t)$ and $\tau(t)$ tend to a common end as $t \to \infty$. Denote by V the connected component of $\overline{M \setminus C}$ corresponding to that end. Without loss of generality it may be assumed that $D(\sigma, \tau) =: H$ is defined (when V is a Riemannian half-plane). Since there are no straight lines γ in M such that $\gamma(-\infty) = \sigma(\infty)$ and $\gamma(\infty) = \tau(\infty)$, we have (cf. the proof of Theorem 3.8.3) for all sufficiently large $t \geq 0$,

$$
F_\sigma(\tau(t)) =
\begin{cases}
F_\sigma^H(\tau(t)) & \text{if V is a Riemannian half-plane,} \\[2mm]
\min\left\{F_\sigma^H(\tau(t)), F_\sigma^{V \setminus H}(\tau(t))\right\} & \text{if V is a Riemannian half-cylinder.}
\end{cases}
$$

Thus, Lemma 3.8.4 proves the theorem. $\qquad\square$

As a direct consequence of Theorems 3.8.5 and 3.8.6 we have:

Theorem 3.8.7. *Assume that there exists a compact subset K of M such that $G \geq 0$ on $M \setminus K$ and $\lambda(\partial H \setminus K) \geq 0$. Then, the following hold.*

(1) $\mathrm{diam}(M(\infty), d_\infty) < \pi/2$ *iff any Busemann function on M is an exhaustion function.*

(2) $\mathrm{rad}(M(\infty), d_\infty) \geq \pi/2$ *iff any Busemann function on M is a nonexhaustion function.*

(3) *The following, (a)–(c), are equivalent to each other.*

    (a) $\mathrm{diam}(M(\infty), d_\infty) \geq \pi/2$ *and* $\mathrm{rad}(M(\infty), d_\infty) < \pi/2$.

    (b) *M admits both an exhaustion Busemann function and a nonexhaustion Busemann function.*

    (c) *$(M(\infty), d_\infty)$ is isometric to a compact arc with length ℓ, $\pi/2 \leq \ell < \pi$.*

4

The cut loci of complete open surfaces

The definition of the cut locus was introduced by Poincaré ([**68**]), and he first investigated the structure of the cut locus of a point on a complete, simply connected and real analytic Riemannian 2-manifold. Myers ([**60, 61**]) determined the structure of the cut locus of a point in a 2-sphere and Whitehead [**107**] proved that the cut locus of a point on a complete two-dimensional Riemannian manifold carries the structure of a local tree. In this chapter we will determine the structure of the cut locus and distance circles of a Jordan curve in a complete Riemannian 2-manifold and will prove the absolute continuity of the distance function of the cut locus.

4.1 Preliminaries

Throughout this chapter (M, g) always denotes a complete connected smooth two-dimensional Riemannian manifold without boundary. Let $\gamma : [0, \infty) \to M$ be a unit-speed geodesic. Note that any geodesic segment $\gamma : [0, a) \to M$ is extensible to $[0, \infty)$ according to Theorem 1.7.3. If, for some positive number b, $\gamma|_{[0,b]}$ is not a minimizing geodesic joining its endpoints, let t_0 be the largest positive number t such that $\gamma|_{[0,t]}$ is minimizing. Note that there always exists such a positive number t_0, by Lemma 1.2.2 . The point $\gamma(t_0)$ is called a *cut point* of $\gamma(0)$ along the geodesic γ. For each point p on M, let $C(p)$ denote the set of all cut points along the geodesics emanating from p.

Let $\mathcal{C}$ be a smooth Jordan curve in M parameterized by a smooth map $z_0 : [0, L_0] \to M$ with arc length parameter s. The map z_0 and other functions of s will be considered periodic with period L_0, for convenience. Throughout this chapter we will assume that $\mathcal{C}$ admits a unit normal smooth field N of period L_0 along itself. For example, the curve admits such a normal field if it lies in a convex ball $B(p, \text{conv}(p))$. A map $z : \mathbf{R} \times [0, L_0] \to M$ is defined by

$$z(t, s) := \exp_{z_0(s)} t N_s.$$

Definition 4.1.1. Let $\pi : \mathcal{N}C \to C$ be the normal bundle of C with projection π.

For every $s \in [0, L_0]$ let $\gamma_s : \mathbf{R} \to M$ be a geodesic with $\gamma_s(t) = z(t, s)$ and $e_s(t)$ a unit parallel vector field along γ_s with $e_s(0) = (\partial z/\partial s)(0, s)$. Since the normal exponential map on the normal bundle $\mathcal{N}C$ maps diffeomorphically around the zero section, each γ_s is a minimizing geodesic from $\gamma_s(t)$ to C if $|t|$ is sufficiently small. Thus if this is the case then it follows from the Gauss lemma 1.2.1 that the map z gives a coordinate system (t, s), that $g(\partial z/\partial t, \partial z/\partial t) = 1$ holds around C and that $g(\partial z/\partial t, \partial z/\partial s) = 0$. For each s let $Y_s(t)$ denote the Jacobi field along γ_s with $Y_s(0) = e_s(0)$, $g(Y_s(t), \gamma'_s(t)) = 0$. By setting

$$f(t, s) := g(Y_s(t), e_s(t)),$$

we have $f(0, s) = 1$, $f_t(0, s) = k(s)$ and $g(\partial z/\partial s, \partial z/\partial s) = f^2(t, s)$, where $k(s)$ denotes the geodesic curvature of C at $z_0(s)$ and $f_t = \partial f/\partial t$. Since Y_s is a Jacobi field we have

$$f_{tt}(t, s) + G(z(t, s))f(t, s) = 0,$$

where $f_{tt} = (\partial/\partial t)f_t$ and G denotes the Gaussian curvature of M.

Exercise 4.1.1. Prove that $f_t(0, s) = k(s)$ holds for each s.

Let $P(s)$ (resp. $N(s)$) denote the least positive t (resp. the largest negative t) with $f(s, t) = 0$, or let $P(s) = +\infty$ (resp. $N(s) = -\infty$) if there is no such zero. If $P(s_0) < +\infty$ (resp. $N(s_0) > -\infty$) then P (resp. N) is smooth around s_0, and $z(P(s_0), s_0)$ (resp. $z(N(s_0), s_0)$) is called *the first positive* (resp. *negative*) *focal point of* C along γ_{s_0}. A unit-speed geodesic $\gamma : [0, a] \to M$ is called a C-*segment* (or *minimizing geodesic from* $\gamma(a)$ *to* C) iff $d(\gamma(t), C) = t$ holds for all $t \in [0, a]$. It follows from the short-cut principle, Lemma 1.8.1, that no two distinct C-segments intersect in its interior. Any C-segment γ issues from C orthogonally, by the first variation formula (cf. Theorem 1.5.1). Therefore every C-segment is a subarc of some γ_s. The set of all terminal points of all C-segments is called the *cut locus of* C and denoted by $C(C)$.

We shall need tools from measure theory to prove some theorems in this chapter. Readers unfamiliar with the theory could consult, for example, [**106**].

Let $h : [a, b] \to \mathbf{R}$ be a continuous function with bounded variation. Then the function defines a Lebesgue–Stieltjes measure Λ_h such that $\Lambda_h((x, y])$ for each subinterval $(x, y]$ of $[a, b]$ equals the total variation of h on $[x, y]$. It is known that any Borel subset B in $[a, b]$ is Λ_h-measurable. The measure $\Lambda_h(B)$ will be called the *variation of h over B*.

Lemma 4.1.1 (Banach [8]). *Let h be a continuous function on $[a, b]$ and $n(t)$ the number of elements of $h^{-1}(t)$. Then h is of bounded variation iff $n(t)$ is Lebesgue integrable on* **R**, *in which case*

$$\int_{-\infty}^{\infty} n(t)\,dt = V(h) \tag{4.1.1}$$

holds, where $V(h) = \Lambda_h(a, b]$. In particular, $n(t)$ is finite for almost all t if h is of bounded variation.

Proof. For each positive integer p, define $2^p + 1$ numbers a_j $(j = 0, 1, 2, \ldots, 2^p)$ by

$$a_j := a + j(b - a)2^{-p}.$$

For each positive integer $j \leq 2^p$, define intervals I_j by

$$I_j := \begin{cases} [a_{j-1}, a_j) & \text{if } j < 2^p, \\ [a_{j-1}, a_j] & \text{if } j = 2^p. \end{cases}$$

Let $L_j(t)$, $j = 1, 2, \ldots, 2^p$, denote the characteristic function of $h(I_j)$, i.e., $L_j(t) = 1$ if t is an element of $h(I_j)$ and $L_j(t) = 0$ otherwise. Let $n_p(t)$ be the function defined by

$$n_p(t) = \sum_{j=1}^{2^p} L_j(t). \tag{4.1.2}$$

It follows from the construction that $n_p(t) \geq n_q(t)$ if p is not less than q. Hence we have a measurable function

$$\bar{n}(t) := \lim_{p \to \infty} n_p(t). \tag{4.1.3}$$

Since $n(t) \geq n_p(t)$ for any positive integer p, we have

$$n(t) \geq \bar{n}(t). \tag{4.1.4}$$

If $n(t) \geq m$ for some integer m and for some real t then, for any sufficiently large integer p, $n_p(t) \geq m$. Thus if, for some t, $n(t) \geq m$ then $\bar{n}(t) \geq m$. It follows from (4.1.4) and this property that $n(t) = \bar{n}(t)$ for any t. It follows from the construction of $L_j(t)$ that

$$\int_{-\infty}^{\infty} L_j(t)\,dt = M_j - m_j, \qquad (4.1.5)$$

where $M_j := \sup h|_{I_j}$ and $m_j := \inf h|_{I_j}$. Then we have

$$\int_{-\infty}^{\infty} n_p(t)\,dt = \sum_{j=1}^{2^p} (M_j - m_j). \qquad (4.1.6)$$

Note that the right-hand side of (4.1.6) is not greater than $V(h)$. By (4.1.6),

$$\int_{-\infty}^{\infty} n_p(t)\,dt \le V(h) \qquad (4.1.7)$$

for any positive integer p. It follows from Fatou's lemma and (4.1.7) that

$$\int_{-\infty}^{\infty} n(t)\,dt \le V(h). \qquad (4.1.8)$$

If h is of bounded variation then by (4.1.8) $n(t)$ is Lebesgue integrable. Conversely, if $n(t)$ is Lebesgue integrable then by (4.1.6) h is of bounded variation, since

$$\lim_{p\to\infty} \sum_{j=1}^{2^p} (M_j - m_j) = V(h) \qquad (4.1.9)$$

holds. $\qquad\qquad\qquad\qquad\qquad\qquad\qquad\qquad\qquad\qquad\qquad\qquad\square$

Corollary 4.1.1. *Let h be continuous and of bounded variation. Then*

$$\Lambda_h(h^{-1}(-\infty, r)) = \Lambda_h(h^{-1}(-\infty, r]) = \int_{-\infty}^{r} n(t)\,dt. \qquad (4.1.10)$$

In particular the function $\mathbf{R} \ni r \mapsto \Lambda_f(f^{-1}(-\infty, r])$ *is absolutely continuous on each compact interval.*

Proof. Suppose that r is a number with $n(r) < +\infty$. Then we can choose a finite number of disjoint intervals I_j such that

$$h^{-1}(-\infty, r] = I_1 \cup \cdots \cup I_N.$$

If we define a function h_r by

$$h_r(t) := \min\{h(t), r\}$$

then h_r is continuous and of bounded variation. Since Λ_h and Λ_{h_r} are measures, we have

$$\Lambda_h(h^{-1}(-\infty, r)) = \sum_j \Lambda_h(I_j) = \sum_j \Lambda_{h_r}(I_j) = \Lambda_{h_r}(\bigcup_j I_j) \quad (4.1.11)$$

where Λ_{h_r} denotes the Lebesgue–Stieltjes measure defined by h_r. However, $\Lambda_{h_r}(\bigcup_j I_j)$ equals the total variation in h_r. Thus by Lemma 4.1.1 we obtain

$$\Lambda_{h_r}\left(\bigcup_j I_j\right) = \int_{-\infty}^{\infty} n_r(t)\, dt \quad (4.1.12)$$

where $n_r(t)$ is the number of elements of $h_r^{-1}(t)$. Clearly $n_r(t) = 0$ for any $t > r$ and $n_r(t) = n(t)$ for any $t \le r$. By (4.1.11) and (4.1.12) we have (4.1.10) for any r with $n(r) < \infty$. The remaining case is easily proved by obtaining an infinite sequence $\{r_i\}$ such that $n(r_i) < +\infty$ for each i and $\lim_{i \to \infty} r_i = r$. Since $h^{-1}(-\infty, r)$ is a countable union of disjoint open intervals, it is easy to prove that $\Lambda_h(h^{-1}(-\infty, r)) = \int_{-\infty}^{r} n(t)\, dt$. $\qquad \square$

Exercise 4.1.2. Give a complete proof of Corollary 4.1.1.

The following lemma is a special case of the Sard lemma.

Lemma 4.1.2 (Sard). *If h is a C^1-function on (a, b) then it follows that $h(\{t \in (a, b)|h'(t) = 0\})$ is of Lebesgue measure zero.*

Exercise 4.1.3. Prove the lemma above.

Remark 4.1.1. The lemma is true for a continuous function (cf. [**87**]), i.e., *the set*

$$h(\{t \in (a, b); h'(t) \text{ exists and equals zero}\})$$

is of Lebesgue measure zero for any continuous function h.

Lemma 4.1.3. *Let $\{I_n\}$ be a collection of mutually disjoint closed subintervals of $[0, 1]$. Then there exists a continuous monotone nondecreasing function f : $[0, 1] \to [0, 1]$ such that $f(0) = 0$, $f(1) = 1$ and such that $f(t_1) = f(t_2)$ for t_1, t_2 iff t_1 and t_2 lie in a common I_n.*

Proof. The function f with the required properties can be constructed in the same way as the Lebesgue–Cantor function ([**106**]). For technical reasons we add a countable family of closed intervals $\{J_k\}$ to $\{I_n\}$. For each component (a, b) of $(0, 1) \setminus \overline{\bigcup_n I_n}$, where $\overline{\bigcup_n I_n}$ denotes the closure of $\bigcup_n I_n$, we add a closed interval $[a, (2a + b)/3]$ (resp. $[(a + 2b)/3, b]$) to $\{I_n\}$ if $a \notin \bigcup_n I_n$ (resp. if $b \notin \bigcup_n I_n$); otherwise, we do not make this addition. Thus we obtain a countable family of mutually disjoint closed intervals $\{I_n, J_k; n, k\}$. First we rename each I_n, J_k as follows. Choose closed intervals $J(0, 0)$ and $J(0, 1)$ from $\{I_n, J_k\}$ such that $0 \in J(0, 0)$, $1 \in J(0, 1)$, if these intervals exist. Otherwise, choose an empty set. Now choose an interval $J(1, 1)$ from $\{I_n, J_k\}$ such that $J(1, 1) \subset [0, 1] \setminus (J(0, 1) \cup J(0, 0))$, $|J(1, 1)| = \max\{|I_n|, |J_k|; I_n \cup J_k \subset [0, 1] \setminus (J(0, 1) \cup J(0, 0))\}$, where $|J(1, 1)|, |I_n|, |J_k|$ denote the respective lengths of the intervals. Suppose that closed intervals $J(n, j)$, $j = 1, 2, \ldots, 2^{n-1}$, have been defined for some positive integer n. Then for each component J of

$$[0, 1] \setminus \bigcup_{\substack{1 \le k \le n \\ 1 \le j \le 2^{n-1}}} J(k, j) \cup J(0, 0) \cup J(0, 1)$$

choose an interval I from $\{I_n, J_k\}$ such that $I \subset J$ and

$$|I| = \max\{|I_n|, |J_k|; I_n \subset J, J_k \subset J\}.$$

If there is no such I for some J, choose an empty set. So, we have chosen closed intervals $J(n+1, j)$, $1 \le j \le 2^n$, with $\max J(n+1, j) < \min J(n+1, j+1)$. From the construction, for each positive integer n we have that

$$|J(n, i)| \ge |J(n+1, j)|$$

holds for any $i \in \{1, 2, \ldots, 2^{n-1}\}$ and $j \in \{1, 2, \ldots, 2^n\}$. Thus it is trivial that

$$\{I_n, J_k\} = \{J(n, i); n = 1, 2, \ldots, 1 \le i \le 2^{n-1}\} \cup \{J(0, 0), J(0, 1)\}.$$

For each positive integer n, let f_n be a continuous monotone nondecreasing function on $[0, 1]$ that satisfies

$$f_n(0) = 0, \qquad f_n(1) = 1, \qquad f_n|_{J(0,0)} = 0,$$

$$f_n|_{J(0,1)} = 1, \qquad f_n|_{J(k,j)} = \frac{2j - 1}{2^k}$$

for each k, j with $1 \le k \le n$, $j = 1, 2, \ldots, 2^{k-1}$, and which is linear on each component of

$$[0, 1] \setminus \left(\bigcup_{k=1}^{n} \bigcup_{j=1}^{2^{k-1}} J(k, j) \cup J(0, 0) \cup J(0, 1) \right).$$

Clearly each f_n is monotone nondecreasing and has the property that $t_1, t_2 \in [0, 1]$ satisfies $f_n(t_1) = f_n(t_2)$ iff t_1 and t_2 are contained in a common $J(k, j)$.

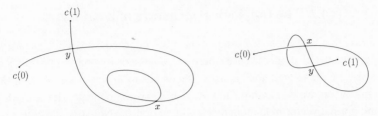

Figure 4.1.1 On the left, $I_1 = I_y \supset I_x$; on the right, $I_1 = I_x$, $I_y \not\subset [0, 1] \setminus I_1$.

It follows from the construction that, for any positive integers $m > n$,

$$|f_n(t) - f_m(t)| \le \frac{1}{2^{n-1}}$$

holds on $[0, 1]$. Therefore, there exists a unique limit function $\tilde{f}(t) = \lim_{n \to \infty} f_n(t)$. Since the convergence is uniform, the function $\tilde{f}$ is monotone nondecreasing and continuous. If (a_n, b_n), $k = 1, 2, \ldots$, denotes all the components of $(0, 1) \setminus \overline{\bigcup_n I_n}$ such that $a_k \notin \bigcup_n I_n$ or $b_k \notin \bigcup_n I_n$, let f be a monotone nondecreasing continuous function on $[0, 1]$ which satisfies $f = \tilde{f}$ on $[0, 1] \setminus \bigcup(a_k, b_k)$ and which is linear on each closed interval $[a_k, b_k]$ with $f(a_k) = \tilde{f}(a_k)$, $f(b_k) = \tilde{f}(b_k)$. Then it is easy to check that f has the required properties. $\qquad\square$

Definition 4.1.2. An injective continuous map from an open or closed interval into M is called a *Jordan arc*. An injective continuous map from a circle into M is called a *Jordan curve*.

Proposition 4.1.1. *Any continuous curve* $c : [0, 1] \to M$ *contains a Jordan subarc joining* $c(0)$ *to* $c(1)$ *if* $c(0) \ne c(1)$ (see Figure 4.1.1).

Proof. For each multiple point x in $c([0, 1])$ let I_x denote the closed subinterval $[\min c^{-1}(x), \max c^{-1}(x)]$ of $[0, 1]$. Since c is continuous, we may choose a closed interval $I_1 := [a_1, b_1]$ from the intervals I_x that has maximal length. Inductively we may choose closed intervals $I_{n+1} := [a_{n+1}, b_{n+1}]$, from the intervals I_x having maximal length and multiple points $x \in c([0, 1] \setminus \bigcup_{k=1}^n I_k)$, which admit distinct elements a, b in $[0, 1] \setminus \bigcup_{k=1}^n I_k$ such that $c(a) = c(b) = x$. Thus we get a sequence $\{I_n\}$ of mutually disjoint closed intervals such that $c(a_n) = c(b_n)$ for each $I_n = [a_n, b_n]$ and such that if two distinct elements x, y in $[0, 1] \setminus \bigcup_n \text{int } I_n$ satisfy $c(x) = c(y)$ then x and y are the endpoints of a common I_n. It follows from Lemma 4.1.3 that there exists a continuous monotone nondecreasing function f on $[0, 1]$ such that $f(0) = 0$, $f(1) = 1$ and such that $f(t_1) = f(t_2)$ for t_1, t_2 iff t_1 and t_2 lie in a common I_n. Then $\bar{c}(u) := c(\max f^{-1}(u))$ on $[0, 1]$ is a continuous Jordan arc joining $c(0)$ and $c(1)$. $\qquad\square$

4.2 The topological structure of a cut locus

It is easy to observe that $M \setminus C$ has at most two connected components. We will not go into detail about the case where $M \setminus C$ has exactly one component, since the case can be treated analogously. The details of this case are given in [**86**]. From now on, we will assume that $M \setminus C$ has two components. Hence C admits a unit normal vector field along itself with period L_0. Note that the case where this is not so would reduce to the case where $M \setminus C$ has two components, for most theorems, propositions and lemmas. In fact for a small positive number ϵ, the normal ϵ-sphere bundle $\mathcal{N}_\epsilon C$ of C is mapped by an exponential map diffeomorphically onto a smooth Jordan curve C_ϵ. It is easy to check that C_ϵ bounds a relatively compact domain and that

$$S(C, t + \epsilon) = S(C_\epsilon, t), \qquad B(C, t + \epsilon) = B(C_\epsilon, t)$$

for any $t > \epsilon$, where

$$B(C, t) = \{q \in M; d(q, C) < t\}.$$

Let

$$\rho(s) := \sup\{t; d(\gamma_s(t), C) = t\}$$

and

$$\nu(s) := \inf\{t < 0; d(\gamma_s(t), C) = -t\}.$$

As we noted in Section 4.1, these two functions are strictly positive valued. If $\rho(s)$ (resp. $\nu(s)$) is finite then $z(\rho(s), s)$ (resp. $z(\nu(s), s)$) is called a *cut point of* C. Thus $\gamma_s|_{[0,\rho(s)]}$ (resp. $\gamma_s|_{[\nu(s),0]}$) is a maximal C-segment contained in $\gamma_s[0, \infty)$ (resp. $\gamma_s|_{(-\infty,0]}$). The functions ρ, ν are called the *distance functions of the cut locus*. By Remark 1.8.1, we have

Lemma 4.2.1. *No two distinct C-segments intersect in their interior.*

From now on, our arguments will be framed in terms of the function ρ, but exactly the same arguments hold for ν.

Proposition 4.2.1. *The function $\rho : [0, L_0] \to (0, \infty]$ is continuous.*

Proof. Let s_0 be any real number in $[0, L_0]$ and $\{s_n\}$ any sequence converging to s_0. Since any limiting geodesic of a sequence of C-segments is itself a C-segment, we obtain

$$\limsup_{n \to \infty} \rho(s_n) \leq \rho(s_0). \tag{4.2.1}$$

Supposing that $\rho(s_0) > b$ for a constant b, we shall prove that $\liminf_{n\to\infty} \rho(s_n)$ $\geq b$. This property implies that

$$\liminf_{n\to\infty} \rho(s_n) \geq \rho(s_0). \tag{4.2.2}$$

Thus, combining inequalities (4.2.1) and (4.2.2), we obtain the claimed proposition. By the above assumption, $\gamma_{s_0}|_{[0,b]}$ is a C-segment and $\gamma_{s_0}(b)$ is not a focal point of C along γ_{s_0}. Therefore the normal exponential map on NC maps a neighborhood U around $b\dot\gamma_{s_0}(0)$ diffeomorphically onto a neighborhood V around $\gamma_{s_0}(b)$. It follows from Lemma 4.2.1 that if x is a point sufficiently close to $\gamma_{s_0}(b)$ then, for any C-segment γ_s from x to C, $d(x, C)\dot\gamma_s(0)$ is an element of U. Therefore, for sufficiently large n we have that $\gamma_{s_n}|_{[0,b]}$ is a C-segment. Thus we get (4.2.2). □

Since the set $\{s \in [0, L_0]; \rho(s) < +\infty\}$ is a union of countably many intervals, it follows from the proposition above that the cut locus is a union of countably many continuous curves. If the cut locus contains a Jordan curve σ that bounds a domain then we get a C-segment from a point in the domain. Thus the segment meets a cut point on σ in its interior. This is a contradiction. Therefore we have

Lemma 4.2.2. *A cut locus does not contain a Jordan curve that lies in a convex neighborhood. Furthermore, if M is simply connected then the cut locus does not contain a Jordan curve.*

For each cut point x of C, let $\Gamma(x)$ denote the set of all C-segments from x to C. The following definition of a sector at a point $x \in C(C)$ plays an important role in investigating the structure of the cut locus.

Definition 4.2.1. In the case where a positive ϵ $(< d(x, C))$ is chosen sufficiently small that $B(x, \epsilon)$ is a convex ball, each component of $B(x, \epsilon) \setminus \bigcup_{\gamma \in \Gamma(x)} \gamma(0, \epsilon)$ is called an ϵ-*sector* (or simply a *sector*) *at* x.

Each sector $\Sigma_\epsilon(x)$ at $x \in C(C)$ has the following four properties.

S1 If $y \in \Sigma_\epsilon(x)$ then the unique minimal geodesic joining y to x lies in $\Sigma_\epsilon(x) \cup \{x\}$. If the inner angle of $\Sigma_\epsilon(x)$ at x is less than π then every minimizing geodesic joining two points in $\Sigma_\epsilon(x)$ lies in $\Sigma_\epsilon(x)$.

S2 There is no element in $\Gamma(x)$ that passes through points in $\Sigma_\epsilon(x)$.

S3 There exists a sequence of cut points of C in $\Sigma_\epsilon(x)$ converging to x.

S4 If $\{q_j\}$ is a sequence of points in $\Sigma_\epsilon(x)$ converging to x then every converging subsequence of C-segments in $\Gamma(q_j)$ has as its limit either γ or σ, where $\gamma, \sigma \in \Gamma(x)$ denote the geodesics whose subarcs, together with the subarc of $S(x, \epsilon)$ with endpoints $\gamma(d(\mathcal{C}, x) - \epsilon)$ and $\sigma(d(\mathcal{C}, x) - \epsilon)$, form the boundary of $\Sigma_\epsilon(x)$.

The properties above would be trivial were it not for the inclusion amongst them of S3. Suppose that there is no cut point near x in $\Sigma_\epsilon(x)$. Let $S(x, t)$ denote the geodesic circle at x with radius t. Then, for each $t < \epsilon$, the arc $S(x, t) \cap \Sigma_\epsilon(x)$ is simply covered by $\mathcal{C}$-segments. If $t_1 < t_2 < \epsilon$ then each $\mathcal{C}$-segment through $S(x, t_1) \cap \Sigma_\epsilon(x)$ is the continuation of some $\mathcal{C}$-segment through $S(x, t_2) \cap \Sigma_\epsilon(x)$. Therefore we get an element of $\Gamma(x)$ through $\Sigma_\epsilon(x)$. This contradicts S2.

Lemma 4.2.3. *For each* $s \in [0, L_0]$ *the inequality* $\rho(s) \leq P(s)$ *holds, and if* $\rho(s_0) = P(s_0) < +\infty$ *for some* s_0 *then* $P'(s_0) = 0$.

Proof. Without loss of generality, we may assume that $P(s_0) < \infty$ in order to prove $\rho(s_0) \leq P(s_0)$. Thus $z(P(s_0), s_0)$ is the first focal point of $\mathcal{C}$. Take any $l > P(s_0)$. Then, by imitating the proof of Lemma 1.6.1, we may construct a piecewise-smooth variation $v : (-\epsilon_0, \epsilon_0) \times [0, l] \to M$ of γ_{s_0}, with $v(\epsilon, 0) = z_0(s_0 + \epsilon)$, such that $L(\epsilon) < L(0)$ for any sufficiently small $\epsilon > 0$, where $L(\epsilon)$ denotes the length of the curve $s \mapsto v(s, \epsilon)$. Thus we have proved that $\rho(s) \leq P(s)$ for any s. Suppose that $P'(s_0) \neq 0$ for some $s_0 \in (0, L_0]$ satisfying $P(s_0) = \rho(s_0) < +\infty$. Again without loss of generality, we may assume that $P'(s_0) > 0$. Take any $s_1 \in [0, s_0)$ such that $P'(s) > 0$ on $[s_1, s_0]$. The length of a subarc $\{z(P(s), s); s \in [s_1, s_0]\}$ equals

$$\int_{s_1}^{s_0} \left\| \frac{\mathrm{d}}{\mathrm{d}s} z(P(s), s) \right\| \mathrm{d}s = P(s_0) - P(s_1).$$

Thus on the one hand, by the definition of the distance function, we obtain

$$d(z(P(s_1), s_1), z(P(s_0), s_0)) \leq P(s_0) - P(s_1), \tag{4.2.3}$$

$$d(z(P(s_1), s_1), \mathcal{C}) \leq L(\gamma_{s_1}|_{[0, P(s_1)]}) = P(s_1).$$

On the other hand, from the triangle inequality,

$$d(z(P(s_1), s_1), \mathcal{C}) + d(z(P(s_1), s_1), z(P(s_0), s_0)) \tag{4.2.4}$$
$$\geq d(z(P(s_0), s_0), \mathcal{C}) = P(s_0).$$

Note that $P(s_0) = \rho(s_0)$. By (4.2.3), the equality holds in the relationship (4.2.4). This implies that the arc combining $\gamma_{s_1}|_{[0, P(s_1)]}$ and $\{z(P(s), s); s \in [s_1, s_0]\}$ is a $\mathcal{C}$-segment from $z(P(s_0), s_0)$ for any $s_1(< s_0)$ sufficiently close to s_0. This contradicts the assumption $P'(s_0) \neq 0$. $\qquad\square$

A cut point of $\mathcal{C}$ is said to be *normal* if it is the endpoint of exactly two $\mathcal{C}$-segments and is not a first focal point along either of them. A cut point of $\mathcal{C}$

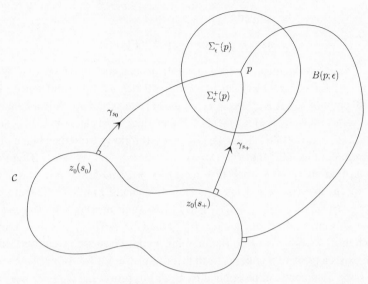

Figure 4.2.1

that is not normal is said to be *anormal*. An anormal cut point x is said to be *totally nondegenerate* iff x is not a first focal point of C along any C-segment from itself. An anormal cut point is said to be *degenerate* iff it is not totally nondegenerate. By Lemmas 4.2.3, 4.2.4 and Propositions 4.2.2, 4.2.3, we will prove that the set of all normal cut points is open and dense in a cut locus with the relative topology.

Lemma 4.2.4. *The set* $F := \{s \in [0, L_0]; \rho(s) < P(s) \text{ and } z(\rho(s), s) \text{ is a degenerate cut point}\}$ *is of Lebesgue measure zero.*

Proof. Fix $s_0 \in F$ and set $p = z(\rho(s_0), s_0)$. Let $\Sigma_\epsilon^+(p)$ and $\Sigma_\epsilon^-(p)$ be the two sectors at p that have the common boundary $\gamma_{s_0}[\rho(s_0) - \epsilon, \rho(s_0)]$; see Figure 4.2.1. Since p is not a focal point of C along γ_{s_0}, there exists a neighborhood V of $\rho(s_0)\dot{\gamma}_{s_0}(0)$ in $\mathcal{N}C$ and a neighborhood U of p in M such that the restriction $\exp_V$ of the normal exponential map to V is a diffeomorphism of V onto U. Let s_+ be the number in $[0, L_0]$ such that $\gamma_{s_+}[\rho(s_0) - \epsilon, \rho(s_0)]$ is the boundary of $\Sigma_\epsilon^+(p)$ distinct from $\gamma_{s_0}[\rho(s_0) - \epsilon, \rho(s_0)]$. Suppose that $P(s_+) = \rho(s_+)$. Choose a positive number ϵ_1 such that U contains $z(\rho(s), s)$ and $z(P(s), s)$ for all $s \in [s_+ - \epsilon_1, s_+ + \epsilon_1]$. Without loss of generality we may assume that $\{z(\rho(s), s); s \in [s_0, s_0 + \epsilon_1]\}$ is a curve lying in $\Sigma_\epsilon^+(p)$. Hence the curve $\{z(\rho(s), s); s \in [s_0 - \epsilon_1, s_0]\}$ lies in $\Sigma_\epsilon^-(p)$. We may choose a positive number $\delta_1 < \epsilon_1$ such that if $z(\rho(s_1), s_1) = z(\rho(s), s)$ for $s_1 \in [0, L_0]$ and $s \in (s_0, s_0 + \delta_1)$ then $s = s_1$ or $s_1 \in (s_+ - \epsilon_1, s_+ + \epsilon_1)$. Let a smooth function v on

$(s_+ - \epsilon_1, s_+ + \epsilon_1)$ be defined by

$$v(s) := z_0^{-1} \circ \pi \circ \exp_V^{-1} z(P(s), s).$$

Recall here that the map $z_0 : [0, L_0] \to M$ denotes the Jordan curve $\mathcal{C}$. Note that if $s \in (s_+ - \epsilon_1, s_+ + \epsilon_1)$ satisfies $P'(s) = 0$ then $v'(s) = 0$ and hence s is a critical point of v. Let $K \subset (s_+ - \epsilon_1, s_+ + \epsilon_1)$ be the set of all critical points of v. If $s \in (s_0, s_0 + \delta_1)$ is an element of F then there exists an $s_1 \in [0, L_0]$ such that $z(\rho(s), s) = z(\rho(s_1), s_1)$, $P(s_1) = \rho(s_1)$. It follows from the choice of δ_1 and from Lemma 4.2.3 that $P'(s_1) = 0$ and $s_1 \in (s_+ - \epsilon_1, s_+ + \epsilon_1)$. Therefore we can find an $s_1 \in K$ such that $z(\rho(s), s) = z(\rho(s_1), s_1) = z(P(s_1), s_1)$. This means that $(s_0, s_0 + \delta_1) \cap F$ is entirely contained in $v(K)$. Therefore if $\rho(s_+) = P(s_+)$ then $(s_0, s_0 + \delta_1) \cap F$ is of Lebesgue measure zero, since $v(K)$ is thus by Lemma 4.1.2. If $\rho(s_+) < P(s_+)$ then there exists a positive number δ such that $(s_0, s_0 + \delta) \cap F = \emptyset$. Summing up this discussion we observe that there exists a positive number δ_1 such that $(s_0, s_0 + \delta_1) \cap F$ is of measure zero. The same argument can be applied to $\Sigma_\epsilon^-(p)$ to prove that $(s_0 - \delta_1, s_0) \cap F$ is of measure zero for some positive number δ_1. Therefore F is covered by countably many sets of measure zero. Hence F is also of measure zero. $\square$

Proposition 4.2.2. *Let p be a normal cut point such that $p = z(\rho(s), s)$ has exactly two solutions, $s = s_1, s_2$, in $[0, L_0)$. Then:*

(1) *$\rho(s)$ is smooth on a neighborhood of $s = s_1, s_2$;*
(2) *there is a strictly monotonic smooth function $v(s)$ in a neighborhood of s_2 with the properties $v(s_2) = s_1$, $\rho(v(s)) = \rho(s)$ and $z(\rho(v(s)), v(s)) = z(\rho(s), s)$ for s near s_2;*
(3) *p has exactly two sectors and the cut locus of $\mathcal{C}$ near p is a smooth curve that bisects the inner angle of each sector at p.*

Proof. Since p is not a focal point along γ_{s_1} or γ_{s_2}, there exist neighborhoods V, U_1 and U_2 around p, $\rho(s_1)\dot{\gamma}_{s_1}(0)$ and $\rho(s_2)\dot{\gamma}_{s_2}(0)$ respectively such that the normal exponential map on $\mathcal{N}C$ is a diffeomorphism of U_i onto V for each $i = 1, 2$. Let f be a smooth function on V defined by

$$f(q) := \| \exp_{U_1}^{-1} q \| - \| \exp_{U_2}^{-1} q \|,$$

where $\exp_{U_i} : U_i \to V$ denotes the restriction of the normal exponential map to U_i. Then for any $w \in T_p M$ the equation

$$df_p(w) = \langle \dot{\gamma}_{s_1}(\rho(s_1)), w \rangle - \langle \dot{\gamma}_{s_2}(\rho(s_2)), w \rangle \tag{4.2.5}$$

follows from the first variation formula. Since the two vectors $\dot{\gamma}_{s_1}(\rho(s_1))$, $\dot{\gamma}_{s_2}(\rho(s_2))$ are distinct, p is not a critical point of f. Thus by the implicit function theorem, $f^{-1}(0)$ is a smooth one-dimensional submanifold. Let v c be a smooth curve in $f^{-1}(0)$ emanating from $p = c(0)$. Put $s_i(t) := z_0^{-1} \circ \pi \circ \exp_{U_i}^{-1} c(t), i = 1, 2$. It is easy to observe that if $|t|$ is sufficiently small then $\gamma_{s_i(t)}|_{[0, \| \exp_{U_i}^{-1} c(t) \|]}$ is a $\mathcal{C}$-segment and $c(t)$ is a normal cut point. Now claim (3) of our proposition is clear by (4.2.5). The other claims also are easily proved. $\qquad \square$

It follows from the first variation formula that, in the above proposition, $\rho'(s) = 0$ at $s = s_1$ (or $s = s_2$) iff the two $\mathcal{C}$-segments $\gamma_{s_1}, \gamma_{s_2}$ make an angle π at $z(\rho(s_1), s_1) = z(\rho(s_2), s_2)$.

Definition 4.2.2. $\gamma_{s_1}|_{[0, 2\rho(s_1)]}$ is called a $\mathcal{C}$-*loop* if the two $\mathcal{C}$-segments $\gamma_{s_1}, \gamma_{s_2}$ make an angle π at $z(\rho(s_1), s_1) = z(\rho(s_2), s_2)$.

The proof of the next proposition is analogous to that of the proposition above.

Proposition 4.2.3. *Let p be a totally nondegenerate anormal cut point. Then:*

(1) *there exists only a finite number $n \geq 3$ of s-values in $[0, L_0)$, say*
    $s = s_1, s_2, \ldots, s_n$, *such that $p = z(\rho(s), s)$;*
(2) *p has exactly n sectors;*
(3) *ρ is piecewise-smooth on all small open intervals including s_j and smooth except at s_j;*
(4) *in each small sector at p, the cut locus is a smooth curve bisecting the inner angle of the sector at p;*
(5) *any cut point in each small sector at p is a normal cut point. In particular, totally nondegenerate cut points are isolated.*

By Lemmas 4.2.3, 4.2.4 and Propositions 4.2.2, 4.2.3 we get

Corollary 4.2.1. *For any $0 \leq a < b < L_0$, the set $\{z(\rho(s), s); a \leq s \leq b, z(\rho(s), s)$ is normal$\}$ is either empty or open dense in $\{z(\rho(s), s); s \in [a, b]\}$. Furthermore, if $\{z(\rho(s), s); a \leq s \leq b, z(\rho(s), s)$ is normal$\}$ is empty then it consists of a single element.*

A topological space T is by definition a *tree* iff any two points on T are joined by a unique Jordan arc in T. A point x on a tree T is by definition an *endpoint* iff $T \setminus x$ is connected. A topological space C is by definition a *local tree* iff, for every point $x \in C$ and for every neighborhood U around x, there exists a smaller neighborhood $T \subset U$ around x that is a tree.

Theorem 4.2.1. *Let M be a complete, connected, smooth Riemannian 2-manifold and $\mathcal{C}$ a smooth Jordan curve in M. Then the cut locus of $\mathcal{C}$ is a local tree. Furthermore, if M is simply connected then the cut locus is a tree.*

Proof. Let x be any cut point of $\mathcal{C}$. Take any sector $\Sigma_\epsilon(x)$ at x. Since ρ is continuous by Proposition 4.2.1, the set

$$\{s \in [0, L_0]; z(\rho(s), s) = x\}$$

is compact. Therefore there exists a finite number of open intervals I_j, $j = 1, 2, \ldots, k$, such that

$$\{s \in [0, L_0]; z(\rho(s), s) = x\} \subset \bigcup_{j=1}^{k} I_j$$

and such that each c_j, where $c_j(s) := z(\rho(s), s)$ for $s \in I_j$, is a curve through x and contained in $B(x, \epsilon)$. By the property S3 of sectors (see Definition 4.2.1), there exists a cut point y in $\Sigma_\epsilon(x)$ sufficiently close to x. Thus $z(\rho(s), s) = y$ for some $s \in I_j$. Since $c_j(I_j)$ contains x, the cut point y can be joined to x by a continuous curve in $\Sigma_\epsilon(x) \cap C(\mathcal{C})$. By Corollary 4.2.1, we may assume that y is a normal cut point, which is the intersection of $\mathcal{C}$-segments σ_y, γ_y of $\Gamma(y)$. Since c_j bisects the two sectors at y, it follows from the property S4 of sectors that, as y tends to x, σ_y and γ_y converge to $\mathcal{C}$-segments σ and γ respectively; $\sigma, \gamma \in \Gamma(x)$ denote the geodesics whose subarcs form the boundary of $\Sigma_\epsilon(x)$, together with the subarc of $S(x, \epsilon)$ with endpoints $\sigma(d(\mathcal{C}, x) - \epsilon)$, $\gamma(d(\mathcal{C}, x) - \epsilon)$.

Thus if y is chosen sufficiently close to x then four $\mathcal{C}$-segments $\sigma, \gamma, \sigma_y, \gamma_y$ and two subarcs $\{z_0(s); s \in J_1\}$, $\{z_0(s); s \in J_2\}$ of $\mathcal{C}$ together bound a disk domain D containing any cut point which lies in $\Sigma_\epsilon(x)$. Here each J_i denotes a closed subinterval in $[0, L_0]$ such that $\{x, y\} = \{z(\rho(s), s); s$ is an endpoint of $J_i\}$. Since a $\mathcal{C}$-segment from a cut point q in D is a subarc of γ_s for some $s \in J_1 \cup J_2$, the cut point q lies on a curve c_j. It follows from Proposition 4.1.1 and Lemma 4.2.2 that any cut point in D (and hence in $\Sigma_\epsilon(x)$) can be joined to x by a unique Jordan arc in $(\Sigma_\epsilon(x) \cup \{x\}) \cap C(\mathcal{C})$. This implies that $C(\mathcal{C})$ is a tree in $(\Sigma_\epsilon(x) \cup \{x\}) \cap \Sigma_{2\epsilon}(y; x)$, where $\Sigma_{2\epsilon}(y; x)$ denotes the 2ϵ-sector at y containing x. If the inner angle of $\Sigma_\epsilon(x)$ at x is sufficiently small then there exists a closed subinterval J of $[0, L_0]$ with $J \subset \bigcup_{j=1}^{k} I_j$ such that the two $\mathcal{C}$-segments σ, γ and the subarc $\{z_0(s); s \in J\}$ of $\mathcal{C}$ cut off by σ and γ bound a disk domain D. Since J is contained in some I_j, any cut point in D lies in the curve c_j. Thus any cut point in D lies in $\Sigma_\epsilon(x)$ and may be joined to x by a Jordan arc in $C(\mathcal{C}) \cap (\Sigma_\epsilon(x) \cup \{x\})$. Summing up the argument, there exist finitely many normal cut points y_i in $B(x, \epsilon)$ such that $U \cap C(\mathcal{C})$ is a tree, where $U := B(x, \epsilon) \cap_i \Sigma_{2\epsilon}(y_i; x)$ is a neighborhood around x. $\square$

In Propositions 4.2.2 and 4.2.3, we observed that each sector at a totally nondegenerate cut point is bisected by a Jordan arc in $C(\mathcal{C})$ emanating from the cut point. This property holds for a sector at any cut point, in a sense. Before stating this property, we need to define some terminology.

A Jordan arc $c : [a, b] \to M$ is said to *have a left tangent* (resp. *a right tangent*) $v \in S_{c(t_0)}M := \{v \in T_{c(t_0)}M; \|v\| = 1\}$ at $c(t_0)$ for some $t_0 \in (a, b]$ (resp. $t_0 \in [a, b)$) iff $\exp^{-1}_{c(t_0)} c(t)/\|\exp^{-1}_{c(t_0)} c(t)\|$ converges to v as $t \to t_0 - 0$ (resp. $t \to t_0 + 0$). Here $\exp^{-1}_{c(t_0)}$ denotes the local inverse of $\exp_{c(t_0)}$ around $\mathbf{0}_{c(t_0)}$.

Proposition 4.2.4. *Let* $c : [0, 1] \to C(\mathcal{C})$ *be a Jordan arc lying in a sector* $\Sigma_{2\epsilon}(p)$ *at* $p = c(0)$. *Then* c *has a right tangent at* $c(0)$ *that bisects the inner angle of* $\Sigma_{2\epsilon}(p)$ *at* p.

Proof. Since the case where $\Gamma(c(0))$ consists of a single $\mathcal{C}$-segment is trivial, by Theorem 1.7.1, we will prove only the other case (see Figure 4.2.2). The inner angle of $\Sigma_{2\epsilon}(p)$ at p is less than 2π, otherwise $\Gamma(p)$ consists of a single element. Let α and β be the $\mathcal{C}$-segments whose subarcs bound $\Sigma_{2\epsilon}(p)$. By choosing a smaller ϵ, we may assume that $c(a) \notin \Sigma_\epsilon(p)$ for some positive a. Then the subarc $c|_{[0,b]}$ ($0 < b < a$) of c divides $\Sigma_\epsilon(p)$ into two components, say D_1, D_2. Since no $\mathcal{C}$-segment meets the curve c in its interior, for each $c(t), t \in (0, b)$, there exist $\mathcal{C}$-segments of $\Gamma(c(t))$ lying in D_1 and D_2 respectively. Thus we get two distinct sectors $\Sigma^+_{2\epsilon}(c(t))$ and $\Sigma^-_{2\epsilon}(c(t))$ for each $c(t), t \in (0, b)$, such that

$$\Sigma^+_{2\epsilon}(c(t)) \supset c(t, b], \qquad \Sigma^-_{2\epsilon}(c(t)) \supset c(0, t].$$

It follows from property S4 of sectors (see Definition 4.2.1) that for each $t \in (0, b]$ we may choose $\mathcal{C}$-segments $\alpha^+_t, \beta^+_t \in \Gamma(c(t))$ (resp. $\alpha^-_t, \beta^-_t \in \Gamma(c(t))$) bounding $\Sigma^+_{2\epsilon}(c(t))$ (resp. $\Sigma^-_{2\epsilon}(c(t))$) such that

$$\lim_{t \to +0} \alpha^+_t = \alpha, \qquad \lim_{t \to +0} \beta^+_t = \beta$$

(resp. $\lim_{t \to +0} \alpha^-_t = \alpha$, $\lim_{t \to +0} \beta^-_t = \beta$).
Assume that there exists a sequence $\{c(t_i)\}$ with $\lim_{i \to \infty} t_i = 0$ such that

$$\lim_{i \to \infty} \angle \left(\exp^{-1}_p \alpha(\epsilon), \exp^{-1}_p c(t_i) \right) =: \theta$$

$$\lim_{i \to \infty} \angle \left(\exp^{-1}_p \beta(\epsilon), \exp^{-1}_p c(t_i) \right) =: \theta'.$$

From the triangle inequality we have

$$d(\beta^-_{t_i}(\epsilon), c(t_i)) - d(\beta^-_{t_i}(\epsilon), p) \le d(\mathcal{C}, c(t_i)) - d(\mathcal{C}, p)$$
$$\le d(\alpha(\epsilon), c(t_i)) - d(\alpha(\epsilon), p).$$

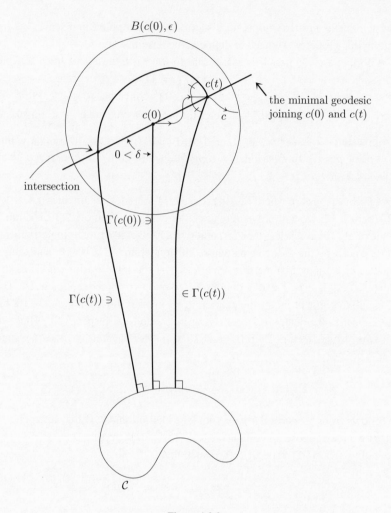

Figure 4.2.2

Applying the first variation formula to each side of the above relation,

$$-\cos\theta' = \lim_{i\to\infty} \frac{d(\beta_{t_i}^-(\epsilon), c(t_i)) - d(\beta_{t_i}^-(\epsilon), p)}{d(c(t_i), p)}$$

$$\leq \lim_{i\to\infty} \frac{d(\alpha(\epsilon), c(t_i)) - d(\alpha(\epsilon), p)}{d(c(t_i), p)} = -\cos\theta.$$

By the symmetry of the above discussion we have $\theta = \theta'$. Thus the claim in the proposition is clear.   □

4.3 Absolute continuity of the distance function of the cut locus

In [34], Hartman tried to prove that ρ is absolutely continuous if it is finite valued. He proved that if the function $\rho_r := \min\{\rho, r\}$ is of bounded variation for any real r then ρ is absolutely continuous where it is finite valued. This property has an interesting application to the Ambrose problem in the two-dimensional case, which was pointed out by Hebda in [35]. Incidentally, this problem in the two-dimensional case can be stated as follows. *Let M and $\bar{M}$ be complete connected Riemannian 2-manifolds with Gaussian curvatures denoted by G and $\bar{G}$ respectively. Suppose that there exists a linear isometry $I : T_p M \to T_{\bar{p}}\bar{M}$ for points $p \in M$ and $\bar{p} \in \bar{M}$ such that $G(\exp_p X) = \bar{G}(\exp_{\bar{p}} I(X))$ for any tangent vector $X \in T_p M$. If M is simply connected, does there exist an isometric immersion $F : M \to \bar{M}$ such that $F(p) = \bar{p}$, $dF = I$ at p?*

In 1993, Hebda in [36] and Itoh in [41] independently and affirmatively solved this problem by proving that ρ_r is of bounded variation for any r in the case where $\mathcal{C} = \{p\}$. Since the function P is smooth if P is finite valued, the following lemma is trivial.

Lemma 4.3.1. *For each real r, the function $P_r : [0, L_0] \to (0, r]$ is Lipschitz continuous, where $P_r := \min\{P, r\}$.*

Let E_0, E_1 be closed subsets of $[0, L_0]$ defined by

$$E_0 := \{s \in [0, L_0]; \rho(s) = P(s)\}$$
$$E_1 := \{s \in [0, L_0]; \rho(s) < +\infty \text{ and } \gamma_s |_{[0, 2\rho(s)]} \text{ is a } \mathcal{C}\text{-loop}\}.$$

See Definition 4.2.2 for the definition of a $\mathcal{C}$-*loop*. For convenience, we can take $0 \in E_0 \cup E_1$ by reparameterizing the curve $\mathcal{C}$. Then we get a countable family of open intervals I_j with $[0, L_0] \setminus E_0 \cup E_1 = \bigcup_j I_j$. By Lemma 4.2.3 the following lemma is trivial.

Lemma 4.3.2. *For any $s_0 \in E_0$ and $s \in [0, L_0]$,*

$$\rho_r(s) - \rho_r(s_0) \le K(P_r) |s - s_0|,$$

where $K(P_r)$ denotes a Lipschitz constant of P_r.

We define positive constants $C_1(r)$, $C_2(r)$, $C_3(r)$ depending on r by

$$C_1(r) := \max\{|f(s, t)|/2; 0 \le s \le L_0, 0 \le t \le r\},$$
$$C_2(r) := C_1(r)(r/\min \rho - 1),$$
$$C_3(r) := \max\{C_1(r), C_2(r)\}.$$

Lemma 4.3.3. *For any $s_0 \in E_1$ and $s_1 \in [0, L_0]$,*

$$\rho_r(s_1) - \rho_r(s_0) \leq C_3(r)|s_1 - s_0|.$$

Proof. We will assume that $\rho(s_0) < \rho(s_1) \leq r$, since the other case is easily proved. Suppose that $\rho(s_1) \leq 2\rho(s_0)$. Define a smooth curve φ on $[0, L_0]$ by $\varphi(s) := z(\rho(s_1), s)$. From the definition of the distance function d,

$$d(\gamma_{s_1}(\rho(s_1)), \gamma_{s_0}(\rho(s_1))) \leq \left| \int_{s_0}^{s_1} |\varphi'(s)| \, ds \right|$$

$$\leq \left| \int_{s_0}^{s_1} |f(\rho(s_1), s)| \, ds \right| \leq 2C_1(r) |s_1 - s_0|. \quad (4.3.1)$$

Since $\gamma_{s_0}|_{[0, 2\rho(s_0)]}$ is a $\mathcal{C}$-loop, we have

$$d(\mathcal{C}, \gamma_{s_0}(\rho(s_1))) = 2\rho(s_0) - \rho(s_1). \quad (4.3.2)$$

Thus by the triangle inequality

$$d(\gamma_{s_1}(\rho(s_1)), \gamma_{s_0}(\rho(s_1))) \geq d(\mathcal{C}, \gamma_{s_1}(\rho(s_1))) - d(\mathcal{C}, \gamma_{s_0}(\rho(s_1)))$$

$$\geq 2(\rho(s_1) - \rho(s_0)). \quad (4.3.3)$$

Therefore by (4.3.1) and (4.3.3) we have proved the claim in the lemma if $\rho(s_1) \leq 2\rho(s_0) \leq 2r$. Now suppose that $\rho(s_1) > 2\rho(s_0)$. Then if we take a positive value $s_2 \in [0, L_0]$ satisfying $\rho(s_2) = 2\rho(s_0)$ we obtain $|s_0 - s_1| \geq |s_0 - s_2|$. By applying the reasoning in the first case to the pair s_0 and s_2, we get

$$\rho(s_0) = \rho(s_2) - \rho(s_0) \leq C_1(r)|s_2 - s_0|. \quad (4.3.4)$$

Thus

$$\rho(s_1) - \rho(s_0) \leq \rho(s_0) \left(\frac{\rho(s_1)}{\min \rho} - 1 \right) \leq (\rho(s_2) - \rho(s_0)) \left(\frac{r}{\min \rho} - 1 \right)$$

$$\leq C_2(r)|s_2 - s_0| \leq C_2(r)|s_1 - s_0|. \quad (4.3.5)$$

$\square$

Lemma 4.3.4. *If ρ attains a local minimum at $s = s_0$ then $s_0 \in E_0 \cup E_1$.*

Proof. Put $p = z(\rho(s_0), s_0)$. Suppose that $s_0 \notin E_0 \cup E_1$. Then p admits at least two sectors. Since $s_0 \notin E_1$, there exists a sector $\Sigma_\epsilon(p)$ at p and adjacent to γ_{s_0} whose inner angle θ at p is less than π. Without loss of generality, we may assume that a Jordan arc $\{z(\rho(s), s); s_0 - \epsilon_1 \leq s \leq s_0\}$ in $C(\mathcal{C})$ lies in the sector $\Sigma_\epsilon(p)$ for some positive $\epsilon_1 > 0$. Since for each $s \in [s_0 - \epsilon, s_0]$

$$\rho(s) = d(\mathcal{C}, \gamma_s(\rho(s))) \leq d(z_0(s_0), \gamma_s(\rho(s)))$$

holds, by the triangle inequality we get

$$\rho(s_0) - \rho(s) \geq d(z_0(s_0), p) - d(z_0(s_0), c(s))$$

$$\geq d(\gamma_{s_0}(\rho(s_0) - \epsilon), p) - d(\gamma_{s_0}(\rho(s_0) - \epsilon), c(s)), \quad (4.3.6)$$

where $c(s) = z(\rho(s), s)$.

It follows from the first variation formula and Proposition 4.2.4 that

$$\lim_{s \to s_0} \frac{d(\gamma_{s_0}(\rho(s_0) - \epsilon)p) - d(\gamma_{s_0}(\rho(s_0) - \epsilon), c(s))}{d(p, c(s))} = \frac{\cos\theta}{2}. \quad (4.3.7)$$

Thus, by (4.3.6) and (4.3.7), ρ cannot attain a local minimum at s_0 since θ is less than π. This is a contradiction. $\qquad\square$

Lemma 4.3.5. *For each interval I_j above,*

$$V(\rho_r|_{I_j}) \leq C(r)|I_j|$$

where $V(\rho_r|_{I_j})$ denotes the total variation of $\rho_r|_{I_j}$, $C(r) := \max\{K(P_r), C_3(r)\}$ and $|I_j|$ denotes the length of I_j.

Proof. Since ρ has no local minimum in the interior of I_j, either $\rho|_{I_j}$ is monotonic or it has a unique maximal value $\rho(t_1)$ such that ρ is monotone nondecreasing on $[a_j, t_1]$ and monotone nonincreasing on $[t_1, b_j]$, where $(a_j, b_j) = I_j$. By Lemmas 4.3.2 and 4.3.3, we have the claim. $\qquad\square$

Proposition 4.3.1. *For any $r > 0$, ρ_r is of bounded variation.*

Proof. Let $0 = t_0 < t_1 < t_2 < \cdots < t_n = L_0$ be any subdivision of $[0, L_0]$. If $(t_i, t_{i+1}) \cap (E_0 \cup E_1) \neq \emptyset$ then we may choose the smallest number a and the largest number b in $[t_i, t_{i+1}] \cap (E_0 \cup E_1)$. Thus

$$(t_i, a) \cap (E_0 \cup E_1) = \emptyset = (b, t_{i+1}) \cap (E_0 \cup E_1).$$

By taking a finer subdivision of $[0, L_0]$, if necessary, we may assume that if $(t_i, t_{i+1}) \cap (E_0 \cup E_1)$ is nonempty then $t_i, t_{i+1} \in E_0 \cup E_1$. Now decompose $\{0, 1, \ldots, n-1\}$ into two disjoint subsets A, B:

$$A := \{i; (t_i, t_{i+1}) \cap (E_0 \cup E_1) = \emptyset\},$$
$$B := \{i; (t_i, t_{i+1}) \cap (E_0 \cup E_1) \neq \emptyset\}.$$

Then it follows from Lemmas 4.3.2, 4.3.3 and 4.3.5 that

$$\sum_{i \in A} |\rho_r(t_i) - \rho_r(t_{i+1})| \leq C(r) \sum_j |I_j| \leq L_0 C(r)$$

and

$$\sum_{i \in B} |\rho_r(t_i) - \rho_r(t_{i+1})| \le C(r) \sum_{i \in B} |t_{i+1} - t_i| \le L_0 C(r).$$

Therefore we get

$$\sum_{i=0}^{n-1} |\rho_r(t_i) - \rho_r(t_{i+1})| \le 2L_0 C(r)$$

for any subdivision $0 = t_0 < t_1 < \cdots < t_n = L_0$ of $[0, L_0]$.   □

Exercise 4.3.1. Generalize Proposition 4.3.1 to the case of a smooth Jordan curve C in M admitting a periodic smooth unit normal vector field along itself.

Definition 4.3.1. A number $t > 0$ is said to be *anormal* (resp. *normal*) if there exists an anormal (resp. no anormal) point in $S(C, t)$. Also, $t > 0$ is said to be *exceptional* if it is anormal or normal but there exists an $s \in \rho^{-1}(t)$ such that $\rho'(s) = 0$. A positive number t is by definition *nonexceptional* iff it is not exceptional.

Lemma 4.3.6. *The set of exceptional values is closed and of Lebesgue measure zero.*

Proof. It follows from Proposition 4.2.2 that the set of normal cut points $z(\rho(s), s)$ with $\rho'(s) \ne 0$ is relatively open in the cut locus to C. Thus the set of exceptional values is closed. By Proposition 4.2.3 the set of ρ-values of totally nondegenerate anormal cut points is discrete. In particular, this set is of Lebesgue measure zero. It follows from Lemmas 4.1.2 and 4.2.3 that the set of ρ-values of all degenerate cut points is of Lebesgue measure zero. Note that $\{s \in [0, L_0]; P(s) < +\infty\}$ is a countable union of intervals. Therefore the set of anormal values is of Lebesgue measure zero. By Proposition 4.2.2, ρ is smooth on

$$\{s \in [0, L_0]; z(\rho(s), s) \text{ is a normal cut point}\}.$$

Thus

$$\{\rho(s); z(\rho(s), s) \text{ is a normal cut point and } \rho'(s) = 0\}$$

is of Lebesgue measure zero, by Lemma 4.1.2. Therefore we have proved that the set of exceptional values is of Lebesgue measure zero.   □

Theorem 4.3.1. *Let M be a complete connected smooth Riemannian 2-manifold and C a smooth Jordan curve separating M into two components.*

Then, for any positive r, ρ_r is absolutely continuous. In particular ρ is absolutely continuous on every compact interval where it is finite valued.

Proof. Without loss of generality, we may assume that r is nonexceptional. Let ϵ be an arbitrary positive number. It follows from Proposition 4.3.1 and Lemma 4.1.1 that there exists a positive number $\delta = \delta(\epsilon, r)$ such that

$$\int_B n_r(t)\,dt < \epsilon \tag{4.3.8}$$

for any Borel subset B of $[0, r]$ with $|B| < \delta$, where $|B|$ denotes the Lebesgue measure of B and $n_r(t)$ denotes the number of elements of $\rho^{-1}(t)$. It follows from Lemma 4.3.6 that there exists a finite number of nonoverlapping open subintervals $I_1, \ldots, I_k$ of $[0, r]$, any of whose endpoints are nonexceptional, such that the set of exceptional values in $[0, r]$ is covered by the union of I_j, $j = 1, \ldots, k$, and such that

$$\sum_{j=1}^{k} |I_j| < \delta. \tag{4.3.9}$$

If Σ denotes the union of I_j, $j = 1, \ldots, k$, then by (4.3.8)

$$\int_\Sigma n_r(t)\,dt < \epsilon. \tag{4.3.10}$$

Since any endpoints of I_j are nonexceptional, $S_1 := \rho^{-1}(\Sigma)$ is a finite union of nonoverlapping open intervals $I_1', \ldots, I_l'$. By Corollary 4.1.1 we have

$$\Lambda_{\rho_r}(S_1) = \int_\Sigma n_r(t)dt. \tag{4.3.11}$$

Since S_1 is open, the set

$$S_2 := \{s \in [0, L_0]; \rho(s) \le r, \ s \notin S_1\}$$

is closed, hence compact. Since ρ is smooth at each point of S_2, by Proposition 4.2.2, there exists a positive constant c such that

$$|\rho'(s)| \le c \tag{4.3.12}$$

on S_2.

Let $J_1, \ldots, J_q$ be arbitrary nonoverlapping closed subintervals of $[0, L_0]$ such that

$$\sum_{j=1}^{q} |J_j| < \frac{\epsilon}{c}. \tag{4.3.13}$$

Since S_1 is a union of finite number of open intervals, we may take each int J_i to be either disjoint from S_1 or contained in S_1 to prove that

$$\sum_{j=1}^{q} |\rho_r(b_j) - \rho_r(a_j)| < 2\epsilon, \qquad (4.3.14)$$

where $[a_j, b_j] = J_j$. Furthermore, we may assume that

$$\max \rho|_{I_j} \leq r \qquad (4.3.15)$$

for each j, since r is nonexceptional and the total variation of ρ_r is zero on a closed interval I with $\min \rho|_I \geq r$. We observe that the inequality (4.3.14) implies the absolute continuity of ρ_r. If int J_j is disjoint from S_1 then I_j is contained in S_2, for we have assumed (4.3.15). Thus by (4.3.12)

$$|\rho_r(b_j) - \rho_r(a_j)| \leq c|J_j|. \qquad (4.3.16)$$

If int J_j is contained in S_1 then

$$|\rho_r(b_j) - \rho_r(a_j)| \leq \Lambda_{\rho_r}(J_j), \qquad (4.3.17)$$

since $V(\rho_r|_{J_j}) = \Lambda_{\rho_r}(J_j)$. It follows from the observation above that we have

$$\sum_{j=1}^{q} |\rho_r(b_j) - \rho_r(a_j)|$$

$$= \sum_{\text{int } J_j \cap S_1 = \emptyset} |\rho_r(b_j) - \rho_r(a_j)| + \sum_{\text{int } J_j \subset S_1} |\rho_r(b_j) - \rho_r(a_j)|$$

$$\leq \sum_{\text{int } J_j \cap S_1 = \emptyset} c|J_j| + \sum_{\text{int } J_j \subset S_1} \Lambda_{\rho_r}(J_j) \leq \epsilon + \Lambda_{\rho_r}(S_1) \leq 2\epsilon.$$

Therefore the proof is complete. $\square$

Exercise 4.3.2. Generalize Theorem 4.3.1 to the case of any smooth Jordan curve C on M admitting a periodic smooth unit normal vector field along itself.

It follows from Proposition 4.3.1 that any two cut points in the same connected component of $C(C)$ can be joined by a rectifiable curve lying in the cut locus. Furthermore, if c denotes a curve in $C(C)$ defined by $c(s) = z(\rho(s), s)$ for $s \in [a, b]$ then it follows from Theorem 4.3.1 that c is absolutely continuous. Thus the length $L(c)$ of the curve can be defined by

$$L(c) := \int_a^b |\dot{c}(s)| \, ds.$$

Therefore an interior metric δ on an each component of $C(C)$ is defined by

$$\delta(p, q) := \inf L(c),$$

where the infimum is taken over all absolutely continuous curves in $C(\mathcal{C})$ joining p to q. For two points p, q not lying in the same component, define

$$\delta(p, q) = +\infty.$$

Theorem 4.3.2. *The interior metric δ is a complete distance function and the topology induced by δ is compatible with the one induced by d.*

Proof. Let $\{p_n\}$ be any sequence of points in $C(\mathcal{C})$ convergent to a point p with respect to d. For each p_n, take a point $s_n \in [0, L_0]$ such that $p_n = z(\rho(s_n), s_n)$. Let $b \in [0, L_0]$ be a limit point of $\{s_n\}$. Then, by Proposition 4.2.1, $p = z(\rho(b), b)$ and hence p is a cut point of $\mathcal{C}$. From the definition of δ, we have

$$\delta(p_n, p) \leq \left| \int_b^{s_n} |\dot{c}(s)| \, ds \right|,$$

where $c(s) := z(\rho(s), s)$. Since c is absolutely continuous, $|\dot{c}(s)|$ is Lebesgue integrable. This implies that $\lim_{n \to \infty} \delta(p_n, p) = 0$. Therefore the latter claim of the theorem is proven. The former claim is an easy consequence from the latter.   □

Exercise 4.3.3. Prove Theorem 4.3.2 for any smooth Jordan curve $\mathcal{C}$ on M admitting a periodic unit normal vector field along itself.

The next lemma is a consequence of the property S3 of sectors (see Section 4.2).

Lemma 4.3.7. *A point p in $C(\mathcal{C})$ is an endpoint of $C(\mathcal{C})$ iff p admits exactly one sector.*

Theorem 4.3.3. *Let M be a complete connected smooth Riemannian 2-manifold and $\mathcal{C}$ a smooth Jordan curve separating M into two components. Then the cut locus of $\mathcal{C}$ consists of countably many rectifiable Jordan arcs and the endpoints of the cut locus.*

Proof. For each $s \in [0, L_0]$ with $\rho(s) < +\infty$, choose a closed subinterval I_s of $[0, L_0]$ such that $c(I_s)$ lies in a convex neighborhood, where $c(u) := z(\rho(u), u)$. We may choose countably many closed intervals $\{I_n\}_{n=1,2,\dots}$ from $\{I_s\}_s$ such that $\bigcup_n I_n = \{s \in [0, L_0]; \rho(s) < +\infty\}$. Since $c|_{I_n}$ lies in a convex neighborhood, $c(I_n)$ is a tree. It is sufficient to prove that, for each I_n, $c(I_n)$ is a union of a countable number of Jordan arcs and the set of endpoints. Fix a point p in $c(I_n)$. Choose a point $q_1 \in c(I_n)$ such that $\delta(q_1, p) = \max\{\delta(p, q); q \in c(I_n)\}$. Let J_1 be the unique rectifiable Jordan arc in $c(I_n)$ joining p to q_1. Suppose that curves $J_1, \dots, J_k$ have been defined. Choose the point $q_{k+1} \in c(I_n)$ that is farthest from the set $\bigcup_{i=1}^k J_i$. Then define J_{k+1} by the unique rectifiable Jordan arcs joining q_{k+1} to $\bigcup_{i=1}^{k} J_i$. If a_k, the length of J_k, is zero for some k then $c(I_n)$ is a finite union of the J_i. If a_k is positive for all k then $\lim_{k\to\infty} a_k = 0$, since $\sum_{k=1}^{\infty} a_k$ does not exceed the length of $c(I_n)$, which is finite by Theorem 4.3.1. Thus the set $\bigcup_{i=1}^{\infty} J_i$ is a dense subset of $c(I_n)$. This implies that $c(I_n)$ is a union of J_i and the set of endpoints, since $c(I_n)$ is a tree.   □

Exercise 4.3.4. Prove Theorem 4.3.3 for any smooth Jordan curve $\mathcal{C}$ on M.

A cut point p of $C(\mathcal{C})$ is called a *branch point* if $T \setminus \{p\}$ has at least three connected components for an open tree T containing p. The number of components of $T \setminus \{p\}$ is called the *order* of the cut point.

Corollary 4.3.1. *The cut locus $C(\mathcal{C})$ admits at most countably many branch points.*

Exercise 4.3.5. Show that a point $p \in C(\mathcal{C})$ is a branch point iff p admits at least three sectors.

Remark 4.3.1. Gluck and Singer constructed a compact convex surface of revolution such that the cut locus of a point on the surface admits a branch point with infinite order (cf. [**28**]).

4.4 The structure of geodesic circles

For each nonpositive t, let $S_+(\mathcal{C}, t)$ denote the set

$$S_+(\mathcal{C}, t) := \{x \in M_+; \ d(x, \mathcal{C}) = t\}.$$

Here M_+ denotes the component of $M \setminus \mathcal{C}$ containing $\{z(\rho(s), s); \rho(s) < \infty\}$. In this section the structure of $S_+(\mathcal{C}, t)$ will be stated for nonexceptional values t.

Definition 4.4.1. Let $\mathcal{E} \subset [0, \infty)$ be the set of all exceptional values.

Note that the set $\mathcal{E}$ is closed and of Lebesgue measure zero by Lemma 4.3.6.

Lemma 4.4.1. *Let t be nonexceptional. Then $\rho^{-1}(t)$ has at most finitely many elements. The number $n_\rho(t)$ of elements is even and constant on every interval of nonexceptional values.*

Proof. From Proposition 4.2.2 and from the definition of a nonexceptional value, ρ is smooth and $\rho' \neq 0$ at each point in $\rho^{-1}(t)$. Thus $\rho^{-1}(t)$ has at most finitely many elements and $n_\rho(t)$ is constant on every interval of nonexceptional values. For each $s_1 \in \rho^{-1}(t)$, there exists a unique element $s_2 \in \rho^{-1}(t) \setminus \{s_1\}$ such that $z(\rho(s_1), s_1) = z(\rho(s_2), s_2)$, since the cut point is normal. Therefore $n_\rho(t)$ is even. □

The following theorem states the structure of $S_+(\mathcal{C}, t)$ for nonexceptional values t in the case where $M \setminus \mathcal{C}$ has two components. The analogous claim holds in the other case. See Proposition 4.3 in [86] for more details.

Theorem 4.4.1. *Let M be a complete connected smooth Riemannian 2-manifold and $\mathcal{C}$ a smooth Jordan curve bounding a domain of M. If $t > 0$ is nonexceptional then $\rho(s) = t$ has an even number of solutions $2m$ mod L_0, which (if $m > 0$) can be enumerated as $\alpha_1 < \beta_1 < \alpha_2 < \cdots < \alpha_m < \beta_m (< \alpha_1 + L_0)$, such that $\rho > t$ on (α_k, β_k) and $\rho < t$ on (β_k, α_{k+1}) for $k = 1, \ldots, m$ with $\alpha_{m+1} = \alpha_1 + L_0$. $S_+(\mathcal{C}, t)$ consists of the set of smooth curves $z(t, s)$, $\alpha_k \leq s \leq \beta_k$, for $k = 1, \ldots, m$ and forms a set of simply closed piecewise-smooth curves whose corners are at $z(t, \alpha_k), z(t, \beta_k), k = 1, \ldots, m$. The length $L_+(t)$ of $S_+(\mathcal{C}, t)$ is*

$$L_+(t) = \sum_{k=1}^{m} \int_{\alpha_k}^{\beta_k} f(t, s) \, ds. \qquad (4.4.1)$$

Furthermore $L_+(t)$ is smooth on the set $(0, \infty) \setminus \mathcal{E}$, and its derivative $L'_+(t)$ with respect to t is given by

$$L'_+(t) = \int_{S_+(\mathcal{C}, t)} \kappa(\bar{s}; t) \, d\bar{s} - \sum_k 2 \tan \frac{\theta_k}{2} \qquad (4.4.2)$$

where $d\bar{s}$ is the line element of $S_+(\mathcal{C}, t)$, $\kappa(\bar{s}; t)$ is the geodesic curvature of the curve and the θ_k are the inner angles of the sectors at nondifferential points of it.

Remark 4.4.1. Fiala [26] proved the above theorem for a real analytic Riemannian plane in connection with an isoperimetric inequality. Hartman extended Fiala's results to a Riemannian plane with C^2-metric. These results were extended to complete open smooth Riemannian 2-manifolds in [81, 82, 85, 86] and to an Alexandrov surfaces in [87].

Proof. The existence of the α_k and β_k is trivial by Lemma 4.4.1. Hence it is easy to check equation (4.4.1). Since t is nonexceptional, the derivative ρ' of ρ exists and is nonzero on a neighborhood of $\rho^{-1}(t)$. Thus α_k, β_k are smooth on

$(0, \infty) \setminus \mathcal{E}$. By differentiating the equation (4.4.1) with respect to t, we get

$$L'_+(t) = \sum_k \int_{\alpha_k}^{\beta_k} f_t(t, s) \, ds + f(t, \beta_k) \frac{d\beta_k}{dt} - f(t, \alpha_k) \frac{d\alpha_k}{dt}. \quad (4.4.3)$$

Since $d(z_0(\beta_k), z(t, \beta_k)) = t$ for any $t \in (0, \infty) \setminus \mathcal{E}$, it follows from the first variation formula that we have

$$\cos \frac{\theta_k}{2} \left\{ 1 + f(t, \beta_k)^2 \left(\frac{d\beta_k}{dt} \right)^2 \right\}^{1/2} = \left\| \frac{d}{dt} z(t, \beta_k(t)) \right\| \cos \frac{\theta_k}{2} = 1. \quad (4.4.4)$$

Since $d\beta_k/dt$ is negative, by (4.4.4) we get

$$f(t, \beta_k) \frac{d\beta_k}{dt} = -\tan \frac{\theta_k}{2}. \quad (4.4.5)$$

A similar argument for the equation $d(z_0(\alpha_k), z(t, \alpha_k)) = t$ leads us to

$$f(t, \alpha_k) \frac{d\alpha_k}{dt} = \tan \frac{\theta_k}{2}. \quad (4.4.6)$$

The geodesic curvature $\kappa(s; t)$ at $z(t, s)$ of $S_+(\mathcal{C}, t)$ is

$$\kappa(s; t) = \frac{f_t(t, s)}{f(t, s)}. \quad (4.4.7)$$

Since the line element of $S_+(\mathcal{C}, t)$ is $f(t, s) \, ds$, we get

$$\sum_k \int_{\alpha_k}^{\beta_k} f_t(t, s) \, ds = \int_{S_+(\mathcal{C}, t)} \kappa(\bar{s}; t) \, d\bar{s}.$$

Therefore by (4.4.3), (4.4.5) and (4.4.6), we get (4.4.2). □

We shall introduce some notation in order to extend the function $L_+(t)$ for exceptional values t. Let

$$D_+ := \{(t, s); 0 \le t < \rho(s), 0 \le s \le L_0\}$$

and $\chi_+(t, s)$ the characteristic function of D_+ such that $\chi_+(s, t) = 1$ or 0 according to whether $(t, s) \in D_+$ or not. For any $t \ge 0$, set

$$L_+(t) := \int_0^{L_0} \chi_+(t, s) f(t, s) \, ds. \quad (4.4.8)$$

It is easy to check that the function $L_+(t)$ defined by (4.4.8) equals the length of $S_+(\mathcal{C}, t)$ if t is nonexceptional. We define for $t \ge 0$ the set $Q_+(t)$ as follows:

$$Q_+(t) := \{s \in \rho^{-1}(t); z(s, t) \text{ is normal and } \rho'(s) = 0\}.$$

For each positive integer n, the set

$$\{t \in [0, \infty); |Q_+(t)| \geq 1/n\}$$

is finite, since $Q_+(t_1)$ and $Q_+(t_2)$ are disjoint subsets of $[0, L_0]$ if t_1 and t_2 are distinct. Thus $Q_+(t)$ is of Lebesgue measure zero except for at most countably many t. We define the function $J_+(t)$ on $[0, \infty)$ by

$$J_+(t) := \sum_{0 \leq u \leq t} \int_{Q_+(u)} f(u, s) \, ds. \tag{4.4.9}$$

Note that L_+ and J_+ are discontinuous at $t = t_0$ iff $|Q_+(t_0)|$ is nonzero. The function $L_+(t)$ has no geometrical meaning at an exceptional value t. Here we shall give a sufficient condition for $L_+(t)$ to be equal to the length of $S_+(\mathcal{C}, t)$.

Definition 4.4.2 (The length of a continuous curve). The length $L(c)$ of a continuous curve $c : [a, b] \to M$ is defined by the following number:

$$L(c) := \sup \left\{ \sum_{i=1}^{k} d(c(t_i), c(t_{i-1})); \ a = t_0 < t_1 < t_2 < \cdots < b = t_k \right\}.$$

Notice that if the curve c is smooth then its length is equal to $\int_a^b \|\dot{c}(t)\| \, dt$.

Lemma 4.4.2. *Let* $f : [a, b] \to \mathbf{R}^n$ *be a C^1-function. Then the one-dimensional Hausdorff measure of the set*

$$f(\{t \in [a, b]; \ df_t = 0\})$$

is zero.

Proof. For simplicity we set

$$A := \{t \in [a, b]; df_t = 0\}, \qquad B := f(A).$$

Let ϵ be any positive number. For each $t_0 \in A$ there exists an open interval $I_\epsilon(t_0) \ (\ni t_0)$ such that $f|_{I_\epsilon(t_0)}$ is a function with Lipschitz constant ϵ. Thus we have $\mathcal{H}^1(f(I_\epsilon(t_0))) \leq \epsilon \, \mathcal{H}^1(I_\epsilon(t_0))$. Here $\mathcal{H}^1$ denotes the one-dimensional Hausdorff measure. Since A is compact, there exist finite elements $t_1, \ldots, t_k$ in A such that

$$A \subset \bigcup_{i=1}^{k} I_\epsilon(t_i).$$

If $I_\epsilon(t_i) \cap I_\epsilon(t_j)$ is nonempty then $f|_{I_\epsilon(t_i) \cup I_\epsilon(t_j)}$ is a function on the interval with Lipschitz constant ϵ. Therefore we may assume that the members of $\{I_\epsilon(t_i)\}$ are mutually disjoint. Thus we get

$$\mathcal{H}^1(B) \leq \epsilon \sum_i \mathcal{H}^1(I_\epsilon(t_i)) \leq \epsilon \mathcal{H}^1([a, b]).$$

This implies that $\mathcal{H}^1(B) = 0$. $\square$

Definition 4.4.3. A point $x \in M$ is called a *critical point of the distance function to a compact subset C of M* if for any unit vector $v \in T_x M$ there exists a minimal geodesic segment σ joining x to C such that $\langle \dot{\sigma}(0), v \rangle \geq 0$.

Remark 4.4.2. If a cut point does not admit a sector whose inner angle at p greater than π then it is a critical point of the distance function $d(C, \cdot)$. Compare [15] for the theory of critical points of distance functions.

Lemma 4.4.3. *Assume that for a compact subset C of M there exist two numbers $t_1 < t_2$ such that $A(C, t_1, t_2) := \overline{B(C, t_2) - B(C, t_1)}$ contains no critical points of the distance function to C. Then $A(C, t_1, t_2)$ is homeomorphic to $S(C, t_1) \times [t_1, t_2]$.*

Proof. For each point q of $A(C, t_1, t_2)$ there exists a local smooth vector field X_U on an open neighborhood N around q such that for any C-segment γ through a point $p \in U$ the angle made by X_U and $\dot{\gamma}(d(C, p))$ at p is greater than $\pi/2$. Since $A(C, t_1, t_2)$ is compact, there exist finitely many local smooth vector fields $X_i := X_{U_i}$ on an open neighborhood U_i, $i = 1, \ldots, k$, such that $A(C, t_1, t_2) \subset \bigcup_{i=1}^k U_i$. Let $\{\psi_i\}$ denote a partition of unity associated with $\{U_i\}$. Clearly $X := \sum_i \phi_i X_i$ is nonvanishing on $A(C, t_1, t_2)$. Furthermore, for each point $q \in A(C, t_1, t_2)$ and C-segment γ through q, the angle made by X_q and $\dot{\gamma}(d(C, q))$ is greater than $\pi/2 + \delta$ for some positive δ. By normalizing X, we may assume that the length $\|X\|$ of X is equal to unity on $A(C, t_1, t_2)$. Let q be an element of $A(C, t_1, t_2)$ and let $\sigma_q : [0, \infty) \to M$ be an integral curve of X with $\sigma_q(0) = q$. Since $d(C, \sigma_q(s))$ is a Lipschitz function in s, this function is differentiable almost everywhere. It follows from the first variation formula that

$$\frac{d}{ds} d(C, \sigma_q(s)) < -\sin \delta$$

if $(d/ds)d(C, \sigma_q(s))$ exists and $\sigma_q(s) \in A(C, t_1, t_2)$. Therefore we obtain the inequality

$$d(C, \sigma_q(0)) - d(C, \sigma_q(s)) < -s \sin \delta$$

if $\sigma_q(s) \in A(C, t_1, t_2)$. This inequality implies that there exists a positive number $s(q)$ depending continuously on q such that $\sigma_q(s(q)) \in \partial B(C, t_1)$. Hence we may construct an injective and continuous map f from $A(C, t_1, t_2)$ into

$\partial B(C, t_1) \times [t_1, t_2]$. Since its inverse map may be constructed in the same manner, the map f is a homeomorphism of $A(C, t_1, t_2)$ onto $\partial B(C, t_1) \times [t_1, t_2]$. □

Lemma 4.4.4. *If any point $q \in S_+(C, t_2)$ admits a sector whose inner angle at q is greater than π for a positive number t_2 then $S_+(C, t_2)$ is a union of finitely many circles and its total length is equal to $L_+(t_2)$.*

Proof. Let $t_1 (< t_2)$ be a nonexceptional value sufficiently close to t_2 such that any cut point q of C admits a sector whose inner angle at q is greater than π, if $t_1 \le d(C, q) \le t_2$. It follows from Theorem 4.4.1 that $S_+(C, t_1)$ is a union of finitely many circles. Therefore, by Lemma 4.4.3, $S_+(C, t_2)$ is also a union of finitely many circles. If $\rho(s)$ is greater than t_2 for any $s \in [0, L_0]$ then the latter claim is trivial. Thus we will suppose that $\rho(s)$ is not greater than t_2 for some $s \in [0, L_0]$. By reparameterizing C we may assume that $\rho(0) \le t_2$. Hence the set $\{s \in [0, L_0]; \rho(s) > t_2\}$ consists of at most countably many open intervals $I_n, n = 1, 2, \ldots$, If $q \in S_+(C, t_2)$ is a totally nondegenerate cut point then, by Propositions 4.2.2 and 4.2.3, q is an element of the set $\{z(t_2, s); s \in \bar{I}_n\}$ for some n. Therefore we get

$$S_+(C, t_2) \setminus F \subset \bigcup_n \{z(t_2, s); s \in \bar{I}_n\} \subset S_+(C, t_2),$$

where F denotes the set of all degenerate cut points of C. Since $\mathcal{H}^1(F) = 0$, by Lemma 4.4.2, we have

$$\mathcal{H}^1(S_+(C, t_2)) = \sum_n \mathcal{H}^1(\{z(t_2, s); s \in \bar{I}_n\})$$

$$= \sum_n \int_{I_n} f(t_2, s) ds = L_+(t_2).$$

Notice that, for any continuous Jordan arc $\sigma : [a, b] \to M, \mathcal{H}^1(\sigma(I)) = L(\sigma)$ holds. □

Theorem 4.4.2. *The function $H_+(t) := L_+(t) + J_+(t)$ is absolutely continuous on any compact subinterval of $[0, \infty)$.*

Proof. Let $[a, b]$ be a compact subinterval of $[0, \infty)$. In order to prove the theorem we shall show that for any positive ϵ there exists a positive $\delta = \delta(\epsilon, a, b)$ such that if $I_1, \ldots, I_k$ are overlapping subintervals of $[a, b]$ then

$$\sum_{i=1}^{k} |I_i H| < (L_0 + 2)\epsilon \tag{4.4.10}$$

whenever $\sum_{i=1}^{k} |I_i| < \delta$, where $I_i = (\sigma, \tau]$ and $I_i H_+ = H_+(\tau) - H_+(\sigma)$. Let

ϵ be fixed. It follows from Proposition 4.2.3 that the set

$$T_b := \{s \in [0, L_0];\ \rho(s) \leq b,\ \text{and } z(\rho(s), s) \text{ is a totally}$$
$$\text{nondegenerate anormal point}\}$$

is finite. Let $c = c(b)$ be a constant satisfying

$$|f(t, s)| \leq c, \qquad |f_t(t, s)| \leq c$$

on $[0, b] \times [0, L_0]$. It follows from Lemma 4.2.4 that the set F^ϵ defined by

$$F^\epsilon := \{s \in [0, L_0];\ \rho(s) \leq b,\ s \in F,\ f(\rho(s), s) \geq \epsilon/2\}$$

is compact and of Lebesgue measure zero, where F is the set defined in Lemma 4.2.4.

Let V^ϵ be a finite union of open subintervals of $[0, L_0]$ such that $|V^\epsilon| < \epsilon/c$ and $V^\epsilon \supset T_b \cup F^\epsilon$. Let Q^ϵ be a set defined as follows:

$$Q^\epsilon := \{s \in [0, L_0];\ \rho(s) \leq b,\ f(\rho(s), s) \leq \epsilon/2\}.$$

Since Q^ϵ is compact, it can be covered by a set S^ϵ consisting of a finite number of open subintervals of $[0, L_0]$ on which $f(\rho(s), s) < 3\epsilon/4$. Then the set

$$R^\epsilon := [0, L_0] \setminus (S^\epsilon \cup V^\epsilon)$$

consists of a finite number of closed subintervals $J_1, \dots, J_p$ of $[0, L_0]$. By the definition of R^ϵ and by Proposition 4.2.2, ρ is smooth at each point $s \in R^\epsilon$ if $\rho(s) \leq b$. Hence the function ρ_b is Lipschitz continuous on each interval $J_j,\ j = 1, \dots, p$. In particular the restriction ρ_j of ρ_b to J_j is of bounded variation. If Λ_j denotes the Lebesgue–Stieltjes measure defined by ρ_j then we observe from Corollary 4.1.1 that, for each I_i,

$$\sum_{j=1}^{k} \Lambda_j\big(\rho_j^{-1}(I_i)\big) = \int_\sigma^\tau n(r)\,\mathrm{d}r, \tag{4.4.11}$$

where $n(r)$ is the Lebesgue, integrable function defined as the number of elements of the set

$$\{s \in R^\epsilon;\ \rho(s) = r\}.$$

Let $O(i)$ be an open set containing $R(i) = \bigcup_{\sigma < t \leq \tau} Q_+(t)$ such that $|O(i) \setminus R(i)| < |I_i|$. Setting $S(i) = \rho^{-1}(I_i)$, we define

$$S_1 := (S(i) \setminus R(i)) \cap O(i),$$
$$S_2 := (S(i) \setminus R(i)) \cap (\{s;\ f(\rho(s), s) < \epsilon\} \cup V^\epsilon),$$
$$S_3 := (S(i) \setminus R(i)) \setminus (S_1 \cup S_2).$$

By (4.4.8), we get

$$I_i L_+ = - \int_{S(i)} f(\rho(s), s) \, ds + \int_0^{L_0} (f(t, s) - f(\sigma, s)) \chi(t, s) \, ds. \quad (4.4.12)$$

By the definition of $Q_+(t)$,

$$I_i J_+ = \sum_{\sigma < t \le \tau} \int_{Q_+(t)} f(t, s) \, ds. \quad (4.4.13)$$

Since $s \in Q_+(t)$ implies that $\rho(s) = t$, we get

$$I_i J_+ = \int_{R(i)} f(\rho(s), s) \, ds. \quad (4.4.14)$$

Combining (4.4.12) and (4.4.14), we obtain

$$|I_i H_+| \le \sum_{j=1}^3 \int_{S_j} f(\rho(s), s) \, ds + 2c L_0 |I_i|$$
$$\le c|I_i| + \epsilon|S(i)| + c|V^\epsilon \cap S(i)| + c|S_3| + 2c L_0 |I_i|. \quad (4.4.15)$$

Since $S_3 \subset R^\epsilon$ and $S_3 \cap O(i) = \emptyset$, ρ is smooth at each point of S_3 and $|\rho'| \ge c_1$ on S_3 holds for some positive constant $c_1 = c_1(\epsilon, a, b)$. From the property of the Lebesgue–Stieltjes measure Λ_j we obtain

$$\sum_{j=1}^p \Lambda_j(J_j \cap S_3) \ge c_1 \sum_{j=1}^p |J_j \cap S_3| = c_1 |R^\epsilon \cap S_3| = c_1 |S_3|. \quad (4.4.16)$$

From (4.4.11) and (4.4.16),

$$|S_3| \le c_1^{-1} \sum_{j=1}^p \Lambda_j(J_j \cap S_3) \le c_1^{-1} \int_\sigma^\tau n(t) \, dt. \quad (4.4.17)$$

From (4.4.15) and (4.4.17),

$$\sum_{i=1}^k |I_i H_+| \le c(1 + 2L_0) \sum_{i=1}^k |I_i| + (L_0 + 1)\epsilon + c c_1^{-1} \sum_{i=1}^k \int_{I_i} n(r) \, dr. \quad (4.4.18)$$

The inequality (4.4.18) implies that we can find a positive $\delta = \delta(\epsilon, a, b)$ satisfying (4.4.10). Note that the function $n(t)$ is Lebesgue integrable. $\quad \square$

Remark 4.4.3. In the case where $\mathcal{C}$ does not bound any domain, a similar claim to that in the above theorem holds if the functions $L_+(t)$, $J_+(t)$ are defined in a modified way. Compare Section 3 in [**86**] for this case. Note that the case where $\mathcal{C}$ does not admit a unit normal vector field along itself is reduced to that where $M - \mathcal{C}$ has two components, as observed in the first paragraph in Section 4.2.

5

Isoperimetric inequalities

In this chapter the structures of distance circles and the cut locus of a Jordan curve in a complete open finitely connected Riemannian 2-manifold will be investigated under the assumption that the manifold admits a total curvature. The existence of a total curvature imposes some strong restrictions on these structures.

5.1 The structures of $S(\mathcal{C}, t)$ and the cut locus of $\mathcal{C}$

In this chapter the notation and definitions used in the last chapter will be used implicitly.

Throughout this section and the next, M always denotes a connected, finitely connected, smooth complete Riemannian 2-manifold without boundary admitting a total curvature, and $\mathcal{C}$ denotes a smooth Jordan curve in M. Take a sufficiently large nonexceptional value t_1 and a core B of M such that the boundary ∂B of B is contained in $S(\mathcal{C}, t_1)$ and such that $M \setminus B$ is homeomorphic to a disjoint union of k tubes, $1 \leq k < \infty$. Hence we are assuming that M has exactly k ends. For each $x \in M \setminus B$ admitting at least two $\mathcal{C}$-segments, let $E(x)$ be the maximal compact disk domain of ones bounded by a subarc of ∂B cut off by two distinct $\mathcal{C}$-segments from x and by the two subarcs of these $\mathcal{C}$-segments cut off by ∂B. Let $\beta(x)$ for each $x \in M \setminus B$ be defined as the inner angle of $E(x)$ at x if x admits at least two $\mathcal{C}$-segments, and let $\beta(x) = 0$ otherwise. Recall that a unit-speed geodesic $\gamma : [0, \infty) \to M$ is called a *ray from* $\mathcal{C}$ if $\gamma(0) \in \mathcal{C}$ and if $\gamma|_{[0,t]}$ is a $\mathcal{C}$-segment for all t. Let F be the set of all points on rays from $\mathcal{C}$. Then it is easy to check that F is nonempty and closed, if M is noncompact. Hence $M \setminus (F \cup B)$ consists of the union $\bigcup_i D_i$ of countably many disjoint and unbounded domains D_i bounded by two subrays γ_i^+, γ_i^- of rays from $\mathcal{C}$

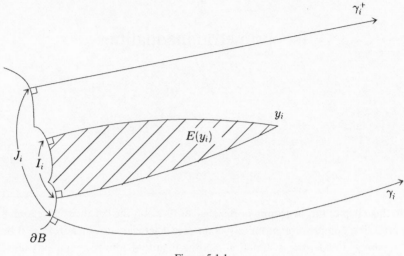

Figure 5.1.1

(or possibly by one subray, if there is a unique ray from $\mathcal{C}$) and a subarc J_i of ∂B (see Figure 5.1.1).

Lemma 5.1.1. *For each D_i, the total curvature $c(D_i)$ of D_i equals*

$$c(D_i) = - \int_{J_i} \kappa(s)\, ds - \sum_k \left(\pi - \omega_k^{(i)}\right), \qquad (5.1.1)$$

where $\kappa(s)$ denotes the geodesic curvature of ∂B with respect to the outward pointing unit normal vector field at B and the $\omega_k^{(i)}$ denote the inner angles of D_i at its nondifferentiable points on J_i. Furthermore, $\lim_{j \to \infty} \beta(x_j) = 0$ for any divergent sequence $\{x_j\}$ of points in M.

We remark that Lemma 5.1.1 is essentially contained in Lemmas 3.4.3 and 3.2.2 under more general conditions.

Proof. We shall prove this theorem in the case where $\gamma_i^+ \neq \gamma_i^-$, since the other case is reduced to this case by considering the universal cover of $\bar{D}_i$, which is a cylinder. Let $\{a_j\}$ be a monotone increasing divergent sequence of nonexceptional values a_j with $a_1 > t_1$. Then there exists a sequence $\{y_j\}$ of cut points $y_j \in S(\mathcal{C}, a_j)$ of $\mathcal{C}$ such that $\{E(y_j)\}$ is a monotone sequence with $\bigcup_{j=1}^{\infty} E(y_j) = D_i$. If I_j denotes the subarc of J_i cut off by $E(y_j)$ then $\{I_j\}$ forms a monotone increasing sequence with $\bigcup_{j=1}^{\infty} I_j = J_i$. Note that there exists no ray from $\mathcal{C}$ lying in D_i. It follows from the Gauss–Bonnet theorem that, for all sufficiently large j,

$$c(E(y_j)) = \beta(y_j) - \int_{I_j} \kappa(s)\,ds - \sum_k \left(\pi - \omega_k^{(i)}\right). \tag{5.1.2}$$

Since $\lim_{j\to\infty} I_j = J_i$ and $\lim_{j\to\infty} E(y_j) = D_i$, the limits of $c(E(y_j))$ and $\int_{I_j} \kappa(s)\,ds$ exist as j goes to infinity and are equal to $c(D_i)$ and $\int_{J_i} \kappa(s)\,ds$ respectively. Therefore by (5.1.2) $\lim_{j\to\infty} \beta(y_j)$ also exists. Thus we get

$$c(D_i) = \lim_{j\to\infty} \beta(y_j) - \int_{J_i} \kappa(s)\,ds - \sum_k \left(\pi - \omega_k^{(i)}\right). \tag{5.1.3}$$

It follows from Lemma 1.2.2 and (5.1.3) that $\lim_{j\to\infty} \beta(y_j) = 0$ and (5.1.1) holds. Suppose that there exists a divergent sequence of points x_j in M and a positive constant c such that $\beta(x_j) \geq c$ for any j. If $\bigcup_j E(x_j)$ contains some D_i then from the argument above it is easy to prove that $\liminf_{j\to\infty} \beta(x_j) = 0$. Thus we may assume that $E(x_k)$ and $E(x_j)$ are disjoint if k and j are distinct. If, for each j, I_j denotes the subarc of ∂B cut off by $E(x_j)$ then $\{I_j\}$ are mutually disjoint subarcs of ∂B. Therefore the length of I_j tends to zero as j goes to infinity. It follows from the Gauss–Bonnet theorem that

$$c(E(x_j)) = \beta(x_j) - \int_{I_j} \kappa(s)\,ds \tag{5.1.4}$$

for any sufficiently large j. In particular, for all sufficiently large j,

$$c(E(x_j)) \geq \frac{c}{2}, \tag{5.1.5}$$

because we have assumed that $c \leq \beta(x_j)$ for any j. The inequality (5.1.5) implies that $c(M) = \infty$. This is impossible, since $c(M) \leq 2\pi\chi(M) < \infty$ (cf. Theorem 2.2.1). $\qquad\square$

Theorem 5.1.1. *Let M be a finitely connected, connected, smooth complete Riemannian 2-manifold without boundary admitting a total curvature and let C be a smooth Jordan curve in M. Then there exists a constant $R_1(> t_1)$ such that, for any $t > R_1$, each tube of $M \setminus B$ has a unique component of $S(C, t)$ homeomorphic to a circle.*

Proof. It follows from Lemma 5.1.1 that there exists a number $R_1(> t_1)$ such that, for any $x \in M \setminus B(C, R_1)$, $\beta(x)$ is less than π. Thus it follows from Lemma 4.4.4 that, for any $t \geq R_1$, $S(C, t)$ is a union of mutually disjoint circles. Suppose that there exist at least two components of $S(C, a)$ in a tube U of M for some $a > R_1$ (see Figure 5.1.2). Since $S(C, a) \cap U$ has a unique component freely homotopic to the boundary of U, there exists a component of $S(C, a) \cap U$ that bounds a compact domain D_1 in U. Thus the function $d(C, \cdot)$

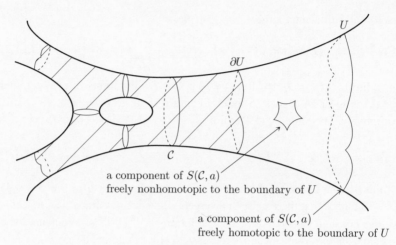

a component of $S(\mathcal{C}, a)$
freely nonhomotopic to the boundary of U

a component of $S(\mathcal{C}, a)$
freely homotopic to the boundary of U

Figure 5.1.2

attains a maximal value $t_2 > a$ at a point x on D_1. This implies that $\beta(x) \geq \pi$ for a point $x \in M \setminus B(\mathcal{C}, R_1)$. This contradicts the choice of R_1. Thus for any $t > R_1$ there exists a unique component of $S(\mathcal{C}, t)$ in each tube of M. □

Corollary 5.1.1. *The length of $S(\mathcal{C}, t)$ is an absolutely continuous function on any compact subinterval of $[R_1, \infty)$.*

Proof. It follows from Theorem 4.4.1 that $H_+(t)$ is absolutely continuous on any compact subinterval of $[0, \infty)$. Furthermore by the theorem above $J_+(t)$ is constant on $[R_1, \infty)$. Thus the assertion is clear. □

Proposition 5.1.1. *For any positive ϵ, there exists a number $t(\epsilon)$ such that if $t > t(\epsilon)$ then*

$$\sum_{x \in S(\mathcal{C}, t)} \beta(x) < \epsilon. \tag{5.1.6}$$

Proof. For each positive integer n, $\{x \in S(\mathcal{C}, t); \ \beta(x) > 1/n\}$ is finite. Thus the sum in (5.1.6) is countable. For each $t > R_1$, let $E(t)$ be the set

$$E(t) := \bigcup_{y \in S(\mathcal{C}, t)} E(y),$$

where $E(y) = \emptyset$ if there exists a unique $\mathcal{C}$-segment from y. Take any element y in $S(\mathcal{C}, t)$, where $t > R_1$, with $E(y) \neq \emptyset$. Then $y \in M \setminus (F \cup B) = \bigcup_i D_i$, and hence $y \in D_i$ for some i. It follows from Theorem 5.1.1 that, for any $t' > t$, $S(\mathcal{C}, t') \cap D_i$ is a Jordan arc bounding a 2-disk, relatively compact, domain that contains $E(y)$. Therefore for each $t' > t$ there exists $y \in S(\mathcal{C}, t')$ such that $E(y') \supset E(y)$. This implies that $E(t)$ is strictly monotone increasing with respect to t. Put $I(t) := E(t) \cap \partial B$ for each $t > R_1$, where B denotes the core of M introduced in Section 5.1. Then $I(t)$ increases monotonically with t and $\lim_{t \to \infty} I(t) = \bigcup_i J_i$, since $\lim_{t \to \infty} E(t) = \bigcup_i D_i$. Therefore there exists a number $t_\epsilon (> R_1)$ such that, for each $t > t_\epsilon$,

$$\int_{\bigcup_i J_i - I(t)} |\kappa(s)| \, ds < \frac{\epsilon}{2} \tag{5.1.7}$$

and

$$\int_{\bigcup_i D_i - E(t)} |G| \, dM < \frac{\epsilon}{2}. \tag{5.1.8}$$

Note that (5.1.8) is a consequence of the fact that $\sum_i c(D_i)$ is bounded, which is due to Lemma 5.1.1. Fix any $t > t_\epsilon$. Applying the Gauss–Bonnet theorem for each $E(x)$ with $x \in S(\mathcal{C}, t)$ and summing up over

$$\{x \in S(\mathcal{C}, t); \ \beta(x) > 0\}$$

one obtains

$$\sum_{x \in S(\mathcal{C}, t)} \beta(x) \le c(E(t)) + \int_{I(t)} \kappa(s) \, ds + \sum_{k, i} \left(\pi - \omega_k^{(i)}\right). \tag{5.1.9}$$

By (5.1.7), (5.1.8) and Lemma 5.1.1, the right-hand side of (5.1.9) is less than ϵ. Thus we obtain (5.1.6). $\qquad \square$

5.2 The case where M is finitely connected

The following lemma is a generalization of l'Hôpital's theorem and will be useful in proving Theorem 5.2.1 below. We omit the proof of the lemma, because it is standard.

Lemma 5.2.1 (Generalization of l'Hôpital's theorem). *Let f, g be absolutely continuous functions with $\lim_{t \to \infty} g(t) = \infty$. If $g(t)$ is monotone nondecreasing then*

$$\limsup_{t \to \infty} \frac{f(t)}{g(t)} \le \limsup_{t \to \infty} \frac{f'(t)}{g'(t)}, \qquad \liminf_{t \to \infty} \frac{f(t)}{g(t)} \ge \liminf_{t \to \infty} \frac{f'(t)}{g'(t)}. \tag{5.2.1}$$

Exercise 5.2.1. Prove the lemma above.

Theorem 5.2.1. *Let M be a connected smooth complete noncompact Riemannian 2-manifold and let $\mathcal{C}$ be a smooth Jordan curve on M. If $L(t)$ and $A(t)$ are* *the length of $S(C, t)$ and the area of $B(C, t)$ then*

$$\lim_{t \to \infty} \frac{L(t)}{t} = \lim_{t \to \infty} \frac{2A(t)}{t^2} = \lambda_\infty(M), \tag{5.2.2}$$

$$\lim_{t \to \infty} \frac{L^2(t)}{2A(t)} = \lambda_\infty(M). \tag{5.2.3}$$

Proof. Suppose that $M - C$ has two components. The other case would be proved analogously. It follows from the Gauss–Bonnet theorem and Theorem 4.4.1 that

$$L'(t) = 2\pi \chi(B(C, t)) - c(B(C, t)) - \sum_k \left(2 \tan \frac{\theta_k}{2} - \theta_k \right) \tag{5.2.4}$$

for any nonexceptional value t. Here θ_k denotes the inner angle of a sector at a cut point of $S(C, t)$. It follows from Proposition 5.1.1 that for any positive ϵ there exists a number $t(\epsilon)$ such that

$$\sum_k \left(2 \tan \frac{\theta_k}{2} - \theta_k \right) < \epsilon \tag{5.2.5}$$

for any nonexceptional $t > t(\epsilon)$. On the one hand, by Theorem 5.1.1, (5.2.4) and (5.2.5) we obtain

$$2\pi \chi(M) - c(B(C, t)) - \epsilon \leq L'(t) \leq 2\pi \chi(M) - c(B(C, t)) \tag{5.2.6}$$

for any nonexceptional $t > t_0(\epsilon) := \max\{R_1, t(\epsilon)\}$. On the other hand, the area $A(t)$ of $B(C, t)$ is given by

$$A(t) = \int_0^t L(u) \, du \tag{5.2.7}$$

for any $t > 0$. From (5.2.7) and Lemma 5.2.1, we get (5.2.2). Note that $L(t)$ is absolutely continuous on $[R_1, \infty)$ by Corollary 5.1.1 and that $A(t)$ is C^1 on $[R_1, \infty)$ by the relation (5.2.7). If $\lim_{t \to \infty} A(t) = \infty$ then we get (5.2.3) by Lemma 5.2.1. Suppose that $\lim_{t \to \infty} A(t) =: A(\infty)$ is finite. Then by (5.2.2) and (5.2.7) we get

$$0 = \lim_{t \to \infty} \frac{2A(t)}{t^2} = \lambda_\infty(M), \qquad \liminf_{t \to \infty} L(t) = 0. \tag{5.2.8}$$

Hence by (5.2.6) there exists $t_1(\epsilon) > t_0(\epsilon)$ such that, for any nonexceptional $t > t_1(\epsilon)$,

$$|L'(t)| < 2\epsilon. \tag{5.2.9}$$

It follows from (5.2.6) and (5.2.9) that, for any $t_2 > t_1 > t_1(\epsilon)$,

$$|L^2(t_2) - L^2(t_1)| \le 2 \int_{t_1}^{t_2} |L(t)L'(t)| \, dt < 4\epsilon A(\infty). \qquad (5.2.10)$$

Therefore $\lim_{t\to\infty} L^2(t)$ exists and equals 0 by (5.2.8). Thus by (5.2.8)

$$\lim_{t\to\infty} \frac{L^2(t)}{A(t)} = \lambda_\infty(M) = 0. \qquad \Box$$

Corollary 5.2.1. *There exists a positive constant R_2 such that $L(t)$ is locally Lipschitz continuous on $[R_2, \infty)$. Furthermore if $\lambda_\infty(M)$ is finite then $L(t)$ is uniformly Lipschitz continuous on $[R_2, \infty)$.*

Proof. It is trivial by (5.2.6). $\qquad \Box$

From now on let M be a two-dimensional connected oriented noncompact smooth manifold without boundary. Let $\mathcal{M}_0(M)$ be the set of all complete metrics on M such that, for every g in $\mathcal{M}_0(M)$, the Gaussian curvature G_g with respect to g satisfies $|G_g| \le 1$. Gromov proved in [31] that the infimum of areas $A(\mathbf{R}^2, g)$ over all $g \in \mathcal{M}_0(\mathbf{R}^2)$ is greater than $4\pi + 0.01$ and not greater than $(2 + 2\sqrt{2})\pi$. He also proved that if $\chi(M)$ is nonpositive then $\inf_{g\in\mathcal{M}_0(M)} A(M, g) = 2\pi|\chi(M)|$. As an application of Theorem 5.2.1 we shall generalize the results on minimal area in the following theorem.

Theorem 5.2.2. *Let M be a connected oriented finitely connected noncompact smooth 2-manifold without boundary and let $\mathcal{M}(M)$ be the set of all complete Riemannian metrics on M such that, for each $g \in \mathcal{M}(M)$, $G_g \le 1$ if $\chi(M) \ge 0$ and $G_g \ge -1$ if $\chi(M) < 0$. Then*

$$\inf_{g\in\mathcal{M}(M)} A(M, g) = \begin{cases} 4\pi & \text{if} \quad \chi(M) = 1 \\ 2\pi|\chi(M)| & \text{if} \quad \chi(M) \le 0. \end{cases} \qquad (5.2.11)$$

Remark 5.2.1. It is not yet known whether the minimal area for $\mathcal{M}_0(\mathbf{R}^2)$ is $(2 + 2\sqrt{2})\pi$. This problem seems rather intractable.

Proof. The proof in the case $\chi(M) < 0$ is clear from the inequalities

$$c(M, g) \ge \int_M G_g^- \, d(M, g) \ge -A(M, g), \qquad (5.2.12)$$

where $G_g^- := \min\{G_g, 0\}$, and $c(M, g)$ and $d(M, g)$ denote respectively the total curvature and area element of the Riemannian manifold (M, g). It follows from Theorem 5.2.1 that if $A(M, g)$ is finite then

$$c(M, g) = 2\pi \chi(M). \qquad (5.2.13)$$

Therefore by (5.2.12) and (5.2.13) we get

$$\inf_{g \in \mathcal{M}(M)} A(M, g) \geq 2\pi |\chi(M)|. \tag{5.2.14}$$

Since there exists a $g \in \mathcal{M}(M)$ such that $A(M, g) < \infty$ and $G_g \equiv -1$, the inequality (5.2.14) holds.

If $\chi(M) = 0$ then for each positive ϵ we can construct a Riemannian metric $g_\epsilon \in \mathcal{M}(M)$ such that $A(M, g_\epsilon) < \epsilon$, as follows. Let $(M, g := \mathrm{d}t^2 + m(t)^2 \, \mathrm{d}\theta^2)$ be a surface of revolution homeomorphic to a cylinder with finite area whose Gaussian curvature is not greater than unity. Define a new metric

$$g_\epsilon := \mathrm{d}t^2 + m_\epsilon(t)^2 \, \mathrm{d}\theta^2,$$

where

$$m_\epsilon := \frac{\epsilon}{2A(M, g)} m(t).$$

Since

$$A(M, g) = 2\pi \int_{-\infty}^{\infty} m(t) \, \mathrm{d}t \quad \text{and} \quad G_g = -\frac{m''}{m},$$

we have

$$A(M, g_\epsilon) = \frac{\epsilon}{2}, \qquad G_{g_\epsilon} = -\frac{m''}{m} \leq 1. \qquad \square$$

The following two lemmas are useful for the proof of Theorem 5.2.2 for the case where $\chi(M)$ is not less than zero.

Lemma 5.2.2 (The Sturm comparison theorem). *Let $K_1(t)$ and $K_2(t)$ be continuous functions on $[0, \infty)$ such that $K_1(t) \geq K_2(t)$. For each $K_i(t)$, $i = 1, 2$, let $u_i(t)$ be the solution to*

$$u_i''(t) + K_i(t)u_i(t) = 0$$

having $u_i = 0$ and $u_i' = 1$ at $t = 0$. If a_i denotes the first zeros after $t = 0$ of $u_1(t)$ and $u_2(t)$ respectively then $a_2 \geq a_1$ and, for any $t \in [0, a_1]$,

$$u_2(t) \geq u_1(t).$$

Proof. Suppose that $a_2 < a_1$. Then

$$\frac{\mathrm{d}}{\mathrm{d}t}\left(\frac{u_1}{u_2}\right) = \frac{f(t)}{u_2{}^2} \tag{5.2.15}$$

on $(0, a_2)$, where $f(t) = u_1'u_2 - u_1u_2'$. Since $f'(t) = (K_2 - K_1)u_1u_2 \leq 0$ on $[0, a_2]$ and $f(0) = 0$, $f(t)$ is nonpositive on $[0, a_2]$. On the one hand, by

(5.2.15) u_1/u_2 is monotone nonincreasing on $(0, a_2)$. On the other hand,

$$\lim_{t \to +0} \frac{u_1}{u_2} = \lim_{t \to +0} \frac{u_1'}{u_2'} = 1. \tag{5.2.16}$$

Thus we get

$$u_1(t) \leq u_2(t) \tag{5.2.17}$$

for any $t \in [0, a_2]$. In particular $u_1(a_2) \leq u_2(a_2) = 0$. This contradicts the assumption $a_2 < a_1$. Therefore $a_2 \geq a_1$. Then the second claim is trivial from the argument above. $\qquad \square$

Lemma 5.2.3. *For every $g \in \mathcal{M}(\mathbf{R}^2)$ with $A(\mathbf{R}^2, g) < \infty$ there exists a point p_0 and a number R such that the metric R-ball $B(p_0, R)$ around p_0 has area greater than 4π.*

Proof. Since the total curvature $c(\mathbf{R}^2, g)$ exists, $c(\mathbf{R}^2, g) = 2\pi$ by (5.2.2). Thus it follows from Theorem 3.8.5 that every Busemann function is an exhaustion function and in particular, takes a minimum. Let p_0 be a point in the minimum set of a Busemann function. Then for every unit vector v at p_0 there exists a ray σ emanating from p_0 such that $\langle \dot{\sigma}(0), v \rangle \geq 0$. Otherwise there exist a Busemann function F_γ, where γ is a ray, and a geodesic α emanating from p_0 such that the gradient vector $\nabla F_\gamma(\alpha(t))$ exists and $\langle \nabla F_\gamma(\alpha(t)), \dot{\alpha}(t) \rangle < 0$ for almost all sufficiently small $t > 0$. Note that F_γ is Lipschitz continuous. Since $(d/dt)F_\gamma(\alpha(t)) = \langle \nabla F_\gamma(\alpha(t)), \dot{\alpha}(t) \rangle$ is negative for almost all sufficiently small $t > 0$, $F_\gamma(p_0)$ is not a minimum. Thus there exist at least two distinct rays emanating from p_0. Let V be the set of all points on all rays emanating from p_0. Set

$$\bigcup_i U_i = \mathbf{R}^2 \setminus V,$$

where the U_i denote the components of $\mathbf{R}^2 \setminus V$. For each i, ∂U_i consists of two distinct rays, and the inner angle of ∂U_i at p_0 is not greater than π. Each U_i contains no ray emanating from p_0 and but does contain a component of $C(p_0)$, the locus of p_0. Let (r, θ) be a set of geodesic polar coordinates around p_0. Then $g = dr^2 + m^2 \, d\theta^2$ on a neighborhood around p_0, where $m = m(r, \theta)$ is the solution of

$$\frac{\partial^2}{\partial r^2}m + G_g m = 0 \tag{5.2.18}$$

with $m(0, \theta) = 0, (\partial/\partial r)m(0, \theta) = 1$. Since $G_g \leq 1$, it follows from

Lemma 5.2.2 that

$$m(r, \theta) \geq \sin r \tag{5.2.19}$$

for any $r \in (0, \pi)$.

Therefore, if the injectivity radius $i(p_0)$ at p_0 is not less than π then the conclusion of the proof is direct from (5.2.19). If, however, $i(p_0) < \pi$ then there is a point $p' \in C(p_0) \cap U_i$ for some i such that $d(p_0, p') < \pi$. Let $p_1 \in C(p_0) \cap U_i$ be a point with the property that $d(p_0, p_1) = d(p_0, C(p_0) \cap U_i) =: a_0$. Then it follows from Lemma 4.3.4 that there exists a geodesic loop γ_0 at p_0 of length $2a_0$ such that $\gamma_0(a_0) = p_1$ and $\gamma_0(0, 2a_0)$ is contained in U_i. The geodesic loop γ_0 bounds a 2-disk D_0 contained in U_i, and the inner angle α_0 of D_0 at p_0 is less than π. Thus D_0 is convex. It follows from $a_0 < \pi$ and $\alpha_0 < \pi$ that there exists a point $q \in C(p_1) \cap D_0$ with the property that $d(p_1, q) < d(p_1, p_0)$. Therefore there exists a point p_2 on $D_0 \cap C(p_1)$ such that $d(p_1, p_2) = d(p_1, C(p_1) \cap D_0) < d(p_0, p_1)$. Now set $a_1 := d(p_1, p_2)$. There exists a geodesic loop γ_1 of length $2a_1$ lying in D_0 such that γ_1 bounds a 2-disk D_1. The inner angle of D_1 at p_1 is less than π. By iterating this procedure, one finally obtains a simple closed geodesic γ in D_0 whose length is $\lim_{j \to \infty} a_j$ and which bounds a 2-disk D contained entirely in D_0. The Gauss–Bonnet theorem implies that $c(D, g) = 2\pi$ and, in particular, $A(D, g) \geq 2\pi$ follows from the assumption $G_g \leq 1$. The above argument shows that if there is a point q on $C(p_0) \cap U_i$ such that $d(p_0, q) < \pi$ then there is a positive number R such that $A(B(p_0, R) \cap U_i, g) > 2\pi$. Therefore, $A(B(p_0, R), g) > 4\pi$ holds for some positive number R if there exist at least two components U_i admitting $q \in C(p_0) \cap U_i$ with $d(q, p_0) < \pi$. If there is a unique U_i such that $d(U_i \cap C(p_0), p_0) < \pi$ then it follows from (5.2.19) that $A(B(p_0, \pi) - U_i, g) \geq 2(2\pi - \theta)$, where θ denotes the inner angle of U_i at p_0. Thus

$$A(B(p_0, R), g) > 2\pi + 2(2\pi - \theta) \geq 4\pi$$

holds for some $R \geq \pi$. $\square$

The proof of Theorem 5.2.2 for $\chi(M)$ nor less than zero is achieved by showing that for any positive ϵ there exists a $g_\epsilon \in \mathcal{M}(\mathbf{R}^2)$ such that $A(\mathbf{R}^2, g) < 4\pi + \epsilon$. Let $y = f(x), x \geq 0$, be the equation of a tractrix with $f(0) = 1$ (see Figure 5.2.1). For a given positive ϵ there is a small positive η such that the area of the surface of revolution in E^3 around the x-axis whose profile curve is given by $y = \eta f(x)$ is less than $\epsilon/2$. Remove from the unit sphere S^2 in E^3 centered at the origin a small ball around the point $(1, 0, 0)$. Then attach to the hole a portion of the surface of revolution such that the total area of the resulting C^0-surface is less than $4\pi + 2\epsilon/3$. This surface is approximated by

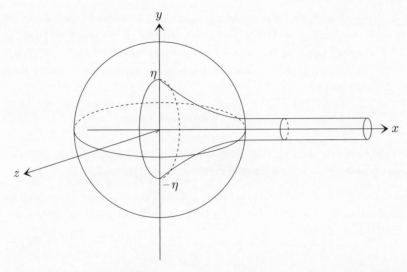

Figure 5.2.1 The unit sphere and the surface of revolution with profile curve $y = \eta f(x)$.

smooth surfaces whose induced metrics have Gaussian curvature not greater than unity, and its area is less than $4\pi + \epsilon$. This completes the proof of Theorem 5.2.2. □

5.3 The case where M is infinitely connected

In this section, we shall generalize Theorem 5.2.1 for an infinitely connected manifold M. Therefore, throughout this section let M be a connected, infinitely connected, complete smooth Riemannian 2-manifold without boundary admitting a total curvature. By Theorem 2.2.2 the total curvature $c(M)$ of M is $-\infty$. The crucial point of our topological observation is to interpret the value $\lambda_\infty(M)$ so as to give a natural meaning to an infinitely connected M. This value should depend only on the Riemannian metric that defines the total curvature. For this purpose we decompose M into a sequence $\{M_j\}$ of submanifolds M_j of M, each of which satisfies the following four properties:

(1) M_j is complete and finitely connected with nonempty compact boundary;
(2) each component of ∂M_j is a nonnull homotopic simply closed geodesic;
(3) $\{M_j\}$ is strictly monotone increasing with j and $\lim_{j\to\infty} M_j = M$;
(4) the sequence $\{\chi(M_j)\}$ of the Euler characteristics of M_j is strictly monotone decreasing.

We shall call such an $\{M_j\}$ a *filtration of finitely connected submanifolds* of M. Such a filtration of M was in fact constructed in the proof of Theorem 2.2.2. We shall prove in Theorem 5.3.1 below that the sequence $\{\lambda_\infty(M_j)\}$ is a monotone nondecreasing sequence of nonnegative numbers and that the limit of the sequence,

$$s(M) := \lim_{j\to\infty} \lambda_\infty(M_j), \tag{5.3.1}$$

is independent of the choice of filtration. The limit $s(M)$ takes a value in $[0, \infty]$ and depends only on the Riemannian metric around the ends.

Theorem 5.3.1. *Let M be a connected, infinitely connected, complete smooth Riemannian 2-manifold and let $\{M_j\}$ be a filtration of finitely connected submanifolds of M. Set*

$$s_j := \lambda_\infty(M_j)$$

for each M_j. Then $\{s_j\}$ is a nondecreasing sequence of nonnegative numbers and $s(M) := \lim_{j\to\infty} s_j$ is independent of the choice of filtration of M and depends only on the Riemannian metric around the ends.

Proof. Take any smooth Jordan curve $\mathcal{C}$ on M with $\mathcal{C} \subset M_1$. For each j, let $L_j(t)$ be the length of $S(\mathcal{C}, t) \cap M_j$ for each nonexceptional value t. It follows from Theorem 5.2.1 that

$$s_{j+1} = \lim_{t\to\infty} \frac{L_{j+1}(t)}{t} \geq s_j = \lim_{t\to\infty} \frac{L_j(t)}{t}.$$

Therefore the sequence $\{s_j\}$ is monotone nondecreasing. Let $\{\tilde{M}_h\}$ be another filtration of finitely connected submanifolds of M and set

$$\tilde{s}_h := \lambda_\infty(M_h).$$

Take any $\tilde{M}_h$ from $\{\tilde{M}_h\}$. By the properties of a filtration of M there exists an M_j such that $\tilde{M}_h \subset M_j$. Thus $\lim_{h\to\infty} \tilde{s}_h \leq \lim_{j\to\infty} s_j$. This shows that $\lim_{h\to\infty} \tilde{s}_h = \lim_{j\to\infty} s_j$. $\square$

Suppose that the function $L(t)$ is extended for any $t > 0$ according to equation (4.4.8). Then the area $A(t)$ of $B(\mathcal{C}, t)$ is given by

$$A(t) = \int_0^t L(t)\, \mathrm{d}t \tag{5.3.2}$$

Thus the following lemma is trivial.

Lemma 5.3.1. *By setting*

$$\bar{\alpha} := \limsup_{t \to \infty} \frac{L(t)}{t}, \qquad \underline{\alpha} := \liminf_{t \to \infty} \frac{L(t)}{t},$$

$$\bar{\beta} := \limsup_{t \to \infty} \frac{2A(t)}{t^2}, \qquad \underline{\beta} := \liminf_{t \to \infty} \frac{2A(t)}{t^2},$$

we have

$$\bar{\alpha} \geq \bar{\beta} \geq \underline{\beta} \geq \underline{\alpha} \geq s(M).$$

Theorem 5.3.2. *Let M be an infinitely connected complete smooth Riemannian 2-manifold admitting a total curvature. For a smooth Jordan curve C and for a nonexceptional value t, let $L(t)$ and $A(t)$ be the length of $S(C, t)$ and the area of $B(C, t)$. Then we have*

$$\liminf_{t \to \infty} \frac{L(t)}{t} \geq s(M), \tag{5.3.3}$$

$$\liminf_{t \to \infty} \frac{2A(t)}{t^2} \geq s(M) \tag{5.3.4}$$

and

$$\limsup_{t \to \infty} \frac{L^2(t)}{A(t)} \geq 2s(M). \tag{5.3.5}$$

Proof. The inequalities (5.3.3) and (5.3.4) are trivial. To prove (5.3.5), we may assume that $s(M) > 0$. Thus $\bar{\beta}$ is positive by Lemma 5.3.1. Suppose that $\bar{\alpha}$ is finite. Let $\{t_n\}$ be a monotone divergent sequence such that $\lim_{n \to \infty} L(t_n)/t_n = \bar{\alpha}$. Then Lemma 5.3.1 implies that

$$\liminf_{n \to \infty} \frac{L(t_n)^2}{2A(t_n)} \geq \frac{\bar{\alpha}^2}{\bar{\beta}} \geq \bar{\alpha} \geq s(M).$$

Hence (5.3.4) is true if $\bar{\alpha}$ is finite. Suppose that $\bar{\alpha}$ is infinite. Then for every positive integer n there exists a number t_n such that

$$t_n := \inf \left\{ t > 0; \; \frac{L(t)}{t} \geq n \right\}.$$

The sequence $\{t_n\}$ is monotone divergent. By the choice of t_n we see that

$$\frac{L(t)}{t} < n \tag{5.3.6}$$

for all $t < t_n$. Since $L(t)$ is right continuous at each t, we get

$$\frac{L(t_n)}{t_n} \geq n. \tag{5.3.7}$$

Therefore by (5.3.2), (5.3.6) and (5.3.7), we obtain

$$A(t_n) < n \int_0^{t_n} t \, dt \le \frac{n}{2} \left(\frac{L(t_n)}{n} \right)^2 = \frac{L(t_n)^2}{2n}. \qquad (5.3.8)$$

The inequality (5.3.8) implies (5.3.5).   □

We will provide three examples, M_1, M_2 and M_3, of complete infinitely connected surfaces having total curvature $-\infty$ that are embedded in $\mathbf{R}^3$. From these examples we will observe that the relations (5.3.3), (5.3.4) and (5.3.5) are optimal. A point $p \in \mathbf{R}^3$ is expressed by a canonical set of coordinates (x, y, z) as $p = (x(p), y(p), z(p))$. For constants $a, b, c \in \mathbf{R}$ let $\Pi_{ax+by+cz} := \{(x, y, z) \in \mathbf{R}^3; \ ax + by + cz = 0\}$.

Example 5.3.1. We fix numbers $l > 3$ and $\theta \in (0, \pi/4)$. A set $W \subset \Pi_z$, which is symmetric with respect to Π_x, is defined as the union of the following three sets (see Figure 5.3.1):

$$\{(0, y, 0); \ -1 \le y \le 0\}; \qquad \{(x, y, 0); \ y = |x| \cot \theta\};$$

$$\bigcup_{n=1}^{\infty} \{p \in \Pi_z; \ l^n \cos \theta \le y(p) \le 2l^n \cos \theta, \ |x(p)| \le y(p) \tan \theta\}.$$

For a sufficiently small $\epsilon > 0$ let $B(W, \epsilon)$ be an ϵ-ball around W and $\partial B(W, \epsilon)$ the boundary of this ball in $\mathbf{R}^3$. The boundary $\partial B(W, \epsilon)$ is an infinitely connected topological surface, with a unique end, which is smooth almost everywhere. The set of nonsmooth points on $\partial B(W, \epsilon)$, which forms a union of portions of ellipses, has a neighborhood $U \subset \partial B(W, \epsilon)$ that can be approximated to obtain a complete smooth surface $M_1 \subset \mathbf{R}^3$ with the following properties:

(i) $\partial B(W, \epsilon) \setminus U$ is contained entirely in M_1 and if $U_1 := M_1 \setminus (\partial B(W, \epsilon) \setminus U)$ then the Gaussian curvature G does not change sign in any component of U_1,

(ii) if $D := \{p \in M_1; y(p) \le -\frac{1}{2}\}$ then D is a disk and a surface of revolution obtained by attaching a convex cap to a flat cylinder and $\partial D =: \mathcal{C}$ is a closed geodesic;

(iii) $G \le 0$ on $M_1 \setminus D$ and $G < 0$ in an open set contained in $U_1 \setminus D$;

(iv) M_1 is symmetric with respect to both Π_z and Π_x.

The cut locus of $\mathcal{C}$ is described as follows. $\mathcal{C}$ is symmetric with respect to Π_z and Π_x and hence so is the cut locus of $\mathcal{C}$. Let $p := (0, -1 - \epsilon, 0) \in \Pi_x \cap \Pi_z \cap M_1$ and let N be a unit normal field along $\mathcal{C}$ such that $N = \partial/\partial y$. If $\gamma_0 : \mathbf{R} \to M_1$ is the geodesic with $\gamma_0(0) = p$ and $\gamma(\mathbf{R})$ lies in Π_z and is symmetric with respect to Π_x then $\gamma_0|_{[0,\infty)}$ and $\gamma(-\infty, 0]$ are both rays from p and their subrays

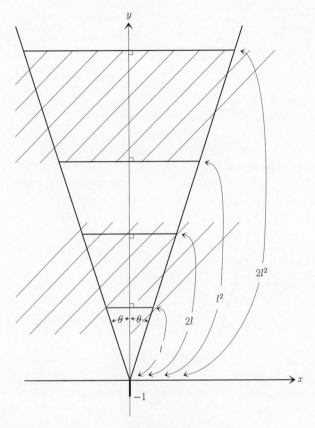

Figure 5.3.1

emanating from points on $\mathcal{C}$ lying in $M_1 \setminus D$ are $\mathcal{C}$-rays. There is no $\mathcal{C}$-ray other than the subrays of γ_0. The set of all cut points of $\mathcal{C}$ is

$$\{p\} \cup (M_1 \cap (\Pi_x \cup \Pi_z) \setminus (D \cup \gamma_0(\mathbf{R}))) .$$

Proposition 5.3.1. *Let $M_1 \subset \mathbf{R}^3$ and $\mathcal{C} \subset M_1$ be as above. Then the following hold:*

(1) *M_1 admits a total curvature and $s(M_1) = 0$;*

(2) *$\liminf_{t \to \infty} \dfrac{L(t)}{t} = s(M_1) = 0$;*

(3) *$\liminf_{t \to \infty} \dfrac{2A(t)}{t^2} > s(M_1)$.*

Proof. (1) is clear from the construction of M_1. Since $\mathcal{C}$ is symmetric in Π_x and Π_z, the cut locus of $\mathcal{C}$ and $S(\mathcal{C}, t)$ are also symmetric in Π_x and Π_z. Note

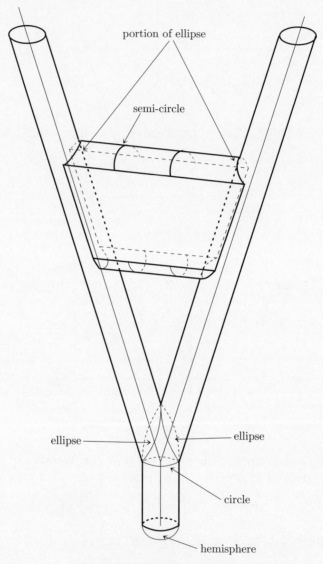

Figure 5.3.2 The region $\partial B(W, \epsilon)$.

that the distance function on M_1 can be approximated by the distance function on $W \subset \Pi_z$. For $n = 1, 2, \ldots$, let

$$X_{2n-1} := (l^n \sin\theta, l^n \cos\theta, 0), \qquad X_{2n} := (2l^n \sin\theta, 2l^n \cos\theta, 0),$$
$$Q_{2n-1} := (0, l^n \cos\theta, 0), \qquad Q_{2n} := (0, 2l^n \cos\theta, 0)$$

The points on $\mathcal{C}$ are expressed as

$$\left(\epsilon \cos \frac{s}{\epsilon}, -\frac{1}{2}, \epsilon \sin \frac{s}{\epsilon}\right), \qquad 0 \le s \le L_0 = 2\pi\epsilon.$$

$L(t)$ and $A(t)$ are estimated as follows. If $t \in (l^n + \frac{1}{2}, l^n(1 + \sin\theta) + \frac{1}{2})$ then $S(\mathcal{C}, t)$ has two components and the right-hand half of it is approximated by two copies of a portion of circle in Π_z centered at X_{2n-1} with radius $t - l^n - \frac{1}{2}$ making an angle $\angle(X_{2n}, X_{2n-1}, Q_{2n-1}) = \frac{1}{2}\pi + \theta$ at X_{2n-1}. Therefore

$$L(t) = 4\left(\frac{\pi}{2} + \theta\right)\left(t - l^n - \frac{1}{2}\right)$$

and

$$A(t) > 2\left(\frac{\pi}{2} + \theta\right)\left(t - l^n - \frac{1}{2}\right)^2 + \sum_{i=1}^{n-1}(3\sin 2\theta)\, l^{2i}.$$

If $t \in (l^n(1+\sin\theta)+\frac{1}{2}, l^n(1+\cos\theta)+\frac{1}{2})$ then $S(\mathcal{C}, t)$ is connected to nonsmooth points on Π_x near the segment $\overline{Q_{2n-1}Q_{2n}}$. Therefore

$$L(t) = 4\left(\frac{\pi}{2} + \theta - \phi_t\right)\left(t - l^n - \frac{1}{2}\right)$$

and

$$A(t) > 2\left(\frac{\pi}{2} + \theta\right)(l^n \sin\theta)^2 + \sum_{i=1}^{n-1}(3\sin 2\theta)\, l^{2i},$$

where we set

$$\cos\phi_t := l^n \frac{\sin\theta}{t - l^n - \frac{1}{2}} \in [\tan\theta, 1].$$

If $t \in (l^n(1 + \cos\theta) + \frac{1}{2}, 2l^n + \frac{1}{2})$ then $S(\mathcal{C}, t)$ has three components, and setting

$$\cos\frac{\psi_t}{2} := l^n \frac{\cos\theta}{t - l^n - \frac{1}{2}},$$

we see that

$$L(t) = 4\left(\frac{\pi}{2} + \theta - \psi_t - \phi_t\right)\left(t - l^n - \frac{1}{2}\right)$$

and

$$A(t) > 2\left(\frac{\pi}{2} + \theta\right)(l^n \sin\theta)^2 + \sum_{i=1}^{n-1}(3\sin 2\theta)\, l^{2i}.$$

If $t \in (2l^n + \frac{1}{2}, l^{n+1} + \frac{1}{2})$ then $S(\mathcal{C}, t)$ has two components lying in flat cylinders of radius ϵ, $L(t) = 4\pi\epsilon$ and $A(t) > \sum_{i=1}^{n}(3\sin 2\theta)l^{2i}$. The claim (2) in the proposition is easily seen from

$$\liminf_{t\to\infty} \frac{L(t)}{t} \le \lim_{n\to\infty} \frac{L(2l^n + 1)}{2l^n + 1} = 0.$$

For every $n = 1, 2, \ldots$, and for $t \in (l^n + \frac{1}{2}, l^{n+1} + \frac{1}{2})$ we have a function that bounds $A(t)/t^2$ from below. Thus we have

$$\liminf_{t\to\infty} \frac{A(t)}{t^2} \ge \lim_{n\to\infty} \frac{\sum_{i=1}^{n-1}(3\sin 2\theta)l^{2i}}{\left(l^{n+1} + \frac{1}{2}\right)^2} = \frac{3\sin 2\theta}{l^2(l^2 - 1)} > 0.$$

This completes the proof of Proposition 5.3.1. $\qquad\qquad\qquad\square$

Example 5.3.2. We shall construct an example $M_2 \subset \mathbf{R}^3$ that shows that the relation (5.3.5), if $\liminf_{t\to\infty} L(t)^2/A(t)$ is replaced by $\limsup_{t\to\infty} L(t)^2/A(t)$, does not hold. Such an M_2 is obtained by joining together $M_1 \setminus D$ and a tube V along $\mathcal{C}$. The tube V is constructed as follows. Let $\beta := \liminf_{t\to\infty} A(t)/t^2$ for the $A(t)$ in Proposition 5.3.1(3). For a positive number c let $\alpha \in (0, 2\pi)$ be a number satisfying

$$\frac{2\alpha}{\alpha + \beta} \le c.$$

V is a surface of revolution around the y-axis such that $\{q \in V;\ y(q) \ge -1\}$ is a flat cylinder with radius ϵ and such that $c(V) = -\alpha$ (see Figure 5.3.3). Then M_2 is infinitely connected, with two ends, and $s(M_2) = \alpha$. If $L_V(t)$ and $L_M(t)$ are the lengths of $S(\mathcal{C}, t) \cap V$ and $S(\mathcal{C}, t) \cap (M_1 \setminus D)$ and $A_V(t)$ and $A_M(t)$ are the areas of $B(\mathcal{C}, t) \cap V$ and $B(\mathcal{C}, t) \cap (M_1 \setminus D)$ respectively, then we have $A(t) = A_V(t) + A_M(t)$ and

$$\lim_{t\to\infty} \frac{L_V(t)}{t} = \lim_{t\to\infty} \frac{2A_V(t)}{t^2} = \alpha.$$

Therefore

$$\liminf_{t\to\infty} \frac{L(t)^2}{A(t)} = \liminf_{t\to\infty} \frac{(L_V(t) + L_M(t))^2}{A_V(t) + A_M(t)} \le \frac{2\alpha^2}{\alpha + \beta} \le cs(M_2).$$

Thus we have proved

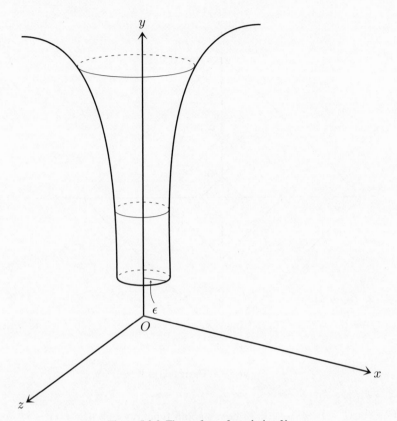

Figure 5.3.3 The surface of revolution V.

Proposition 5.3.2. *For every $c > 0$ there exists a complete infinitely connected surface $M_2 \subset \mathbf{R}^3$ having a total curvature with $s(M_2) \in (0, 2\pi)$ and a simply closed curve $\mathcal{C}$ in M_2 such that $\liminf_{t \to \infty} L(t)^2/A(t) \le cs(M_2)$.*

Example 5.3.3. The previous example shows that there exists an $M_3 \subset \mathbf{R}^3$ and a simply closed curve $\mathcal{C}$ in M_3 such that in the relations (5.3.3), (5.3.4) and (5.3.5) the equalities hold. Fix an $l > 3$ and an $\epsilon \in (0, 1/2)$ such that

$$A := \sum_{n=1}^{\infty} (\epsilon \, l)^n < \infty.$$

Let $W \subset \Pi_z$ (see Figure 5.3.4) be defined as follows:

$$W := \{p \in \Pi_z \cap \Pi_y; |x(p)| \ge l\} \cup \{p \in \Pi_z \cap \Pi_x; |y(p)| \ge l\}$$

$$\bigcup_{n=1}^{\infty} \{p \in \Pi_z; |x(p)| + |y(p)| = l^n\}.$$

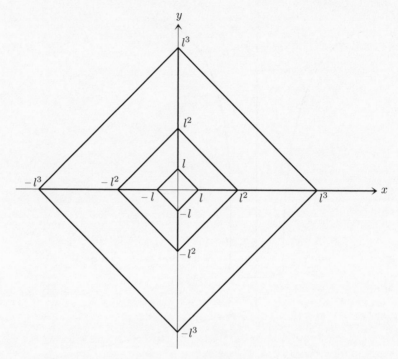

Figure 5.3.4 The region of W.

Let $f : (l - 1, \infty) \to (0, \epsilon]$ be a smooth function such that $f(t) = \epsilon^n$ for $l^n - 1 \le t \le l^n + 1$ and $f'(t) \in (-\epsilon^n/l^n, 0)$ for $l^n + 1 < t < l^{n+1} - 1$. Let $B(W) \subset \mathbf{R}^3$ (Figure 5.3.5) be defined as a union of the following three sets:

$$\{q \in \mathbf{R}^3; |x(q)| \ge l, y(q)^2 + z(q)^2 \le f(x(q))^2\};$$
$$\{q \in \mathbf{R}^3; |y(q)| \ge l, x(q)^2 + z(q)^2 \le f(y(q))^2\};$$
$$\bigcup_{n=1}^{\infty} \{q \in \mathbf{R}^3; d_n(q) \le \epsilon^n\}.$$

Here $d_n(q)$ denotes the distance between q and the rectangle $|x| + |y| = l^n$ in Π_z.

By the same manner as in Example 5.3.1, the boundary $\partial B(W)$ of $B(W)$ in $\mathbf{R}^3$ can be approximated by a smooth surface M_3 with the following properties:

(1) $\partial B(W) \setminus \bigcup_{n=1}^{\infty} \{B_{4\epsilon^n}(0, \pm l^n, 0) \cup B_{4\epsilon^n}(\pm l^n, 0, 0)\}$ is contained entirely in M_3;

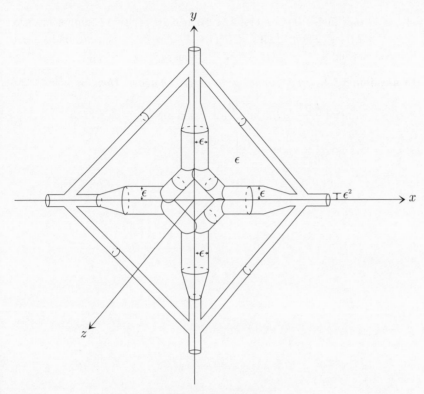

Figure 5.3.5 The region $B(w)$.

(2) the four components $B_{4\epsilon^n}(0, \pm l^n, 0) \cap M_3$ and $B_{4\epsilon^n}(\pm l^n, 0, 0) \cap M_3$ are congruent and have nonpositive curvature for all $n = 1, 2, \ldots$;

(3) M_3 is symmetric with respect to $\Pi_x, \Pi_y, \Pi_z, \Pi_{x+y}$ and Π_{x-y}.

It follows from the construction that M_3 has nonpositive Gaussian curvature; hence it admits a total curvature and $s(M_3) = 0$. The total area of M_3 is bounded above by

$$8\pi \left(\sum_{n=1}^{\infty} (l^{n+1} - l^n) \epsilon^n + \sqrt{2}\, l^n \epsilon^n \right) = 8\pi A(l - 1 + \sqrt{2}).$$

Let $\mathcal{C} \subset M_3$ be the closed geodesic of length $L_0 < 4\sqrt{2}l$ lying in Π_z that is symmetric with respect to $\Pi_x, \Pi_y, \Pi_z, \Pi_{x+y}$ and Π_{x-y}. There are exactly eight $\mathcal{C}$-rays whose images are in Π_x and Π_y. The cut locus of $\mathcal{C}$ is $(\Pi_z \cup \Pi_{x-y} \cup$

$\Pi_{x+y}) \cap (M_3 \setminus \mathcal{C})$. If $t \in (l^n, l^n(1 + 1/\sqrt{2}))$ then $S(\mathcal{C}, t)$ has 12 components and $24\pi\epsilon^n \leq L(t) \leq 24\pi\epsilon^{n-1}$. If $t \in (l^n(1 + 1/\sqrt{2}), l^{n+1})$ then $S(\mathcal{C}, t)$ has four components and $8\pi\epsilon^n \leq L(t) \leq 8\pi\epsilon^{n-1}$. Thus we have proved

Proposition 5.3.3. *Let M_3 and $\mathcal{C}$ be as described above. Then we have*

$$\liminf_{t \to \infty} \frac{L(t)}{t} = \liminf_{t \to \infty} \frac{2A(t)}{t^2} = \limsup_{t \to \infty} \frac{L(t)^2}{2A(t)} = s(M_3).$$

6

Mass of rays

We observed in Chapter 5 that the existence of a total curvature imposes some strong restrictions on the structure of distance circles. In this chapter, we shall see that the total curvature of a finitely connected complete open two-dimensional Riemannian manifold imposes strong restrictions on the mass of rays emanating from an arbitrary fixed point. The first result on the relation between the total curvature and the mass of rays was proved by Maeda in [51]. In [76], Shiga extended this result to the case where the sign of the Gaussian curvature changes. Some relations between the mass of rays and the total curvature were investigated, in detail, by Oguchi, Shiohama, Shioya and Tanaka [62, 83, 84, 90]. Also, Shioya investigated the relation between the mass of rays and the ideal boundary of higher-dimensional spaces with nonnegative curvature (cf. [90]).

6.1 Preliminaries; the mass of rays emanating from a fixed point

Let M be a connected, finitely connected, smooth complete Riemannian 2-manifold.

Note that if M contains no straight line (see Definition 2.2.1) then it has exactly one end.

Lemma 6.1.1. *Assume that M contains no straight line. Then, for each compact subset K of M, there exists a number $R(K)$ such that if $q \in M$ satisfies $d(q, K) > R(K)$ then no ray emanating from q passes through any point on K.*

Proof. Suppose that there exists a compact subset K of M and a divergent sequence $\{q_j\}$ of points on M such that for each q_j there exists a ray γ_j emanating from q_j and passing through K. Let p_j be an intersection point of γ_j and K.

Since K is compact, the sequence $\{p_j\}$ has a convergent subsequence $\{p_{j_k}\}$. Since the sequence $\{v_{j_k}\}$ of unit-speed vectors of γ_{j_k} at p_{j_k} has a convergent subsequence, the sequence $\{\gamma_j\}$ has a limiting geodesic γ through K. Since $\{q_j\}$ is divergent, γ is a straight line. This contradicts the hypothesis of our lemma. $\square$

For each Borel subset F of M admitting a total curvature let the total curvature of F be denoted by $c(F)$, i.e.,

$$c(F) := \int_F G \, dM.$$

For each core K of M (see Definition 2.1.2) and each ray or straight line $\gamma : I \to M$, where $I = [0, \infty)$ or $\mathbf{R}$, intersecting K, we define two numbers $t(\gamma), s(\gamma)$ by

$$t(\gamma) := \max\{t \in I;\, \gamma(t) \in K\}, \qquad s(\gamma) := \min\{t \in I;\, \gamma(t) \in K\}.$$

Theorem 3.7.3 implies the following:

Lemma 6.1.2. *Suppose M has exactly one end. If M contains a straight line and if it admits a total curvature then*

$$\lambda_\infty(M) \geq 2\pi.$$

Lemma 6.1.3. *Suppose that there exist two distinct rays α, β emanating from a common point p bounding a domain D which contains no ray emanating from p. If M admits a total curvature then*

$$c(D) = 2\pi(\chi(D) - 1) + \theta,$$

where θ denotes the inner angle of D at p. In particular if D is homeomorphic to a plane then $c(D) = \theta$.

Proof. Take a small positive ϵ less than the injectivity radius at p. Then $\mathcal{C} := \{\exp_p \epsilon v;\, v \in T_p M,\, \|v\| = 1\}$ is a smooth Jordan curve on M and, for any $t > \epsilon$,

$$S(p, t) = S(\mathcal{C}, t - \epsilon).$$

Therefore, by Theorem 4.4.1, there exists a sequence $\{q_j\}$ of normal cut points of p in D such that $\{d(p, q_j)\}$ is a strictly monotone increasing and divergent sequence of nonexceptional values with respect to $S(\mathcal{C}; t)$ and such that the domains D_j bounded by the two distinct minimal geodesics α_j, β_j joining p to q_j form a strictly monotone increasing sequence. Since there is no ray emanating from p in D, $\lim_{j \to \infty} D_j = D$ and we may assume that

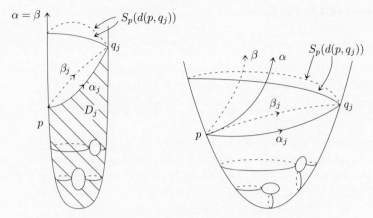

Figure 6.1.1 On the left, the case $\alpha = \beta$; on the right, the case $\alpha \neq \beta$.

$\lim_{j \to \infty} \alpha_j = \alpha$, $\lim_{j \to \infty} \beta_j = \beta$ (Figure 6.1.1). Let θ_j and ξ_j be the inner angles of D_j at p and q_j respectively. Then $\lim_{j \to \infty} \theta_j = \theta$ and $\lim_{j \to \infty} \xi_j = 0$ by Proposition 5.1.1. By applying the Gauss–Bonnet theorem to D_j, we get

$$c(D_j) = 2\pi(\chi(D_j) - 1) + \theta_j + \xi_j. \tag{6.1.1}$$

Since $\lim_{j \to \infty} D_j = D$, we have

$$\lim_{j \to \infty} c(D_j) = c(D), \qquad \chi(D_j) = \chi(D) \tag{6.1.2}$$

for all sufficiently large j. Thus by (6.1.1) and (6.1.2)

$$c(D) = 2\pi(\chi(D) - 1) + \theta. \qquad \square$$

Lemma 6.1.4. *If there exists a unique ray emanating from a point p and if M admits a total curvature then M has exactly one end and $\lambda_\infty(M) = 0$.*

Proof. The lemma is trivial since $M(\infty)$ consists of a single point. $\qquad \square$

Theorem 3.7.4 implies the following:

Lemma 6.1.5. *Suppose that M has exactly one end. If M contains no straight lines and if it admits a total curvature then M satisfies $\lambda_\infty(M) \leq 2\pi$.*

Definition 6.1.1. Let A_p be the set of all unit vectors tangent to the rays emanating from a point p. We denote by μ the Lebesgue measure on the unit circle $S_p M := \{v \in T_p M; \|v\| = 1\}$ with total measure 2π.

Theorem 6.1.1. *The equality*

$$\mu(A_p) = 2\pi \chi(M) - c(M \setminus F_p) \tag{6.1.3}$$

holds for any connected, finitely connected, smooth complete Riemannian 2-manifold M admitting a total curvature c(M). Here

$$F_p := \{\exp_p tu; u \in A_p, t \geq 0\}. \tag{6.1.4}$$

Proof. First of all, we shall observe that the set A_p is measurable. Since the limit of a sequence of rays in M is also a ray, the set A_p is closed and hence a measurable subset of $S_p M$. Take a core $B := \{q \in M; d(p, q) \leq t_1\}$ of M as defined in Section 2.1. Here we shall use some notation from this section. Let $T > t_1$ be a sufficiently large nonexceptional value such that $S(p, T)$ consists of k piecewise-smooth closed curves $C^{(i)}, i = 1, \ldots, k$; the nondifferentiable points $x_m^{(i)}, 1 \leq m \leq m(i)$, of $C^{(i)}$ are joined to p by exactly two distinct minimizing geodesics $\alpha_m^{(i)}, \beta_m^{(i)}, 1 \leq m \leq m(i)$, with $\alpha_m^{(i)}(0) = \beta_m^{(i)}(0) = p$, $\alpha_m^{(i)}(T) = \beta_m^{(i)}(T) = x_m^{(i)}$ (see Figure 6.1.2). This is possible by Theorem 4.4.1. Here we have defined the points $x_m^{(i)}$ and the geodesics $\alpha_m^{(i)}, \beta_m^{(i)}$ in such a way that for each $C^{(i)}$ the two endpoints of each maximal smooth subarc of $C^{(i)}$ are $x_m^{(i)}, x_{m+1}^{(i)}(x_{m(i)+1}^{(i)} := x_1^{(i)})$ and the subarc is orthogonal to $\alpha_m^{(i)}, \beta_{m+1}^{(i)}(\beta_{m(i)+1}^{(i)} := \beta_1^{(i)})$ at $x_m^{(i)}, x_{m+1}^{(i)}$ respectively. We denote this subarc by $[x_m^{(i)}, x_{m+1}^{(i)}]$. Let $F_m^{(i)}, 1 \leq m \leq m(i), 1 \leq i \leq k$, be the disk domain bounded by $\alpha_m^{(i)}, [x_m^{(i)}, x_{m+1}^{(i)}]$ and $\beta_{m+1}^{(i)}$. On the one hand, by the Gauss–Bonnet theorem we get

$$c(B) = 2\pi \chi(M) - \lambda(\partial B) - \theta, \tag{6.1.5}$$

where θ denotes the sum of all outer angles of B at the break points of ∂B and $\lambda(\partial B)$ denotes the geodesic total curvature of ∂B with respect to the inward-pointing unit normal field. On the other hand, by Lemma 4.1.1,

$$c(M \setminus (B \cup F_p)) = \theta + \sum_i \lambda(J_i). \tag{6.1.6}$$

For each $F_m^{(i)}$,

$$c\left(B \cap F_m^{(i)}\right) = \xi_m^{(i)} - \lambda\left(\partial B \cap F_m^{(i)}\right), \tag{6.1.7}$$

where $\xi_m^{(i)}$ denotes the angle between $\dot\alpha_m^{(i)}(0)$ and $\dot\beta_{m+1}^{(i)}(0)$. Thus it follows from (6.1.5) and (6.1.7) that

$$c\left(B \setminus \bigcup_{i,m} F_m^{(i)}\right) = 2\pi \chi(M) - \theta - \sum_{i,m} \xi_m^{(i)} - \lambda\left(\partial B \setminus \bigcup_{i,m} F_m^{(i)}\right). \tag{6.1.8}$$

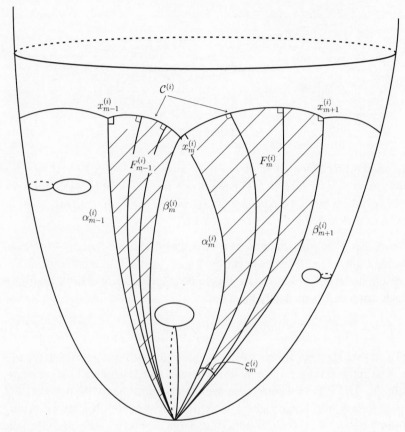

Figure 6.1.2

Since the total arc length $\partial B \setminus \bigcup_{i,m} F_m^{(i)} - \partial B \setminus F_p$ tends to zero as $T \to \infty$, we get

$$\lim_{T \to \infty} \lambda \left(\partial B \setminus \bigcup_{i,m} F_m^{(i)} \right) = \sum_i \lambda(J_i). \tag{6.1.9}$$

Furthermore it is trivial that $\lim_{T \to \infty} \sum_{i,m} \xi_m^{(i)} = \mu(A_p)$. Therefore, by (6.1.8),

$$c(B \setminus F_p) = \lim_{T \to \infty} c \left(B \setminus \bigcup_{i,m} F_m^{(i)} \right) = 2\pi \chi(M) - \theta - \mu(A_p) - \sum_i \lambda(J_i). \tag{6.1.10}$$

Combining (6.1.6) and (6.1.10), we obtain

$$c(M \setminus F_p) = c(M \setminus (B \cup F_p)) + c(B \setminus F_p) = 2\pi \chi(M) - \mu(A_p). \qquad \square$$

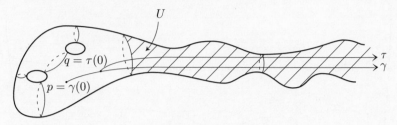

Figure 6.1.3

Lemma 6.1.6. *Suppose that M has finite area. Let* $\gamma : [0, \infty) \to M$ *be a ray that, except for its subarcs with finite length lies in a tube U of M. Then, for any positive b,* $\gamma|_{[b,\infty)}$ *is a unique ray emanating from* $\gamma(b)$ *that, except for its subarcs with finite length, lies in the tube U.*

Proof. Set $q := \gamma(b)$. Suppose that there exists a ray $\tau : [0, \infty) \to M$ emanating from $q = \tau(0)$ and distinct from $\gamma|_{[\gamma(b),\infty)}$, which, except for its subarcs with finite length, lies in U (see Figure 6.1.3). Then there exists a positive δ such that for any sufficiently large t

$$d(\gamma(0), \tau(t)) \leq t + b - \delta. \tag{6.1.11}$$

$S(q, t) \cap U$ does not always have a unique component for a sufficiently large nonexceptional value t, since we do not assume that M admits a total curvature. But $S(q, t) \cap U$ has a unique component freely homotopic to the boundary ∂U of U for any nonexceptional $t > \max\{d(x, q); x \in \partial U\}$. If a ray emanating from q meets $S(q, t) \cap U$ for sufficiently large nonexceptional value t then the ray meets $S(q, t) \cap U$ at a point of the unique component of $S(q, t) \cap U$ freely homotopic to ∂U. Therefore $\gamma(t + b)$ and $\tau(t)$ lie in the same component of $S(q, t)$ for any sufficiently large nonexceptional value t and hence, for such a t,

$$d(\tau(t), \gamma(t + b)) \leq \frac{L(t)}{2} \tag{6.1.12}$$

holds, where $L(t)$ denotes the length of $S(q, t)$ for the nonexceptional value t. By the triangle inequality, (6.1.11) and (6.1.12),

$$\begin{aligned} t + b = d(\gamma(0), \gamma(t + b)) &\leq d(\gamma(0), \tau(t)) + d(\tau(t), \gamma(t + b)) \\ &\leq t + b - \delta + \frac{L(t)}{2}. \end{aligned}$$

Thus we get

$$L(t) \geq 2\delta \tag{6.1.13}$$

for any sufficiently large nonexceptional t. Thus by (5.2.7) or (5.3.2), the area of M is infinite, a contradiction.   □

Theorem 6.1.2. *Suppose that M has finite area. Then there exists a set E of Lebesgue measure zero such that each point $q \in M \setminus E$ has the property that there exist exactly k rays emanating from q, where k is the number of ends of M.*

Proof. Let K be a core of M. Thus $\overline{M \setminus K}$ has k tubes $U_1, \dots, U_k$. Take a point p. For each tube U_i, choose a ray $\gamma_i : [0, \infty) \to M$ emanating from p that, except for its subarcs with finite length, lies in U_i. Let F_i denote the Busemann function with respect to γ_i. Set $E := E_1 \cup \cdots \cup E_k$, where E_i denotes the set of all nondifferentiable points of F_i. Since each set E_i is of Lebesgue measure zero, so is E. Take any point $q \in M \setminus E$. Then it is trivial that there exist at least k rays $\sigma_1, \dots, \sigma_k$ emanating from q such that each σ_i, except for its subarcs with finite length, lies in U_i. We shall prove that there is no ray emanating from q except $\sigma_1, \dots, \sigma_k$. Suppose that there exists a ray $\gamma : [0, \infty) \to M$ emanating from $q = \gamma(0)$ and distinct from $\sigma_1, \dots, \sigma_k$. For sufficiently large a, $\gamma|_{[a,\infty)}$ is contained in a unique tube, say U_1. Take any positive b. Let τ be a coray of σ_1 emanating from $\gamma(b)$. Then by definition τ, except for its subarcs with finite length, lies in U_1. By Lemma 6.1.6, τ is a subray of γ; thus γ is a coray of σ_1. Since q is a differentiable point of F_1, q admits a unique coray of σ_1, a contradiction.   □

Lemma 6.1.7. *Suppose that M admits a total curvature. Let $\{D_n\}$ be a sequence of domains on M. If there exists a Borel subset H of M such that $\lim_{n \to \infty} c(D_n \cap K) = c(H \cap K)$ holds for any compact subset K then*

$$\limsup_{n \to \infty} c(D_n) \le c(H).$$

Proof. Let ϵ be any positive number. It follows from Theorem 2.2.1 that there exists a compact subset $K(\epsilon)$ of M such that

$$\int_{M \setminus K(\epsilon)} G_+ \, \mathrm{d}M < \epsilon. \tag{6.1.14}$$

Here $G_+ = \max\{G, 0\}$. Let $\{K_j\}$ be a monotone increasing sequence of compact subsets satisfying

$$\bigcup_j K_j = M, \qquad K_1 \supset K(\epsilon).$$

Then it follows from (6.1.14) that

$$c(D_n) \le c(D_n \cap K_j) + \epsilon \tag{6.1.15}$$

for any j, n. Hence we get

$$\limsup_{n \to \infty} c(D_n) \leq c(H \cap K_j) + \epsilon \qquad (6.1.16)$$

for any j. Since $c(H) = \lim_{j \to \infty} c(H \cap K_j)$, we have $\limsup_{n \to \infty} c(D_n) \leq c(H)$. $\qquad \square$

Exercise 6.1.1. Find a sequence $\{D_n\}$ and an H as in the above lemma satisfying

$$\limsup_{n \to \infty} c(D_n) < c(H).$$

Definition 6.1.2. For each point $p \in M$ and $u \in S_p M$, let $\gamma_u(t) := \exp_p tu$, $t \geq 0$, denote the geodesic emanating from p and tangent to u.

Definition 6.1.3. For each core K of M, point $p \notin K$ and distinct tangent vectors $u, v \in \{w \in S_p M; \gamma_w[0, \infty) \cap K \neq \emptyset\}$, let $D(u, v)$ denote the disk domain bounded by $\gamma_u|_{[0, s(\gamma_u)]}$, $\gamma_v|_{[0, s(\gamma_v)]}$ and the subarc of ∂K cut off by γ_u and γ_v.

Theorem 6.1.3. *Suppose that M admits a total curvature. Let K be any core of M and ϵ any positive number. Then there exists a number $R(\epsilon)$ such that if $p \in M$ satisfies $d(p, K) > R(\epsilon)$ then $\theta_K(p) < \epsilon$, where $\theta_K(p)$ denotes the maximal inner angle of the domains $D(u, v)$, where $u, v \in \{u \in A_p; \gamma_u[0, \infty) \cap K \neq \emptyset\}$, at p.*

Proof. It is sufficient to prove that any divergent sequence $\{p_n\}$ of points on M has a subsequence $\{p_{n_j}\}$ such that $\lim_{j \to \infty} \theta_K(p_{n_j}) = 0$. Without loss of generality we may assume that all p_n are contained in a tube U that is a connected component of $M \setminus \text{int } K$. Consider the universal Riemannian covering space $\tilde{U}$ of U. Let $\tau : [0, \infty) \to M$ be a *ray from the boundary* $\partial \tilde{U}$ of U lying on U, i.e., $d(\tau(t), \partial U) = t$ for any $t > 0$. Cut open U along $\tau[0, \infty)$ and let, $\ldots, \tilde{U}_{-1}, \tilde{U}_0, \tilde{U}_1, \ldots$, be the fundamental domains of U lying in this order in $\tilde{U}$. Let $\tilde{\tau}_i : [0, \infty) \to \tilde{U}$ be the lifted ray of τ whose image coincides with $\partial \tilde{U}_{i-1} \cap \partial \tilde{U}_i$ and

$$\tilde{W} := \tilde{U}_0 \cup \tilde{U}_1 \cup \tilde{U}_2 \bigcup_{i=0}^{3} \tilde{\tau}_i[0, \infty).$$

Then $\partial \tilde{W}$ consists of two rays, $\tilde{\tau}_0[0, \infty)$ and $\tilde{\tau}_3[0, \infty)$, and the subarc of $\partial \tilde{U}$ cut off by $\tilde{\tau}_0$ and $\tilde{\tau}_3$. For each n, choose $u_n, v_n \in \{u \in A_p; \gamma_u[0, \infty) \cap K \neq \emptyset\}$ in such a way that $\theta_K(p_n)$ is equal to the inner angle of $D(u_n, v_n)$ at p_n. For simplicity set $\alpha_n := \gamma_{u_n}$, $\beta_n := \gamma_{v_n}$. It follows from Lemma 3.2.2 that we may assume that no p_n lies on τ. Therefore for each n, we may choose a lifted point $\tilde{p}_n$ in $\tilde{U}_1$ of p_n. Let $\tilde{\alpha}_n : [0, t_n] \to \tilde{U}$, $\tilde{\beta}_n : [0, s_n] \to \tilde{U}$, where $t_n := s(\alpha_n)$,

$s_n := s(\beta_n)$, denote the lifted geodesics of $\alpha_n|_{[0,t_n]}$, $\beta_n|_{[0,s_n]}$ emanating from $\tilde{p}_n$. It follows from the minimality of rays that, for each n, $\tilde{\alpha}_n$ and $\tilde{\beta}_n$ respectively intersect $\bigcup_{i=-\infty}^{\infty} \tilde{\tau}_i[0, \infty)$ at most once. This fact means that these geodesics lie in $\tilde{W}$ and, in particular, that $\tilde{x}_n := \tilde{\alpha}_n(t_n)$, $\tilde{y}_n := \tilde{\beta}_n(s_n)$ are on $\partial\tilde{W} \cap \partial\tilde{U}$. By choosing a subsequence, if necessary, we may demonstrate that $\{\tilde{x}_n\}$ and $\{\tilde{y}_n\}$ converge to $\tilde{x}$, and $\tilde{y}$ respectively and that $\{\tilde{\alpha}_n\}$ and $\{\tilde{\beta}_n\}$ converge to rays $\tilde{\alpha}$ and $\tilde{\beta}$ respectively. By Lemma 3.2.2, we get $\lim_{n\to\infty} \theta_K(p_n) = 0$. $\qquad\square$

6.2 Asymptotic behavior of the mass of rays

Throughout this section, M will always denote a connected, finitely connected, smooth complete Riemannian 2-manifold admitting a total curvature $c(M)$.

Theorem 6.2.1. *Suppose that M has a unique end and $\lambda_\infty(M) < 2\pi$. Then for each positive ϵ there exists a compact subset $K(\epsilon)$ of M such that*

$$| \mu(A_p) - \lambda_\infty(M) | < \epsilon$$

for any $p \in M \setminus K(\epsilon)$.

Proof. By Theorem 2.2.1 and the hypothesis of our theorem,

$$-\infty < 2\pi(\chi(M) - 1) < c(M) \leq 2\pi\chi(M).$$

Thus $\int_M |G|\, dM$ is finite. Fix an arbitrary positive number ϵ and choose a core K of M such that

$$\int_{M\setminus K} |G|\, dM < \frac{\epsilon}{2}. \tag{6.2.1}$$

Take any point p with $d(p, K) > R(K)$, where $R(K)$ is the number defined in Lemma 6.1.1. Let D be the component of $M \setminus F_p$ such that $K \subset D$, where F_p is the set defined by (6.1.4). It follows from Lemma 6.1.3 and (6.2.1) that

$$c(M) - 2\pi\chi(M) + 2\pi + \frac{\epsilon}{2} > \theta = c(D) - 2\pi\chi(M) + 2\pi$$

$$> c(M) - 2\pi\chi(M) + 2\pi - \frac{\epsilon}{2}. \tag{6.2.2}$$

The set $M \setminus F_p$ is expressed as a disjoint union $\bigcup_i D_i \cup D$ of at most countably many disjoint open sets D_i and D. Note that each D_i is bounded by two rays emanating from p and homeomorphic to a plane. It follows from Lemma 6.1.3,

(6.2.1) and (6.2.2) that

$$c(M) - 2\pi\chi(M) + 2\pi - \epsilon < 2\pi - \mu(A_p) = \theta + \sum_i c(D_i)$$

$$< c(M) - 2\pi\chi(M) + 2\pi + \epsilon.$$

This inequality implies that

$$2\pi\chi(M) - c(M) + \epsilon > \mu(A_p) > 2\pi\chi(M) - c(M) - \epsilon$$

for any $p \in M \setminus K(\epsilon)$, where $K(\epsilon) = \{q \in M; d(q, K) \le R(K)\}$. □

From Theorems 6.1.1 and 6.2.1 we obtain the following two corollaries.

Corollary 6.2.1. *If M is a Riemannian plane with nonnegative Gaussian curvature then*

$$\inf_{p \in M} \mu(A_p) = \lambda_\infty(M).$$

Corollary 6.2.2. *If M is a Riemannian plane then*

$$2\pi - \int_M G_+ \, dM \le \inf_{p \in M} \mu(A_p) \le 2\pi - c(M) = \lambda_\infty(M),$$

where $G_+ = \max\{G, 0\}$.

Theorem 6.2.2. *Suppose that M has exactly one end and $\lambda_\infty(M) < 2\pi$. If $\{K_j\}$ is a monotone increasing sequence of compact subsets of M with $\lim K_j = M$ then we have*

$$\lim_{j \to \infty} \frac{\int_{p \in K_j} \mu(A_p) \, dM}{\int_{K_j} dM} = \lambda_\infty(M).$$

Proof. First of all, we shall observe that the function $M \ni p \mapsto \mu(A_p)$ is measurable. Since the limit of a sequence of rays in M is a ray, the function $M \ni p \mapsto \mu(A_p) \in \mathbf{R}$ is upper semicontinuous. Therefore this function is locally integrable. If M has infinite area then the integral formula above follows trivially from Theorem 6.2.1. If it has finite area then, by Theorem 6.1.2, $\mu(A_p)$ is zero for almost all $p \in M$. Thus the left-hand side of the equation is zero. However, by Theorem 5.2.1 $\lambda_\infty(M) = 0$. Thus the proof is complete. □

It is trivial that the theorem above is still true in the case where $\lambda_\infty(M) = 2\pi$, if M does not admit straight lines. Making use of Theorem 6.1.1, in [**84**] the present authors extended Theorem 6.2.1 for any M with $\lambda_\infty(M) = 2\pi$ and hence the theorem above. The existence of straight lines makes the proof difficult. Furthermore, in [**90**] Shioya generalized these results in the case where $\lambda_\infty(M) \ge 2\pi$, as follows:

Theorem 6.2.3. *Let U_i be a tube of M. Then for each positive ϵ there exists a compact subset $K(\epsilon)$ of M such that*

$$| \min\{2\pi, \lambda_\infty(U_i)\} - \mu(A_p) | \le \epsilon$$

for any $p \in U_i \setminus K(\epsilon)$.

Therefore we get

Theorem 6.2.4. *Suppose that M has exactly one end. If $\{K_j\}$ is a monotone increasing sequence of compact subsets of M with $\lim K_j = M$ then we have*

$$\lim_{j \to \infty} \frac{\int_{p \in K_j} \mu(A_p) \, dM}{\int_{K_j} dM} = \min\{2\pi, \lambda_\infty(M)\}.$$

Note that the limit of the left-hand side of the integral formula of Theorem 6.2.4 does not always exist in the case where M has more than one end. But the limit exists for a sequence of balls about a Jordan curve.

Theorem 6.2.5. *Let $\mathcal{C}$ be a smooth Jordan curve in M and $B(t) := B(\mathcal{C}, t)$. Then we have*

$$\lim_{t \to \infty} \frac{\int_{p \in B(t)} \mu(A_p) \, dM}{\int_{B(t)} dM} = \begin{cases} (\lambda_\infty(M))^{-1} \sum_{i=1}^k \lambda_\infty(U_i) \\ \times \min\{2\pi, \lambda_\infty(U_i)\} & \text{if } \lambda_\infty(M) > 0, \\ 0 & \text{if } \lambda_\infty(M) = 0, \end{cases}$$

where U_i is a family of tubes such that $M \setminus \text{int} \bigcup_i U_i$ is a core.

Proof by assuming Theorem 6.2.3. Take any positive ϵ. By Theorem 6.2.3 we may choose a sufficiently large nonexceptional value t_1 such that $M \setminus \text{int } B$, where $B := B(\mathcal{C}, t_1)$ is a union of k mutually disjoint tubes $U_1, \ldots, U_k$, and such that for any $p \in U_i$

$$| \min\{2\pi, \lambda_\infty(U_i)\} - \mu(A_p) | \le \epsilon. \tag{6.2.3}$$

For each i let M_i denote a complete open Riemannian 2-manifold with the properties that there exists an isometric embedding ι of $B \cup U_i$ into M_i and that $M_i \setminus \iota(B \cup U_i)$ consists of $k - 1$ mutually disjoint disks. Then it is trivial from the Gauss–Bonnet theorem that $\lambda_\infty(M_i) = \lambda_\infty(U_i)$ and that the difference of $\int_{U_i \cap B(t)} dM$ and the area $A_i(t)$ of $B(\mathcal{C}, t)$ is bounded. Therefore it follows from

Theorem 5.2.1 that

$$\lim_{t\to\infty}\frac{1}{t^2}\int_{U_i\cap B(t)}\mathrm{d}M = \lim_{t\to\infty}\frac{1}{t^2}A_i(t) = \frac{\lambda_\infty(U_i)}{2}, \qquad (6.2.4)$$

$$\lim_{t\to\infty}\frac{1}{t^2}\int_{B(t)}\mathrm{d}M = \frac{\lambda_\infty(M)}{2}.$$

Thus by (6.2.3),

$$\frac{1}{2}\sum_i \lambda_\infty(U_i)(\min\{2\pi,\lambda_\infty(U_i)\} - \epsilon)$$

$$\leq \lim_{t\to\infty}\frac{1}{t^2}\int_{p\in B(t)}\mu(A_p)\,\mathrm{d}M$$

$$\leq \frac{1}{2}\sum_i \lambda_\infty(U_i)(\min\{2\pi,\lambda_\infty(U_i)\} + \epsilon).$$

Since ϵ is arbitrary, we get

$$\lim_{t\to\infty}\frac{1}{t^2}\int_{p\in B(t)}\mu(A_p)\,\mathrm{d}M = \frac{1}{2}\sum_i \lambda_\infty(U_i)\min\{2\pi,\lambda_\infty(U_i)\} \qquad (6.2.5)$$

If $2\pi\chi(M) - c(M)$ is positive then it follows from (6.2.4) and (6.2.5) that

$$\lim_{t\to\infty}\frac{\int_{p\in B(t)}\mu(A_p)\,\mathrm{d}M}{\int_{B(t)}\mathrm{d}M} = \frac{1}{\lambda_\infty(M)}\sum_{i=1}^k \lambda_\infty(U_i)\min\{2\pi,\lambda_\infty(U_i)\}.$$

Suppose that $\lambda_\infty(M) = 0$. Then $\lambda_\infty(U_i) = 0$ for each i, by Lemma 3.1.1. Thus by (6.2.3)

$$\mu(A_p) \leq \epsilon \qquad (6.2.6)$$

for any $p \in M \setminus B$. If the area of M is infinite then it follows from (6.2.6) that

$$\lim_{t\to\infty}\frac{\int_{p\in B(t)}\mu(A_p)\,\mathrm{d}M}{\int_{B(t)}\mathrm{d}M} = \lim_{t\to\infty}\frac{1}{\int_{B(t)}\mathrm{d}M}\int_{p\in B(t)\setminus B}\mu(A_p)\,\mathrm{d}M$$

$$\leq \epsilon\lim_{t\to\infty}\frac{1}{\int_{B(t)}\mathrm{d}M}\int_{B(t)\setminus B}\mathrm{d}M = \epsilon.$$

Thus $\lim_{t\to\infty}(\int_{p\in B(t)}\mu(A_p)\,\mathrm{d}M / \int_{B(t)}\mathrm{d}M) = 0$. The case where the area of M is finite is trivial, from Theorem 6.1.2. $\qquad\square$

Definition 6.2.1. For each core K of M let the $U_i(K), i = 1,\ldots,k$, denote the tubes of M with $M \setminus \mathrm{int}\,K = \bigcup_{i=1}^k U_i(K)$.

Definition 6.2.2. If U_i denotes a tube of M then for each $p \in \operatorname{int} U_i$ we define three subsets of A_p by

$$A_p^c(U_i) := \{u \in A_p; \gamma_u[0, \infty) \not\subset U_i \text{ but } \gamma_u[t_1, \infty) \subset U_i$$
$$\text{for sufficiently large } t_1\},$$
$$A_p(U_i) := \{u \in A_p; \gamma_u[0, \infty) \subset U_i\},$$
$$A_p(\operatorname{int} U_i) := \{u \in A_p; \gamma_u[0, \infty) \subset \operatorname{int} U_i\}.$$

Note that $A_p^c(U_i)$ is the complement of $A_p(U_i)$ in the set

$$\{v \in A_p; \gamma_v[t_1, \infty) \subset U_i \text{ for sufficiently large } t_1\}.$$

We need the following three lemmas to assist us in the long proof of Theorem 6.2.3. Before proving the lemmas, we shall prove the theorem.

Lemma 6.2.1. *Let $\{p_j\}$ be a divergent sequence of points in a tube $U_i(K_0)$ of M. Suppose that, for each core $K \supset K_0$ of M, $A_{p_j}^c(U_i(K))$, where $U_i(K)$ denotes the subtube of $U_i(K_0)$, is empty for all sufficiently large j. Then $\lambda_\infty(U_i(K)) \leq 2\pi$ and*

$$\lim_{j \to \infty} \mu(A_{p_j}) = \lambda_\infty(U_i(K)) = \min\{2\pi, \lambda_\infty(U_i(K))\}.$$

Lemma 6.2.2. *For each core K of M there exists a compact subset $K_1 \supset K$ of M such that $A_p(U_i(K))$ is nonempty for all $p \in M \setminus K_1$.*

Definition 6.2.3. If $A_p(U_i(K)) \neq \emptyset$ for a point $p \in \operatorname{int} U_i(K)$, then $\xi_K(p)$ denotes the inner angle of $D_p(K)$ at p and $D_p(K)$ denotes the unique component of $M \setminus \{\gamma_v(t); t \geq 0, v \in A_p(U_i(K))\}$ that contains K; see Figure 6.2.1 .

Lemma 6.2.3. *Let K be a core of M. For any divergent sequence $\{p_j\}$ of points within a tube $U_i(K)$ such that $A_{p_j}^c(U_i(K))$ is nonempty for each j, we have*

$$\lim_{j \to \infty} \xi_K(p_j) = 0.$$

Let us give the proof of Theorem 6.2.3 assuming Lemmas 6.2.1, 6.2.2 and 6.2.3.

Proof of Theorem 6.2.3. Let $\{p_j\}$ be any divergent sequence of points within the tube U_i. It is sufficient to prove that the sequence has a subsequence $\{p_{j_n}\}$ such that $\lim_{n \to \infty} \mu(A_{p_{j_n}}) = \min\{2\pi, \lambda_\infty(U_i)\}$. By Lemma 6.2.1 we may assume that there exists a core K_0 such that $A_{p_j}^c(U_i(K_0))$ is nonempty for all j. Take

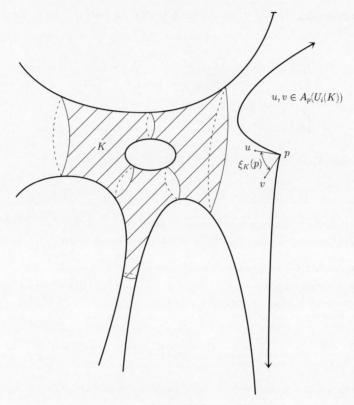

Figure 6.2.1

any positive ϵ. Let $K(\epsilon) \supset K_0$ be a core of M such that

$$\int_{M \setminus K(\epsilon)} G_+ \, dM < \epsilon. \tag{6.2.7}$$

It follows from Lemmas 6.2.2 and 6.2.3 that there exists a number j_0 such that

$$\xi_{K(\epsilon)}(p_j) < \epsilon \tag{6.2.8}$$

for any $j > j_0$. It follows from Lemma 6.1.3 and (6.2.7) that for any $j > j_0$

$$2\pi - \mu(A_{p_j}) - \xi_{K(\epsilon)}(p_j) \leq \sum_m c(D_m) < \epsilon,$$

where D_m denotes a component of $U_i(K(\epsilon)) \setminus (D_{p_j}(K(\epsilon)) \cup \{\gamma_v(t); \, t \geq 0, \, v \in A_{p_j}(U_i(K(\epsilon)))\})$. Therefore by (6.2.8)

$$\mu(A_{p_j}) \geq 2\pi - 2\epsilon$$

for any $j > j_0$. Thus

$$\lim_{j\to\infty} \mu(A_{p_j}) = 2\pi.$$

However, there exists a straight line lying, except for its subarcs with finite length, in $U_i(K_0)$, because there exists a divergent sequence $\{p_j\}$ of points in U_i such that $A^c_{p_j}(U_i(K_0))$ is nonempty for all j. It follows from Theorem 3.7.3 that $\lambda_\infty(U_i(K_0))$ is not less than 2π. Thus

$$\lim_{j\to\infty} \mu(A_{p_j}) = \min\{2\pi, \lambda_\infty(U_i(K_0))\}.$$

This completes the proof. $\qquad\square$

Proof of Lemma 6.2.1. Let j_0 be a positive integer such that, for any $j \geq j_0$, $A^c_{p_j}(U_i(K_0))$ is empty. Using the same method as in the proof of Lemma 6.1.3, we get

$$c(D_{p_j}(K_0) \cap U_i(K_0)) = -\lambda(\partial U_i(K_0)) + \xi_{K_0}(p_j) - 2\pi \qquad (6.2.9)$$

for any $j \geq j_0$. In particular,

$$-(c(D_{p_j}(K_0) \cap U_i(K_0)) + \lambda(\partial U_i(K_0))) \leq 2\pi.$$

Since $\bigcup_{j=j_0}^\infty D_{p_j}(K_0) \cap U_i(K_0) = U_i(K_0)$, by the hypothesis of our lemma, we have

$$\lambda_\infty(U_i(K_0)) \leq 2\pi.$$

In particular, $c(U_i(K_0))$ is finite. Take any positive ϵ and let $K(\epsilon) \supset K_0$ be a core of M such that

$$\int_{U_i(K_0)\backslash K(\epsilon)} |G|\,\mathrm{d}M < \frac{\epsilon}{2}.$$

Since $D_{p_j}(K_0) \supset K(\epsilon)$ for any sufficiently large j, it follows from (6.2.9) that

$$2\pi - \lambda_\infty(U_i(K_0)) - \frac{\epsilon}{2} < \xi_{K_0}(p_j) < 2\pi - \lambda_\infty(U_i(K_0)) + \frac{\epsilon}{2}$$

holds for any sufficiently large j. Therefore from Theorem 6.1.3

$$\lambda_\infty(U_i(K_0)) - \epsilon \leq \mu(A_{p_j}) \leq \lambda_\infty(U_i(K_0)) + \epsilon$$

holds for all sufficiently large j. Thus the proof is complete. $\qquad\square$

For each pair of distinct rays α, β emanating from a point $p \in \mathrm{int}\, U$ (where U denotes a tube) such that $\dot\alpha(0)$, $\dot\beta(0) \in A^c_p(U)$, let $H(\alpha, \beta)$ denote the component containing p of

$$\mathrm{int}\, U \setminus (\alpha[t(\alpha), \infty) \cup \beta[t(\beta), \infty));$$

see Figure 6.2.2.

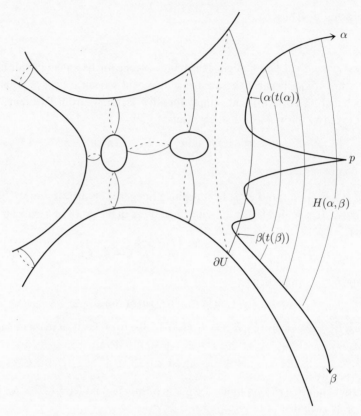

Figure 6.2.2

Proof of Lemma 6.2.2. We shall get a contradiction by supposing that there exists a core K and a divergent sequence $\{p_j\}$ of points in a tube $U := U_i(K)$ such that $A_{p_j}(U)$ is empty for any j. First of all we shall note that for all sufficiently large j, $A^c_{p_j}(U)$ does not consist of a single element. If $A^c_{p_j}(U)$ did consist of a single element then on the one hand U would admit a unique ray emanating from p_j. Thus we would get $\lambda_\infty(U) = 0$. On the other hand, for all sufficiently large j there would exist a ray emanating from p_j that would intersect ∂U. Thus there would exist a straight line lying, except for its subarcs with finite length, in U. This would imply that $\lambda_\infty(U) \geq 2\pi$. We obtain a contradiction. Therefore, without loss of generality we may assume that, for any j, $A^c_{p_j}(U)$ consists of more than one element. For any pair of distinct rays $\tau_i, i = 1, 2$, with $\dot\tau_i(0) \in A^c_{p_j}(U)$ and for any ray γ with $\dot\gamma(0) \in A^c_{p_j}(U), \gamma(t(\gamma))$ is not an element of $\bar{D}(\dot\tau_1(0), \dot\tau_2(0))$; here $\bar{D}(\dot\tau_1(0), \dot\tau_2(0))$ denotes the closure of the domain $D(\dot\tau_1(0), \dot\tau_2(0))$ defined in Definition 6.1.3; see Figure 6.2.3.

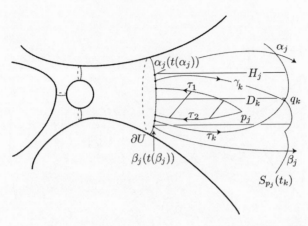

Figure 6.2.3 The region between the rays τ_1 and τ_2 is $D(\tau_1'(0), \tau_2'(0))$.

Therefore we may choose two distinct rays α_j, β_j, with $\dot{\alpha}_j(0)$, $\dot{\beta}_j(0) \in A_{p_j}^c(U)$, such that there is no ray γ with $\gamma(t(\gamma)) \in \bar{H}_j \cap \partial U$ except for α_j and β_j. Here $\bar{H}_j$ denotes the closure of $H_j := H(\alpha_j, \beta_j)$. Then we can prove that the total curvature $c(H_j)$ of H_j satisfies

$$c(H_j) = \pi - \lambda(\partial H_j). \tag{6.2.10}$$

Let $\{t_k\}$ be a divergent monotone increasing sequence of nonexceptional values with respect to $S_{p_j}(t)$ such that there exists a cut point on $S_{p_j}(t_1) \cap H_j$. Then for each k we may choose a normal cut point $q_k \in S_{p_j}(t_k) \cap H_j$ such that the sequence $\{D(-\dot{\gamma}_k(t_k), -\dot{\tau}_k(t_k))\}$ of relatively compact domains in H_j is monotone increasing, where γ_k, τ_k denote the distinct minimal geodesics joining p_j to q_k. Set $D_k := D(-\dot{\gamma}_k(t_k), -\dot{\tau}_k(t_k))$ for simplicity. Since we have assumed that $A_{p_j}(U)$ is empty, the sequences $\{\gamma_k\}$, $\{\tau_k\}$ converge to rays γ_∞, τ_∞ with $\dot{\gamma}_\infty(0)$, $\dot{\tau}_\infty(0) \in A_{p_j}^c(U)$ respectively, as k goes to infinity. These rays γ_∞, τ_∞ satisfy $\gamma_\infty(t(\gamma_\infty))$, $\tau_\infty(t(\tau_\infty)) \in \bar{H}_j \cap \partial U$. It follows from the property of H_j that the sequences $\{\gamma_k\}$, $\{\tau_k\}$ converge to α_j, β_j respectively. Thus $\bigcup_{k=1}^\infty D_k = H_j$ and hence

$$\lim_{k \to \infty} c(D_k) = c(H_j). \tag{6.2.11}$$

It follows from Proposition 5.1.1 that the sequence of the inner angles of D_k at q_k converges to 0. Hence by the Gauss–Bonnet theorem

$$\lim_{k \to \infty} c(D_k) = \pi - \lambda(\partial H_j). \tag{6.2.12}$$

By (6.2.11) and (6.2.12), we get (6.2.10). Without loss of generality we may

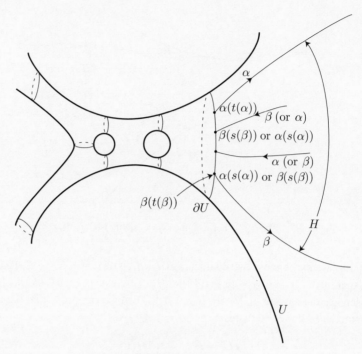

Figure 6.2.4

assume that two limiting straight lines,

$$\alpha := \lim_{k \to \infty} \alpha_k, \qquad \beta := \lim_{k \to \infty} \beta_k,$$

exist; see Figure 6.2.4. Let H denote the component of int $U \setminus (\alpha[t(\alpha), \infty) \cup \beta[t(\beta), \infty))$ such that $\partial H \cap \partial U = \lim_{i \to \infty} \partial H_i \cap \partial U$. For each j, $\bar{H}_j$ contains two geodesic segments $\alpha_j|_{[0, s(\alpha_j)]}$, $\beta_j|_{[0, s(\beta_j)]}$. Hence $\alpha|_{(-\infty, s(\alpha)]}$ divides H into two half-planes H_α, H_β, say $\bar{H}_\alpha \supset \alpha[t(\alpha), \infty)$. Then it follows from Lemma 2.2.2 and Corollary 2.2.1 that

$$c(H_\alpha) \leq -\lambda(\partial H_\alpha), \qquad (6.2.13)$$

$$c(H_\beta) \leq \pi - \lambda(\partial H_\beta). \qquad (6.2.14)$$

Therefore we get

$$c(H) = c(H_\alpha) + c(H_\beta) \leq \pi - (\lambda(\partial H_\alpha) + \lambda(\partial H_\beta)) = -\lambda(\partial H). \qquad (6.2.15)$$

By Lemma 6.1.7, however,

$$\limsup_{i \to \infty} c(H_i) \leq c(H). \qquad (6.2.16)$$

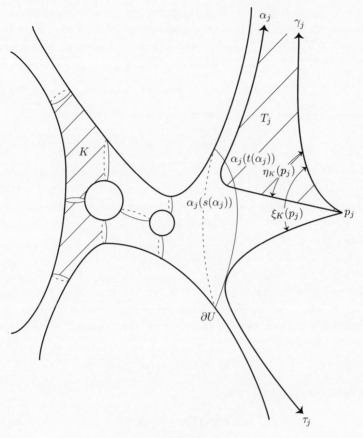

Figure 6.2.5

Thus it follows from (6.2.10) and (6.2.16) that

$$\limsup_{i\to\infty} c(H_i) = \lim_{i\to\infty} c(H_i) = \pi - \lim_{i\to\infty} \lambda(\partial H_i) = \pi - \lambda(\partial H) \le c(H).$$

This contradicts (6.2.15).   □

Proof of Lemma 6.2.3. Set $U := U_i(K)$ for simplicity. It is sufficient to prove
that any divergent sequence $\{p_j\}$ has a subsequence $\{p_{j_n}\}$ such that $\lim_{n\to\infty} \xi_K$
$(p_{j_n}) = 0$. By Lemma 6.2.2 we may assume that for any j, $A_{p_j}(U)$ is nonempty.
For each j let γ_j, τ_j be the rays emanating from p_j that form the boundary of
$D_j := D_{p_j}(K)$, the domain defined in Definition 6.2.3; see Figure 6.2.5. If
$\dot{\gamma}_j(0) \in A_{p_j}(\text{int } U)$ then we may choose a ray α_j with $\dot{\alpha}_j(0) \in A^c_{p_j}(U)$ in
such a way that there exists a component T_j of $\text{int } U \cap D_j \setminus (\alpha_j[0, s(\alpha_j)] \cup$

$\alpha_j[t(\alpha_j), \infty))$ such that there is no ray emanating from p_j whose subarc with infinite length lies in T_j. Let $\eta_K(p_j)$ denote the inner angle of T_j at p_j. Define $\eta_K(p_j) = 0$ if $\dot{\gamma}_j(0) \notin A_{p_j}(\mathrm{int}\, U)$. For each τ_j, we define the corresponding inner angle of τ_j by $\lambda_K(p_j)$. Then it is trivial that

$$\xi_K(p_j) \le \theta_K(p_j) + \eta_K(p_j) + \lambda_K(p_j). \tag{6.2.17}$$

It follows from Theorem 6.1.3 that if we can prove that

$$\lim_{j \to \infty} \eta_K(p_j) = \lim_{j \to \infty} \lambda_K(p_j) = 0,$$

then we obtain $\lim_{j \to \infty} \xi_K(p_j) = 0$, our conclusion. If $\{\gamma_j\}$ has a convergent subsequence $\{\gamma_{j_n}\}$ then by Theorem 6.1.3 we get $\lim_{j \to \infty} \eta_K(p_{j_n}) = 0$. Thus we may assume that $\{\gamma_j\}$ is a divergent sequence. Since T_j admits a cut point of p_j, by an analogous calculation to that for H_j in the proof of Lemma 6.2.2 we arrive at

$$c(T_j) = \pi - \lambda(\partial T_j). \tag{6.2.18}$$

Choosing a subsequence of $\{p_j\}$ if necessary, we may assume that a straight line $\alpha := \lim_{j \to \infty} \alpha_j$ exists and that a component H of $U \setminus (\alpha[t(\alpha), \infty) \cup \alpha(-\infty, s(\alpha)])$ contains each τ_j except for its finite subarcs. It follows from Corollary 2.2.1 that

$$c(H) \le -\lambda(\partial H).$$

Thus on the one hand, by Lemma 6.1.7,

$$\limsup_{j \to \infty} c(T_j) \le -\lambda(\partial H). \tag{6.2.19}$$

On the other hand, by (6.2.18),

$$\limsup_{j \to \infty} c(T_j) = -\lambda(\partial H) + \limsup_{j \to \infty} \eta_K(p_j). \tag{6.2.20}$$

So, combining (6.2.19) and (6.2.20), we obtain

$$\lim_{j \to \infty} \eta_K(p_j) = 0.$$

In the same way,

$$\lim_{j \to \infty} \lambda_K(p_j) = 0.$$

Thus by (6.2.17), we find that

$$\lim_{j \to \infty} \xi_K(p_j) = 0. \qquad \square$$

7

The poles and cut loci of a surface of revolution

It is not easy to find a nontrivial pole even on a surface of revolution, unless the latter has a nonpositive Gaussian curvature. We shall give a necessary and sufficient condition for a surface of revolution to have nontrivial poles. The proof is achieved by obtaining Jacobi fields along any geodesic (see [**101**]). The method is found in a classical work of von Mangoldt [**59**]. We will also determine the cut loci of a certain class of surfaces of revolution containing well-known examples: the two-sheeted hyperboloids of revolution and the paraboloids of revolution (see [**102**]). von Mangoldt proved in [**59**] that any point on a two-sheeted hyperboloid of revolution is a pole if the point is sufficiently close to the vertex. Furthermore, he proved in [**59**] that the two umbilic points of a two-sheeted hyperboloid are poles and that the poles of any elliptic paraboloid are the two umbilic points. These surfaces are typical examples of a Liouville surface. A definition of global Liouville surfaces was introduced by Kiyohara in [**44**]. See also [**40**] for poles on noncompact complete Liouville surfaces.

7.1 Properties of geodesics

A *surface of revolution* means a complete Riemannian manifold (M, g) home-omorphic to $\mathbf{R}^2$ that admits a point p such that the Gaussian curvature of M is constant on $S(p, t)$ for each positive t. The point p is called the *vertex of the surface of revolution*.

Throughout this chapter (M, g) denotes a surface of revolution and p denotes the vertex of the surface. Note that the vertex is unique unless M has a constant Gaussian curvature.

Lemma 7.1.1. *The vertex p is a pole. Furthermore, for any linear isometry F on $T_p M$, $\exp_p \circ F \circ \exp_p^{-1} : M \to M$ is an isometry.*

Proof. Suppose that p is not a pole. Thus we can take a cut point x of p that is an endpoint of the cut locus of p. Then x is a conjugate point of p along any minimizing geodesic $\tau \colon [0, d(p, x)] \to M$ joining p to x. Fix such a geodesic τ. Since M is noncompact and complete, there exists a ray $\gamma \colon [0, \infty) \to M$ emanating from p. Hence

$$G(\gamma(t)) = G(\tau(t))$$

for any $t \in [0, d(p, x)]$, since $d(p, \tau(t)) = d(p, \gamma(t)) = t$ on $[0, d(p, x)]$. Thus the Jacobi fields along $\gamma|_{[0,d(p,x)]}$ and $\tau|_{[0,d(p,x)]}$ orthogonal to the geodesics respectively satisfy the same differential equation. Therefore $\gamma(d(p, x))$ is conjugate to p along γ, which contradicts the fact that γ is a ray. Thus we have proved that p is a pole. Suppose that F is a linear isometry on $T_p M$. Set $f := \exp_p \circ F \circ \exp_p^{-1}$ for simplicity. Since it is trivial that df_p is a linear isometry, we shall show that for any point q distinct from p, df_q is a linear isometry. This means that f is an isometry on M. Let a point $q \, (\neq p)$ be fixed. Let $\gamma \colon [0, \infty) \to M$ be the unit-speed geodesic emanating from $p = \gamma(0)$ and passing through $q = \gamma(t_0)$; here we put $t_0 := d(p, q)$. By definition, $f(\gamma(t)) = \exp_p t F(\dot\gamma(0))$ holds for any t. Hence

$$df_q(\dot\gamma(t_0)) = \dot{\bar\gamma}(t_0) \tag{7.1.1}$$

where $\bar\gamma(t) = \exp_p t F(\dot\gamma(0))$. Let $v(\theta)$ be a unit-speed positively oriented smooth curve in $S_p M$ emanating from $v(0) = \dot\gamma(0)$. Then we get geodesic variations

$$\alpha(t, \theta) := \exp_p t v(\theta) \tag{7.1.2}$$

and

$$\bar\alpha(t, \theta) := f(\exp_p t v(\theta)) = \exp_p t F(v(\theta)) \tag{7.1.3}$$

of $\gamma, \bar\gamma$ respectively. From the variations above we obtain the Jacobi fields $Y(t) := (\partial\alpha/\partial\theta)(t, 0)$ and $\bar Y(t) := (\partial\bar\alpha/\partial\theta)(t, 0)$ along γ and $\bar\gamma$ respectively. It is easy to see that

$$\bar Y(t_0) = df_q(Y(t_0)) \tag{7.1.4}$$

and

$$\begin{aligned} Y(0) &= 0, & Y'(0) &= \dot v(0), \\ \bar Y(0) &= 0, & \bar Y'(0) &= F(\dot v(0)). \end{aligned} \tag{7.1.5}$$

Here $Y'(t)$ and $\bar Y'(t)$ denote the covariant derivatives of $Y(t)$ and $\bar Y(t)$ along γ and $\bar\gamma$ respectively. However, by the Gauss lemma 1.2.1, $Y(t_0)$ and $\bar Y(t_0)$

are orthogonal to $\dot{\gamma}(t_0)$ and $\dot{\bar{\gamma}}(t_0)$ respectively. Since $G(\gamma(t)) = G(\bar{\gamma}(t))$ for any $t \in [0, t_0]$, it follows that Y and $\bar{Y}$ satisfy the same differential equation. Therefore

$$\|Y(t_0)\| = \|\bar{Y}(t_0)\|. \tag{7.1.6}$$

Combining (7.1.1), (7.1.4) and (7.1.6), we observe that df_q is a linear isometry.

$\square$

Definition 7.1.1. For each $q \in M \setminus \{p\}$, let $\mu_q : [0, \infty) \to M$ be the unit-speed geodesic emanating from p through $q = \mu_q(d(p, q))$, where μ_q is called the *meridian* through q.

Let $\mu : [0, \infty) \to M$ be a meridian. Then we can introduce local coordinates (r, θ) on $M \setminus \mu[0, \infty)$, i.e., $r(q) := d(p, q)$ and $\theta(q)$ denotes the oriented angle measured from $-\dot{\mu}(0)$ to $\dot{\mu}_q(0)$. The coordinates (r, θ) are called *geodesic polar coordinates around p on M*. It follows from Lemma 7.1.1 that the function

$$g\left(\frac{\partial}{\partial \theta}, \frac{\partial}{\partial \theta}\right)$$

is constant on $S(p, t)$ for each $t > 0$. Thus a smooth function $m : (0, \infty) \to (0, \infty)$ can be defined by

$$m(r(q)) := \sqrt{g_q\left(\frac{\partial}{\partial \theta}, \frac{\partial}{\partial \theta}\right)}. \tag{7.1.7}$$

From the Gauss lemma 1.2.1, we obtain the relation

$$g = dr^2 + m^2(r)d\theta^2. \tag{7.1.8}$$

Proposition 7.1.1. *The function m is extensible to a smooth odd function around 0 satisfying $m'(0) = 1$.*

Proof. Let $\gamma : \mathbf{R} \to M$ be a unit-speed geodesic with $\gamma(0) = p$ and let $Y(t)$ be the Jacobi field along γ, with $Y(0) = 0$ and $Y'(0) = \dot{v}(0)$, where $v(\theta)$ is the curve defined in the proof of Lemma 7.1.1. Then

$$Y(t) = \left(\frac{\partial}{\partial \theta}\right)_{\gamma(t)}$$

also holds for any positive t. Therefore

$$\left(\frac{\partial}{\partial \theta}\right)_{\gamma(t)} = g(Y(t), E(t))E(t) \tag{7.1.9}$$

also holds for any positive t. Here $E(t)$ denotes the unit parallel vector field along the geodesic γ with $E(0) = Y'(0) (= \dot{v}(0))$, which is orthogonal to $\dot{\gamma}(t)$ for each t. Thus

$$m(t) = g(Y(t), E(t)) \tag{7.1.10}$$

holds for any positive t. Let F be the reflection on $T_p M$, with $F(\dot{\gamma}(0)) = -\dot{\gamma}(0)$. It is easy to check that

$$\mathrm{d}f_{\gamma(t)}Y(t) = -Y(-t), \qquad \mathrm{d}f_{\gamma(t)}E(t) = E(-t) \tag{7.1.11}$$

holds for any real t, where $f := \exp_p \circ F \circ \exp_p^{-1}$. Since f is an isometry on M,

$$g(Y(t), E(t)) = g(\mathrm{d}f_{\gamma(t)}Y(t), \mathrm{d}f_{\gamma(t)}E(t)) = -g(Y(-t), E(-t)). \tag{7.1.12}$$

By equations (7.1.10) and (7.1.12), $m(t)$ is extensible to a smooth odd function and $m'(0) = 1$. $\qquad\square$

Since the Jacobi field $Y(t)$ satisfies (1.4.1), we obtain

Corollary 7.1.1. *The function m satisfies the following differential equation:*

$$m'' + Gm = 0 \tag{7.1.13}$$

with initial condition $m(0) = 0$, $m'(0) = 1$.

Lemma 7.1.2. *Let $f : [a, b] \to \mathbf{R}$ be a smooth function. Then there exists a smooth function h satisfying*

$$f(x) = f(a) + (x - a)h(x).$$

Proof. It is trivial, from the following equation:

$$f(x) - f(a) = \int_0^1 \frac{\mathrm{d}}{\mathrm{d}t} f(tx + (1 - t)a)\,\mathrm{d}t$$

$$= (x - a) \int_0^1 f'(tx + (1 - t)a)\,\mathrm{d}t. \qquad\square$$

Lemma 7.1.3. *Let $f : \mathbf{R} \to \mathbf{R}$ be a smooth even function. Then the function $F : \mathbf{R}^2 \to \mathbf{R}$, $(x, y) \mapsto f(\sqrt{x^2 + y^2})$, is smooth.*

Proof. Let n be any positive fixed integer. Since f is an even function, $f^{(2k-1)}(0) := (\mathrm{d}^{2k-1}/\mathrm{d}x^{2k-1})f(0) = 0$ for any positive integer k. Thus it follows from Lemma 7.1.2 that there exists a smooth function $R_n(t)$ such that

$$f(t) = \sum_{k=0}^{n-1} t^{2k} \frac{f^{(2k)}}{(2k)!}(0) + t^{2n} R_n(t)$$

By induction on n, it is easily proved that the function

$$\mathbf{R}^2 \ni (x, y) \mapsto (x^2 + y^2)^n R_n(\sqrt{x^2 + y^2}) \in \mathbf{R}$$

is C^n. Therefore F is C^n for any positive integer n. $\qquad\square$

The converse of Proposition 7.1.1 is true also.

Theorem 7.1.1. *Let $m : (0, \infty) \to (0, \infty)$ be a smooth function that is extensible to a smooth odd function around 0 with $m'(0) = 1$. Then the Riemannian metric $g = \mathrm{d}r_0^2 + m^2(r_0)\,\mathrm{d}\theta_0^2$ on $\mathbf{R}^2$ defines a surface of revolution whose vertex is the origin of $\mathbf{R}^2$. Here (r_0, θ_0) denotes geodesic polar coordinates around the origin for the Euclidean plane $(\mathbf{R}^2, g_0)$.*

Proof. Let (x, y) be the canonical coordinates for $(\mathbf{R}^2, g_0)$, i.e., $x = r_0 \cos\theta_0$, $y = r_0 \sin\theta_0$. Then it follows from direct evaluation that

$$g\left(\frac{\partial}{\partial x}, \frac{\partial}{\partial x}\right) = 1 + y^2 f(r),$$

$$g\left(\frac{\partial}{\partial x}, \frac{\partial}{\partial y}\right) = -xy f(r), \tag{7.1.14}$$

$$g\left(\frac{\partial}{\partial y}, \frac{\partial}{\partial y}\right) = 1 + x^2 f(r),$$

where $r = \sqrt{x^2 + y^2}$ and $f(r) = (m^2(r) - r^2)/r^4$. From the assumption on m and from Lemma 7.1.2 there exists a smooth even function $k(t)$ satisfying $m(r) = r + r^3 k(r)$. Therefore

$$f(r) = k(r)(2 + r^2 k(r))$$

is a smooth even function. By Lemma 7.1.3 the function

$$\mathbf{R}^2 \ni (x, y) \mapsto f(\sqrt{x^2 + y^2}) \in \mathbf{R}$$

is smooth around the origin. It follows from (7.1.14) that g is a smooth Riemannian metric on $\mathbf{R}^2$. Moreover, each geodesic $\theta_0 = $ constant emanating from the origin can be defined on $0 \le r < \infty$. By the Hopf-Rinow–de Rham theorem 1.7.3, g defines a complete Riemannian metric. Therefore $(\mathbf{R}^2, g)$ is a surface of revolution with the origin as its vertex. $\qquad\square$

In what follows, (r, θ) denotes a set of geodesic polar coordinates around the vertex p of a surface of revolution (M, g). Let $\gamma : I \to M$ be a geodesic such that $\gamma(t) \ne p$ for any $t \in I$, where I denotes an interval. Define

$$r(t) := r(\gamma(t)), \qquad \theta(t) := \theta(\gamma(t)).$$

Proposition 7.1.2. *Let* $\gamma : I \to M$ *be a geodesic such that* $\gamma(t) \neq p$ *for any* $t \in I$. *Then* γ *satisfies the following differential equations:*

$$r'' - mm'(\theta')^2 = 0, \tag{7.1.15}$$

$$\theta'' + 2\frac{m'r'\theta'}{m} = 0. \tag{7.1.16}$$

Conversely, any smooth curve $(r(s), \theta(s))$ *satisfying* (7.1.15) *and* (7.1.16) *is a geodesic on* M.

Proof. In terms of tensor calculus, set

$$g_{11} := g\left(\frac{\partial}{\partial r}, \frac{\partial}{\partial r}\right),$$

$$g_{12} = g_{21} := g\left(\frac{\partial}{\partial r}, \frac{\partial}{\partial \theta}\right),$$

$$g_{22} := g\left(\frac{\partial}{\partial \theta}, \frac{\partial}{\partial \theta}\right).$$

Then it follows from (7.1.8) that

$$g_{11} = 1, \qquad g_{12} = g_{21} = 0, \qquad g_{22} = m^2. \tag{7.1.17}$$

Thus the Christoffel symbols are

$$\Gamma^1_{i1} = \Gamma^1_{1i} = \Gamma^2_{ii} = 0$$

for $i = 1, 2$, and

$$\Gamma^1_{22} = -mm', \qquad \Gamma^2_{12} = \Gamma^2_{21} = \frac{m'}{m}.$$

By the differential equation for geodesics (1.2.1), we get (7.1.15) and (7.1.16). The converse is trivial. □

It follows from (7.1.16) that

$$\frac{d}{dt}(m^2(r(t))\theta'(t)) = 0 \tag{7.1.18}$$

for any $t \in I$. Therefore there exists a constant v such that

$$\theta'(t) = \frac{v}{m^2(r(t))} \tag{7.1.19}$$

for any $t \in I$. If $\xi(t)$ denotes the angle made by the geodesic $\gamma(t)$ and the *parallel* through $\gamma(t)$, i.e., $S(p, d(p, \gamma(t)))$, then we obtain Clairaut's relation

$$m(r(t))\cos\xi(t) = v$$

for any $t \in I$. Summing up, we get

Theorem 7.1.2 (Clairaut). *Let* $\gamma: I \to M$ *be a geodesic such that* $\gamma(t) \neq p$ *for any* $t \in I$. *Then there exists a constant* v *such that*

$$m(r(t)) \cos \xi(t) = v \tag{7.1.20}$$

for any $t \in I$.

Definition 7.1.2. The constant v in (7.1.20) is called *Clairaut's constant* for the geodesic γ.

Note that Clairaut's constant is zero iff the geodesic is tangent to a meridian. Therefore, if Clairaut's constant for a geodesic is nonzero then the geodesic does not pass through the vertex.

Lemma 7.1.4. *A parallel* $S(p, t_0)$ *is a geodesic iff* $m'(t_0) = 0$.

Proof. Suppose that the parallel $S(p, t_0)$ is a geodesic; then $S(p, t_0)$ can be expressed as

$$r(t) = t_0, \qquad \theta(t) = t. \tag{7.1.21}$$

By (7.1.15), we get $m'(t_0) = 0$. Alternatively, the curve defined by (7.1.21) satisfies (7.1.15) and (7.1.16); therefore $S(p, t_0)$ is a geodesic. □

Lemma 7.1.5. *Let* $\gamma: I \to M$ *be a geodesic. If* γ *is not a parallel then the zero points of* r' *are discrete. Furthermore, if* $r' = 0$ *for some* $s_0 \in I$ *then* $m'(r(s_0))$ *is nonzero (see Figure 7.1.1).*

Proof. If γ is tangent to a meridian then the claim is trivial. Thus we may assume that $\gamma(t) \neq p$ for any $t \in I$. Suppose $r'(s_0) = 0$ for some $s_0 \in I$. Then γ is tangent to the parallel $S(p, r(s_0))$. Since γ is not a parallel, $m'(r(s_0))$ is not zero by Lemma 7.1.4. By (7.1.15) and (7.1.19) $r''(s_0)$ is not zero. Therefore the zero points of r' are discrete. The fact $r'(s_0) = 0$ implies that γ is tangent to the parallel $S(p, r(s_0))$. Hence the latter claim is clear from Lemma 7.1.4. □

Lemma 7.1.6. *Let* $\gamma: \mathbf{R} \to M$ *be a geodesic not tangent to any meridian. If* $r' = 0$ *at two distinct parameter values then there exists a rotation* T *around* p *and a nonzero constant* b *such that*

$$T(\gamma(t)) = \gamma(t + b)$$

for any $t \in \mathbf{R}$. *In particular* $\gamma|_{[0,\infty)}$ *is not a ray.*

Proof. Suppose $r' = 0$ at $t = s_0$. Thus γ is tangent to the parallel $S(p, r(s_0))$. Then there exists a reflection f_0 on M (i.e., $\exp_p^{-1} \circ f_0 \circ \exp_p$ is a reflection

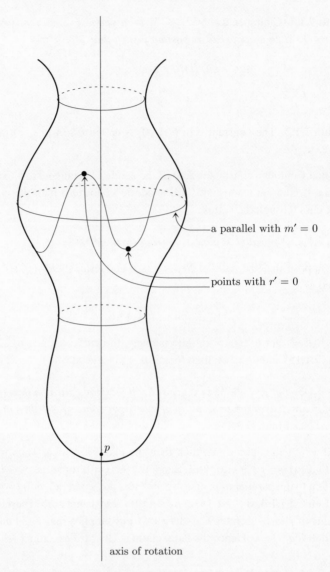

a parallel with $m' = 0$

points with $r' = 0$

axis of rotation

Figure 7.1.1

on T_pM) fixing the meridian through $\gamma(s_0)$ such that $f_0(\gamma(t)) = \gamma(2s_0 - t)$ for any t. Hence if $r' = 0$ at two distinct real numbers s_0 and s_1 then there exist two reflections f_0, f_1 on M such that

$$f_1 \circ f_0(\gamma(t)) = \gamma(t + 2s_1 - 2s_0)$$

for any t; furthermore, $f_1 \circ f_0$ is a rotation around p.   □

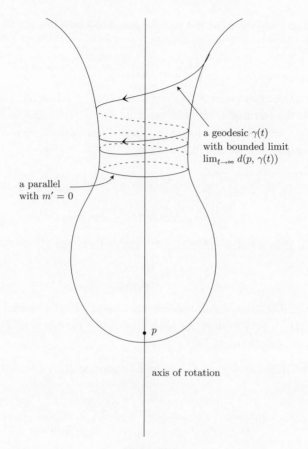

a geodesic $\gamma(t)$
with bounded limit
$\lim_{t\to\infty} d(p, \gamma(t))$

a parallel
with $m' = 0$

p

axis of rotation

Figure 7.1.2

Lemma 7.1.7. *Let $\gamma : [0, \infty) \to M$ be a geodesic whose Clairaut's constant is v. If $r_0 := \lim_{t\to\infty} d(p, \gamma(t))$ is bounded then $m(r_0) = |v|$ and $m'(r_0) = 0$ (see Figure 7.1.2). Hence $S(p, r_0)$ is a geodesic.*

Proof. From the assumption, Clairaut's constant v for γ is nonzero and $\gamma(t) \neq p$ for any $t \in [0, \infty)$. If γ is a parallel, the claim is trivial from Lemma 7.1.4. Thus we can assume that γ is not a parallel. It follows from Lemma 7.1.6 that there exists a positive constant t_0 such that $r' \neq 0$ on $[t_0, \infty)$. Without loss of generality, we may assume that γ is a unit-speed geodesic. Thus

$$(r'(t))^2 + m^2(r(t))(\theta')^2 = 1$$

for any $t \in [0, \infty)$. It follows from (7.1.19) that on $[t_0, \infty)$ either

$$r' = \sqrt{1 - \frac{v^2}{m^2}} \qquad \text{or} \qquad r' = -\sqrt{1 - \frac{v^2}{m^2}}.$$

If $m(r_0) \neq |\nu|$ then there exists a positive constant δ such that

$$\sqrt{1 - \frac{\nu^2}{m^2(r(t))}} \geq \delta$$

on $[t_0, \infty)$. Therefore either $r' \geq \delta$ on $[t_0, \infty)$ or $r' \leq -\delta$ on $[t_0, \infty)$. But this contradicts the fact that $0 \leq r_0 < \infty$. Thus $m(r_0) = |\nu|$ and, in particular, $\lim_{t \to \infty} r'(t) = 0$. Suppose that $m'(r_0) \neq 0$. Since $\lim_{t \to \infty} r(t) = r_0$, there exists a constant $t_1 \geq t_0$ such that

$$\frac{|\nu|}{2} = \frac{m(r_0)}{2} \leq m(r(t)) \leq 2m(r_0) = 2|\nu|$$

$$\frac{|m'(r_0)|}{2} \leq |m'(r(t))| \leq 2|m'(r_0)|$$

on $[t_1, \infty)$. Thus it follows from (7.1.15) and (7.1.19) that

$$|r''| \geq \frac{|m'(r_0)|}{16m(r_0)}$$

on $[t_1, \infty)$. This inequality implies that $\lim_{t \to \infty} r'(t) = \pm\infty$. However, $\lim_{t \to \infty} r'(t) = 0$. Therefore we have proved that $m'(r_0) = 0$. The second claim is clear from Lemma 7.1.4. □

Definition 7.1.3. Let $f(r, \nu)$, $g(r, \nu)$ be functions on $\{(r, \nu); r > 0, m(r) > |\nu|\}$ defined by

$$f(r, \nu) := \frac{\nu}{m(r)\sqrt{m^2(r) - \nu^2}}, \qquad g(r, \nu) := \frac{m(r)}{\sqrt{m^2(r) - \nu^2}}.$$

(7.1.22)

Proposition 7.1.3. Let $\gamma : [a, b) \to M$ be a unit-speed geodesic whose Clairaut's constant ν is nonzero. If $r'(t)$ is nonzero on $[a, b)$ then the geodesic γ parameterized by r satisfies

$$\frac{d\theta}{dr} = \epsilon f(r, \nu), \qquad (7.1.23)$$

and moreover

$$\theta(b) - \theta(a) \equiv \epsilon \int_{r(a)}^{r(b)} f(r, \nu) \, dr \qquad \mod 2\pi, \qquad (7.1.24)$$

$$b - a = \epsilon \int_{r(a)}^{r(b)} g(r, \nu) \, dr, \qquad (7.1.25)$$

where ϵ denotes the sign of $r'(t)$, $t \in [a, b)$.

Proof. Since γ is parameterized by arc length,

$$r'(t)^2 + m^2(r(t))\,\theta'^2 = 1$$

for any $t \in [a, b)$. Thus it follows from (7.1.19) that

$$r'(t) = \frac{\epsilon}{g(r(t), \nu)}. \tag{7.1.26}$$

Therefore

$$\frac{\mathrm{d}\theta}{\mathrm{d}r} = \frac{\theta'}{r'} = \epsilon f(r, \nu).$$

Suppose that $r'(b) \neq 0$. It follows from (7.1.23) and (7.1.26) that

$$b - a = \int_a^b \mathrm{d}t = \int_{r(a)}^{r(b)} \frac{\mathrm{d}r}{r'(t)} = \epsilon \int_{r(a)}^{r(b)} g(r, \nu)\,\mathrm{d}r$$

$$\theta(b) - \theta(a) \equiv \int_a^b \frac{\mathrm{d}\theta}{\mathrm{d}r}\,\mathrm{d}r = \epsilon \int_{r(a)}^{r(b)} f(r, \nu)\,\mathrm{d}r \qquad \mathrm{mod}\ 2\pi.$$

Now suppose that $r'(b) = 0$. By taking a strictly monotone increasing sequence $\{b_i\}$ convergent to b, we get (7.1.24) and (7.1.25). Note that the integrals in (7.1.24) and (7.1.25) are singular, since $m(r(b)) = |\nu|$, but both are finite, since $m'(r(b)) \neq 0$. □

Definition 7.1.4. For each $q \in M \setminus \{p\}$, let $\tau_q : [0, \infty) \to M$ be the unit-speed geodesic emanating from $q = \tau_q(0)$ through $p = \tau_q(d(p, q))$. For each positive t, let $L(t)$ denote the length of $S(p, t)$, which equals $2\pi m(t)$.

Lemma 7.1.8. *If* $\liminf_{t \to \infty} L(t) = 0$ *then for any point* q *distinct from* p *the subray* $\mu_q|_{[d(p,q),\infty)}$ *of* μ_q *is a unique ray emanating from* q.

Proof. Let $\{t_i\}$ be a divergent sequence with $\lim_{i \to \infty} L(t_i) = 0$. Let q be any point distinct from p. For simplicity, put

$$\rho := d(p, q).$$

It follows from the triangle inequality that

$$d(q, \tau_q(t_i + \rho)) \leq d(q, \mu_q(t_i)) + d(\mu_q(t_i), \tau_q(t_i + \rho))$$
$$\leq t_i - \rho + \tfrac{1}{2}L(t_i).$$

Since $\lim_{i \to \infty} L(t_i) = 0$, $d(q, \tau_q(t_i + \rho)) < t_i$ for sufficiently large i. This implies that τ_q is not a ray. Let γ be a geodesic emanating from q that is not tangent to any meridian. Since $\liminf_{t \to \infty} L(t) = \liminf_{t \to \infty} 2\pi m(t) = 0$, it follows from (7.1.20) that the image of γ is bounded. Hence γ is not a ray. Thus the geodesic $\mu_q|_{[\rho,\infty)}$ is a unique ray emanating from q. □

Lemma 7.1.9. *Suppose that $\int_1^\infty L^{-2}(t)\, dt$ is infinite. Let q be any point distinct from the vertex. If a geodesic γ emanating from q is not tangent to the meridian through q then γ is not a ray. In particular, the vertex is a unique pole.*

Proof. Since $\gamma: [0, \infty) \to M$ is not tangent to μ_q, the Clairaut constant v for γ is non-zero. By Lemma 7.1.6, we may assume

$$\lim_{t \to \infty} r(t) = +\infty$$

and the existence of a positive number b such that r' is positive on $[b, \infty)$. It follows from (7.1.24) that

$$\theta(t) - \theta(b) \equiv \int_{r(b)}^{r(t)} f(r, v)\, dr \qquad \mathrm{mod}\ 2\pi$$

for any real $t > b$. Since $L(t) = 2\pi m(t)$, $|\theta(s) - \theta(b)| > \pi$ for some $s > b$. This implies that γ is not a ray. In particular, q is not a pole. $\qquad\square$

7.2 Jacobi fields

Choose a point q distinct from the vertex and let it be fixed throughout this section. In order to derive the Jacobi fields along a geodesic emanating from q, we need some notation. For brevity we put

$$\rho := d(p, q). \qquad (7.2.1)$$

Let $\{e_1, e_2\}$ be an orthonormal basis for $T_q M$ such that

$$e_1 := \left(\frac{\partial}{\partial r}\right)_q, \qquad e_2 := \frac{1}{m(\rho)}\left(\frac{\partial}{\partial \theta}\right)_q. \qquad (7.2.2)$$

For each $v \in [-m(\rho), m(\rho)]$ let $\beta_v, \gamma_v: [0, \infty) \to M$ denote unit-speed geodesics emanating from q whose velocity vectors at $t = 0$ are

$$\dot\beta_v(0) = \frac{1}{g(\rho, v)}e_1 + \frac{v}{m(\rho)}e_2, \qquad \dot\gamma_v(0) = -\frac{1}{g(\rho, v)}e_1 + \frac{v}{m(\rho)}e_2.$$
$$(7.2.3)$$

Here $g(\rho, v)$ is defined in (7.1.22). Then it follows from (7.1.20) that the Clairaut constant of each β_v, γ_v equals v. Since β_v, γ_v are parameterized by arc length, we get

$$\left(\frac{\partial}{\partial t}r(\beta_\nu(t))\right)^2 + m^2(r(\beta_\nu(t)))\left(\frac{\partial}{\partial t}\theta(\beta_\nu(t))\right)^2 = 1$$

$$\left(\frac{\partial}{\partial t}r(\gamma_\nu(t))\right)^2 + m^2(r(\gamma_\nu(t)))\left(\frac{\partial}{\partial t}\theta(\gamma_\nu(t))\right)^2 = 1. \qquad (7.2.4)$$

Since the two geodesics β_ν, γ_ν both depend smoothly on ν if $\nu \in (-m(\rho), m(\rho))$, we obtain the Jacobi fields X_ν, Y_ν:

$$X_\nu(t) := \frac{\partial}{\partial \nu}(\beta_\nu(t)), \qquad Y_\nu(t) := \frac{\partial}{\partial \nu}(\gamma_\nu(t)) \qquad (7.2.5)$$

along β_ν, γ_ν respectively. Since $\beta_\nu(0) = q = \gamma_\nu(0)$ for any $\nu \in [-m(\rho), -m(\rho)]$, it is trivial that

$$X_\nu(0) = Y_\nu(0) = 0. \qquad (7.2.6)$$

If X_ν', Y_ν' denote the covariant derivatives along β_ν, γ_ν respectively, we have

$$X_\nu'(0) = -f(\rho, \nu)e_1 + \frac{1}{m(\rho)}e_2, \qquad Y_\nu'(0) = f(\rho, \nu)e_1 + \frac{1}{m(\rho)}e_2, \quad (7.2.7)$$

where $f(\rho, \nu)$ is defined in (7.1.22).

Proposition 7.2.1. *Let $\beta: [0, s] \to M$ be a geodesic β_c defined by (7.2.3) for some $c \in (-m(\rho), m(\rho))$ such that $p \notin \beta[0, s)$. If $r'(t)$, i.e., $dr(\beta(t))/dt$, is nonzero on $[0, s)$ then the Jacobi field X_c defined by (7.2.5) is given by*

$$X_c(t) = r'(t)\int_\rho^{r(t)} \frac{m(r)}{\sqrt{m^2(r) - c^2}^3}\, dr \left\{-c\left(\frac{\partial}{\partial r}\right)_{\beta(t)} + r'(t)\left(\frac{\partial}{\partial \theta}\right)_{\beta(t)}\right\}$$

$$(7.2.8)$$

on $[0, s)$.

Proof. Fix any positive $s_1 < s$. Then it follows from the assumption that there exists a positive δ such that, for each ν with $|\nu - c| < \delta$, the function

$$\frac{\partial}{\partial t}r(\beta_\nu(t))$$

is nonzero on $[0, s_1]$. Put $r(t, \nu) := r(\beta_\nu(t))$ and $\theta(t, \nu) := \theta(\beta_\nu(t))$. In particular, put $r(t) := r(t, c)$ and $\theta(t) := \theta(t, c)$ for simplicity. Since r' is positive on $[0, s_1]$,

$$\frac{\partial}{\partial t}r(t, \nu)$$

is positive for any $t \in [0, s_1]$ and any v with $|v - c| < \delta$. Thus it follows from (7.1.19) and (7.2.4) that, for each v with $|v - c| < \delta$,

$$\frac{\partial}{\partial t} r(t, v) = \frac{1}{g(r(t, v), v)} \qquad (7.2.9)$$

holds on $[0, s_1]$. By (7.1.25),

$$t = \int_\rho^{r(t,v)} g(r, v) \, dr \qquad (7.2.10)$$

for any $t \in [0, s_1]$ and any v with $|v - c| < \delta$. If we differentiate (7.2.10) with respect to v, then we get

$$\frac{\partial r}{\partial v}(t, c) = -cr'(t) \int_\rho^{r(t)} \frac{m(r)}{\sqrt{m^2(r) - c^2}^3} \, dr. \qquad (7.2.11)$$

However, it follows from Proposition 7.1.3 that

$$\theta(t, v) \equiv \theta(q) + \int_\rho^{r(t,v)} f(r, v) \, dr \qquad \text{mod } 2\pi; \qquad (7.2.12)$$

hence, by (7.2.11),

$$\frac{\partial \theta}{\partial v}(t, c) = r'(t)^2 \int_\rho^{r(t)} \frac{m(r)}{\sqrt{m^2(r) - c^2}^3} \, dr. \qquad (7.2.13)$$

It follows from (7.2.11) and (7.2.13) that for any $t \in [0, s)$ we get (7.2.8). $\quad\square$

By an argument analogous to that in the proof of Proposition 7.2.1, we conclude that the following holds:

Proposition 7.2.2. *Let* $\gamma \colon [0, s] \to M$ *be a geodesic* γ_c *defined by (7.2.3) for some* $c \in (-m(\rho), m(\rho))$ *such that* $p \notin \gamma[0, s)$. *If* $r'(t) \, (= dr(\gamma(t))/dt$ *is nonzero on* $[0, s)$ *then the Jacobi field* Y_c *defined in (7.2.5) is given by*

$$Y_c(t) = -r'(t) \int_\rho^{r(t)} \frac{m(r)}{\sqrt{m^2(r) - c^2}^3} \, dr \left\{ -c \left(\frac{\partial}{\partial r} \right)_{\gamma(t)} + r'(t) \left(\frac{\partial}{\partial \theta} \right)_{\gamma(t)} \right\}$$

$$(7.2.14)$$

on $[0, s)$.

Corollary 7.2.1. *Suppose that* $c \neq 0$. *If* $r'(s)$ *is zero then*

$$Y_c(s) = \frac{c}{|c| m'(r(s))} \left(\frac{\partial}{\partial r} \right)_{\gamma(s)}. \qquad (7.2.15)$$

In particular, for each γ_v *with* $|v| \in (0, m(\rho))$, $\gamma_v|_{[0,s]}$ *has no conjugate points of* q *along* γ_v *if* $\partial r(\gamma_v(t))/\partial t$ *is nonzero on* $[0, s)$.

Remark 7.2.1. It follows from Lemma 7.1.5 that $m'(r(s)) \neq 0$.

Proof. From the definition of γ_v, (7.2.3),

$$r'(t) = -\frac{1}{g(r(t), c)}. \tag{7.2.16}$$

By l'Hôpital's theorem,

$$\lim_{t \to s-0} r'(t) \int_\rho^{r(t)} \frac{m}{(m^2 - c^2)^{3/2}} \, dr = \lim_{t \to s-0} \frac{1}{r'} \int_\rho^{r(t)} \frac{m}{\sqrt{m^2 - c^2}^3} \, dr$$

$$= \lim_{t \to s-0} \frac{-(r')^3 m(r(t))}{r''(m^2 - c^2)^{3/2}}. \tag{7.2.17}$$

Thus by (7.2.16)

$$\lim_{t \to s-0} r'(t) \int_\rho^{r(t)} \frac{m}{(m^2 - c^2)^{3/2}} \, dr = \frac{1}{r''(s) m^2(r(s))}. \tag{7.2.18}$$

It follows from (7.1.15) and (7.1.19) that

$$r''(t) = \frac{c^2 m'(r(t))}{m^3(r(t))}. \tag{7.2.19}$$

By (7.2.18) and (7.2.19),

$$\lim_{t \to s-0} r'(t) \int_\rho^r \frac{m}{(m^2 - c^2)^{3/2}} \, dr = \frac{m(r(s))}{c^2 m'(r(s))}. \tag{7.2.20}$$

It follows from (7.2.16) that $m(r(s)) = |c|$. Therefore by (7.2.14) and (7.2.20) we get (7.2.15). $\qquad\square$

Proposition 7.2.3. *Let $\gamma \colon [0, \infty) \to M$ be a geodesic γ_c defined by (7.2.3) for some $|c| \in (0, m(\rho))$. Suppose that there exists a positive point s_1 such that $s_0 \in (0, s_1)$ is a unique zero point of r' on $[0, s_1]$. Then the Jacobi field Y_c defined in (7.2.5) is given by*

$$Y_c(t) = \frac{\partial \theta}{\partial v}(t, c) \left\{ -c\, g(r(t), c) \left(\frac{\partial}{\partial r} \right)_{\gamma(t)} + \left(\frac{\partial}{\partial \theta} \right)_{\gamma(t)} \right\} \tag{7.2.21}$$

on $(s_0, s_1]$, where

$$\theta(t, v) = \theta(\gamma_v(t)). \tag{7.2.22}$$

Proof. Since $r'(s_0) = 0$, it follows that γ is tangent to the parallel $S(p, r(s_0))$. By Lemma 7.1.4, $m'(r(s_0))$ is nonzero. Thus by (7.1.15) and (7.1.19)

$$\frac{\partial^2}{\partial t^2} r(t, c)$$

is nonzero at $t = s_0$. Here we have set

$$r(t, v) = r(\gamma_v(t)). \tag{7.2.23}$$

It follows from the inverse-function theorem that there exists a smooth function $\bar{s}$ on $(c - \delta, c + \delta)$ for some positive δ with $\delta < |c|$ satisfying

$$\bar{s}(v) = \min\{t > 0; \partial r(t, v)/\partial t = 0\}$$

for each $v \in (c - \delta, c + \delta)$ and $\bar{s}(c) = s_0$. By taking a smaller δ if necessary, we may assume that for each $v \in (c - \delta, c + \delta)$, $\bar{s}(v)$ is the unique zero point of $\partial r(t, v)/\partial t$ on $[0, s_1]$. It follows from (7.1.24) and (7.1.25) that, for each $v \in (c - \delta, c + \delta)$,

$$\theta(t, v) \equiv \theta(q) + \int_{r(\bar{s}(v),v)}^{\rho} f(r, v)\,dr + \int_{r(\bar{s}(v),v)}^{r(t,v)} f(r, v)\,dr \qquad \text{mod } 2\pi \tag{7.2.24}$$

and

$$t = \int_{r(\bar{s}(v),v)}^{\rho} g(r, v)\,dr + \int_{r(\bar{s}(v),v)}^{r(t,v)} g(r, v)\,dr \tag{7.2.25}$$

on $(\bar{s}(v), s_1]$. Since

$$g(r, v) = \frac{1}{g(r, v)} + vf(r, v)$$

holds, it follows from (7.2.25) that

$$t = \int_{r(\bar{s}(v),v)}^{\rho} \frac{1}{g(r, v)}\,dr + \int_{r(\bar{s}(v),v)}^{r(t,v)} \frac{1}{g(r, v)}\,dr$$
$$+ v \left(\int_{r(\bar{s}(v),v)}^{\rho} f(r, v)\,dr + \int_{r(\bar{s}(v),v)}^{r(t,v)} f(r, v)\,dr \right). \tag{7.2.26}$$

If we differentiate equation (7.2.26) with respect to v, we get

$$0 = \frac{1}{g(r(t, v), v)} \frac{\partial r}{\partial v}(t, v) + v \frac{\partial \theta}{\partial v}(t, v). \tag{7.2.27}$$

Note that

$$\frac{1}{g(r, v)} = \frac{\sqrt{m^2(r) - v^2}}{m(r)} = 0$$

at $r = r(\bar{s}(v), v)$ and

$$\frac{\partial}{\partial v} \frac{1}{g(r, v)} = -f(r, v).$$

Hence

$$\frac{\partial r}{\partial v}(t, v) = -v\, g(r(t, v), v)\, \frac{\partial \theta}{\partial v}(t, v) \tag{7.2.28}$$

holds for any $t \in (\bar{s}(v), s_1]$ and $v \in (c - \delta, c + \delta)$. By (7.2.28) we obtain (7.2.21). $\qquad\square$

Let a_1 be a small positive number such that m' is positive on $[0, a_1]$. Note that $m'(0) = 1$ by Proposition 7.1.1. Let $k : [0, m(a_1)] \to \mathbf{R}$ be the inverse function of $m|_{[0,a_1]}$. By Corollary 7.1.1,

$$k'(0) = 1, \qquad k''(0) = 0.$$

By Lemma 7.1.2 there exists a smooth function h on $[0, m(a_1)]$ satisfying

$$k'(t) = 1 + t^2 h(t). \tag{7.2.29}$$

Proposition 7.2.4. *Let Y_0 be the Jacobi field Y_v along $\tau_q = \gamma_v$, defined by (7.2.5), for $v = 0$. Then Y_0 is given by*

$$Y_0(t) = \left\{ -\frac{2}{m(a_1)} + 2m(a_1)h(m(a_1)) - 2\int_0^{m(a_1)} th'(t)\,dt \right.$$
$$\left. + \int_{a_1}^{\rho} \frac{1}{m^2}\,dr + \int_{a_1}^{t-\rho} \frac{1}{m^2}\,dr \right\} \left(\frac{\partial}{\partial \theta} \right)_{\tau_q(t)} \tag{7.2.30}$$

on (ρ, ∞).

Proof. Fix an arbitrary number $t > \rho$. Since Y_v depends smoothly on v, it follows from Proposition 7.2.3 that

$$Y_0(t) = \lim_{v \to +0} Y_v(t) = \lim_{v \to +0} \frac{\partial \theta}{\partial v}(t, v) \left(\frac{\partial}{\partial \theta} \right)_{\tau_q(t)} \tag{7.2.31}$$

on (ρ, ∞). If we set $m(r) = t$ then it follows from integrating by parts that, for any sufficiently small positive number v,

$$\int_{r(\bar{s}(v),v)}^{a_1} f(r, v)\,dr = \int_v^{m(a_1)} \left(\arccos\frac{v}{t} \right)' + v\left(\sqrt{t^2 - v^2} \right)' h(t)\,dt$$
$$= \arccos\frac{v}{m(a_1)} + v\sqrt{m(a_1)^2 - v^2}\, h(m(a_1))$$
$$- v \int_v^{m(a_1)} \sqrt{t^2 - v^2}\, h'(t)\,dt. \tag{7.2.32}$$

Thus

$$\lim_{v \to +0} \frac{\partial}{\partial v} \int_{r(\bar{s}(v),v)}^{a_1} f(r, v)\, \mathrm{d}r = -\frac{1}{m(a_1)} + m(a_1)\, h(m(a_1)) - \int_0^{m(a_1)} t h'(t)\, \mathrm{d}t.$$
(7.2.33)

By (7.2.24),

$$\lim_{v \to +0} \frac{\partial \theta}{\partial v}(t, v) = 2 \lim_{v \to +0} \frac{\partial}{\partial v} \int_{r(\bar{s}(v),v)}^{a_1} f(r, v)\, \mathrm{d}r$$

$$+ \lim_{v \to +0} \frac{\partial}{\partial v} \int_{a_1}^{\rho} f(r, v)\, \mathrm{d}r + \lim_{v \to +0} \frac{\partial}{\partial v} \int_{a_1}^{r(t,v)} f(r, v)\, \mathrm{d}r.$$
(7.2.34)

Therefore, combining (7.2.31), (7.2.33) and (7.2.34), we obtain (7.2.30). □

Corollary 7.2.2. *If $\int_1^\infty L(t)^{-2}\, \mathrm{d}t$ is finite then there exists a positive constant δ such that, for any point q with $0 < d(p, q) \le \delta$, there is no conjugate point of q along τ_q. If $\int_1^\infty L(t)^{-2}\, \mathrm{d}t$ is infinite then, for any point q distinct from p, there exists a conjugate point of q along τ_q. Moreover the geodesic $\mu_q|_{[d(p,q),\infty)}$ is a unique ray emanating from q.*

Proof. Suppose that

$$\int_1^\infty \frac{1}{L(t)^2}\, \mathrm{d}t$$

is finite. Thus, by the relation $L(t) = 2\pi m(t)$,

$$\int_{a_1}^\infty \frac{1}{m^2}\, \mathrm{d}r$$

is finite also. Choose a small positive number δ satisfying

$$-\int_\delta^{a_1} \frac{1}{m^2}\, \mathrm{d}r + \int_{a_1}^\infty \frac{1}{m^2}\, \mathrm{d}r - \frac{2}{m(a_1)}$$

$$+ 2m(a_1) h(m(a_1)) - 2\int_0^{m(a_1)} t h'(t)\, \mathrm{d}t < 0. \qquad (7.2.35)$$

Then, for any point q with $0 < d(p, q) \le \delta$ and any $t > d(p, q)$, $Y_0(t)$ is nonzero by (7.2.30) and (7.2.35). Since p is a pole, it is trivial that Y_0 does not vanish on $(0, \rho]$. Therefore, for any point q with $0 < d(p, q) \le \delta$, there is no conjugate point of q along τ_q. Suppose that

$$\int_1^\infty \frac{1}{L(t)^2}\, \mathrm{d}t$$

is infinite. Then

$$\int_{a_1}^{\infty} \frac{1}{m^2(r)}\, dr$$

is infinite also. Take any point q distinct from p; by (7.2.30) there exists a $t_1 > \rho$ such that $Y_0(t_1) = 0$. Therefore there is a conjugate point of q along τ_q. The last claim of the corollary is clear from Lemma 7.1.9. $\qquad \square$

From now on, we will regard

$$\liminf_{t \to \infty} L(t)$$

as nonzero. Since $L(t) = 2\pi m(t)$, by taking a smaller positive number a_1, if necessary, we may assume that

$$m(t) > m(a_1) \qquad\qquad (7.2.36)$$

for any $t > a_1$. Suppose that the point q satisfies $\rho < a_1$. Then it follows from the choice of a_1 that, for each $v \in (0, m(\rho))$, the function

$$\frac{\partial r}{\partial t}(t, v)$$

has a unique zero point at $t = t(v)$, where $r(t, v)$ is the function defined by (7.2.23) and $t(v)$ is the unique solution $t > 0$ of $r(t, v) = k(v)$. Geometrically, γ_v is tangent to a unique parallel $S(p, k(v))$ at a unique point $\gamma_v(t(v))$ for each $v \in (0, m(\rho))$. By (7.1.25) we have

$$t(v) = \int_{k(v)}^{\rho} g(r, v)\, dr.$$

Definition 7.2.1. For each positive number $\rho < a_1$, let $\phi_\rho(r, v)$ be a function on $\{(r, v); m(\rho) > v > 0, r > k(v)\}$ defined by

$$\phi_\rho(r, v) = \int_{k(v)}^{\rho} f(r, v)\, dr + \int_{k(v)}^{r} f(r, v)\, dr, \qquad (7.2.37)$$

where $f(r, v)$ is defined in (7.1.22).

Proposition 7.2.5. *Let* $\gamma : [0, \infty) \to M$ *be a geodesic* γ_c *defined by* (7.2.3) *for some* $c \in (0, m(\rho))$. *If the number* ρ *is less than* a_1 *then the Jacobi field* Y_c *defined in* (7.2.5) *is given by*

$$Y_c(t) = \frac{1}{g^2(r(t), c)} \frac{\partial \phi_\rho}{\partial v}(r(t), c) \left\{ -cg(r(t), c)\left(\frac{\partial}{\partial r}\right)_{\gamma(t)} + \left(\frac{\partial}{\partial \theta}\right)_{\gamma(t)} \right\}$$

$$(7.2.38)$$

on $(t(c), \infty)$, *where* $r(t) = r(\gamma(t))$.

Proof. From (7.2.24) we have

$$\frac{\partial \theta}{\partial v}(t, v) = \frac{\partial \phi_\rho}{\partial v}(r(t, v), v) + \frac{\partial r}{\partial v}(t, v) f(r(t, v), v), \qquad (7.2.39)$$

where $\theta(t, v)$, $r(t, v)$ are functions defined by (7.2.22) and (7.2.23) respectively. Thus it follows from (7.2.28) that

$$g^2(r(t, v), v) \frac{\partial \theta}{\partial v}(t, v) = \frac{\partial \phi_\rho}{\partial v}(r(t, v), v) \qquad (7.2.40)$$

holds on $\{(t, v); t > t(v), 0 < v < m(\rho)\}$. From Proposition 7.2.3 we obtain (7.2.38). $\qquad \square$

Lemma 7.2.1. *The inequality*

$$\frac{\partial}{\partial v} \int_{k(v)}^{r} f(r, v) \, dr$$

$$\leq \frac{-1}{\sqrt{m^2(r) - v^2}}(1 + v^2 h(m(r))) + (2m^2(a_1) + m(a_1)) C(m) \qquad (7.2.41)$$

holds on $\{(r, v); v \in (0, m(\rho)), k(v) < r \leq a_1\}$, *where*

$$C(m) := \max\{|h(t)|, |h'(t)|; 0 \leq t \leq m(a_1)\} \qquad (7.2.42)$$

is a constant depending on m.

Proof. By the same procedure as for equation (7.2.32), we get

$$\int_{k(v)}^{r} f(r, v) \, dr = \arccos \frac{v}{m(r)} + v \sqrt{m(r)^2 - v^2} \, h(m(r))$$

$$- v \int_{v}^{m(r)} \sqrt{t^2 - v^2} \, h'(t) \, dt. \qquad (7.2.43)$$

Thus

$$\frac{\partial}{\partial v} \int_{k(v)}^{r} f(r, v) \, dr = \frac{-1}{\sqrt{m(r)^2 - v^2}}(1 + v^2 h(m(r)))$$

$$+ \sqrt{m(r)^2 - v^2} \, h(m(r))$$

$$+ \int_{v}^{m(r)} \left(\frac{v^2}{\sqrt{t^2 - v^2}} - \sqrt{t^2 - v^2} \right) h'(t) \, dt. \qquad (7.2.44)$$

Since

$$\left| \int_{v}^{m(r)} \frac{v^2}{\sqrt{t^2 - v^2}} h'(t) \, dt \right| \leq C(m) v \int_{v}^{m(r)} \frac{t}{\sqrt{t^2 - v^2}} \, dt$$

$$= C(m) v \sqrt{m^2(r) - v^2},$$

(7.2.41) follows. $\qquad \square$

Lemma 7.2.2. *Suppose that*

$$\liminf_{t \to \infty} L(t)$$

is nonzero and that

$$\int_1^\infty \frac{1}{L^2(t)} \, dt$$

is finite. Then there exists a positive constant $b < a_1$ such that, for each positive number $\rho \leq b$,

$$\frac{\partial \phi_\rho}{\partial v}(r, v)$$

is strictly negative on $\{(r, v); \; r > k(v), \; m(\rho) > v > 0\}$.

Remark 7.2.2. von Mangoldt [59] proved this lemma for a two-sheeted hyperboloid of revolution.

Proof. Choose a positive number $b < a_1$ satisfying

$$\frac{1 - m^2(b) \, C(m)}{m(b)} > C_1(m) + (2\pi)^2 \, g(a_1, m(b))^3 \int_{a_1}^\infty \frac{1}{L^2(t)} \, dt, \quad (7.2.45)$$

where

$$C_1(m) := 2(2m^2(a_1) + m(a_1)) \, C(m) \qquad (7.2.46)$$

is a constant depending on m. Note that

$$m^2(b) \, C(m) < 1 \qquad (7.2.47)$$

because the right-hand side of the inequality (7.2.45) is positive. Suppose that $r \leq a_1$. By Lemma 7.2.1,

$$\frac{\partial \phi_\rho}{\partial v}(r, v) \leq -\frac{1 + v^2 \, h(m(\rho))}{\sqrt{m^2(\rho) - v^2}} + C_1(m) - \frac{1 + v^2 \, h(m(r))}{\sqrt{m^2(r) - v^2}}. \quad (7.2.48)$$

Since

$$1 + v^2 \, h(m(r)) \geq 1 - m^2(b)C(m) > 0$$

and

$$1 + v^2 \, h(m(\rho)) \geq 1 - m^2(b)C(m) > 0,$$

by (7.2.47), we obtain

$$\frac{\partial \phi_\rho}{\partial v}(r, v) \leq -\frac{1 - m^2(b) \, C(m)}{m(b)} + C_1(m). \qquad (7.2.49)$$

Therefore, again by (7.2.47),

$$\frac{\partial \phi_\rho}{\partial v}(r, v)$$

is negative on $\{(r, v); 0 < v < m(\rho), a_1 \geq r > k(v)\}$ for each positive $\rho \leq b$. Suppose that $r > a_1$. By Lemma 7.2.1,

$$\frac{\partial \phi_\rho}{\partial v}(r, v) \leq -\frac{1 - m^2(b)C(m)}{m(b)} + C_1(m) + \int_{a_1}^r \frac{m}{(m^2 - v^2)^{3/2}}\, dr.$$

Since

$$\frac{m(r)}{(m^2(r) - v^2)^{3/2}} \leq \frac{g^3(a_1, v)}{m^2(r)} < \frac{(2\pi)^2 g(a_1, m(b))^3}{L^2(r)}$$

on $[a_1, \infty)$, we get

$$\frac{\partial \phi_\rho}{\partial v}(r, v) \leq -\frac{1 - m^2(b)\, C(m)}{m(b)} + C_1(m)$$
$$+ (2\pi)^2 g(a_1, m(b))^3 \int_{a_1}^\infty \frac{1}{L^2(t)}\, dt. \qquad (7.2.50)$$

From the choice of b, the right-hand side of (7.2.50) is negative. Therefore we have proved the lemma. $\qquad \square$

Lemma 7.2.3. *Let M be a surface of revolution with vertex p. If a point q ($\neq p$) is a pole then any point x with $d(p, x) \leq d(p, q)$ is also a pole. Hence the set of poles on M forms a closed ball centered at p.*

Proof. It follows from Lemma 7.1.1 that any point on $S(p, \rho)$, where $\rho = d(p, q)$, is a pole. Thus it is sufficient to prove that any point x with $0 < d(p, x) < d(p, q)$ is a pole. Take such a point x and fix it. Let A_x be the set of all unit tangent vectors v such that the geodesic $\gamma_v(t) := \exp tv, t \in [0, \infty)$, is a ray. We shall prove that A_x is open in $S_x M$. Fix any $v \in A_x$. Since γ_v intersects $S(p, \rho)$ at a pole y on $S(p, \rho)$, γ_{-v} is a subray of the ray emanating from y through x. Note that each point on $S(p, \rho)$ is a pole. Hence there exists an open arc I containing $-v$ of $S_x M$ such that, for any $w \in I$, γ_w intersects $S(p, \rho)$. This implies that, for any w with $-w \in I$, γ_w is a subray of a ray emanating from a point on $S(p, \rho)$. Thus A_x is open in $S_x M$. Since A_x is a nonempty closed set, it follows from the connectedness of $S(p, \rho)$ that $A_x = S_x M$. $\qquad \square$

Definition 7.2.2. For a surface of revolution M with vertex p, define a number $r(M)$ by

$$r(M) := \sup\{d(p, q); \ q \text{ is a pole}\}.$$

Theorem 7.2.1. *Let M be a surface of revolution with vertex p. Then the set of poles on M forms the closed ball with radius $r(M)$ centered at p. Moreover, $r(M)$ is nonzero iff M satisfies*

$$\int_1^\infty \frac{1}{L^2(t)} \, dt < \infty \qquad (7.2.51)$$

and

$$\liminf_{t \to \infty} L(t) > 0, \qquad (7.2.52)$$

where $L(t)$ denotes the length of $S(p, t)$.

Proof. The first part of the claim is clear from Lemma 7.2.3. If $r(M)$ is nonzero then it follows from Lemmas 7.1.8 and 7.1.9 that we obtain (7.2.51) and (7.2.52). Suppose that the conditions (7.2.51), (7.2.52) hold. We shall prove that if a point q is sufficiently close to p then, for any geodesic emanating from q, there is no conjugate point of q along the geodesic. Note that any geodesic equals β_v, $|v| \leq m(\rho)$, or γ_v, $|v| < m(\rho)$. If $0 < d(p, q) < a_1$ then for, each $|v| \in (-m(\rho), m(\rho))$,

$$\frac{\partial}{\partial t} r(\beta_v(t))$$

is nonzero on $[0, \infty)$. Therefore it follows from Proposition 7.2.1 that there is no conjugate point of q along β_v. If $v = \pm m(\rho)$ then

$$\frac{\partial}{\partial t} r(\beta_v(t))$$

is zero at $t = 0$ but positive on $(0, \infty)$. Take any positive t_1. If we apply Corollary 7.2.1 to the reversed geodesic of $\beta_v|_{[0,t_1]}$ emanating from $\beta_v(t_1)$ then q is not conjugate to $\beta_v(t_1)$ along β_v. Since t_1 is arbitrary, there is no conjugate point of q along β_v, in the case $v = \pm m(\rho)$. It follows from Corollaries 7.2.1 and 7.2.2, Proposition 7.2.5 and Lemma 7.2.2 that if $d(p, q)$ is sufficiently small then for each $v \in [0, m(\rho))$ there is no conjugate point of q along γ_v. Therefore, by Lemma 7.1.1, for each $v \in (-m(\rho), m(\rho))$ there is no conjugate point of q along γ_v, if $d(p, q)$ is sufficiently small. Thus the proof is complete. $\square$

Corollary 7.2.3. *Let M be a surface of revolution admitting a total curvature $c(M)$. If $c(M)$ is less than 2π then the set of poles on M forms a closed ball centered at the vertex of M, which does not consist of a single pole.*

Proof. It follows from Theorem 5.2.1 that

$$\lim_{t \to \infty} \frac{L(t)}{t} = 2\pi - c(M).$$

Thus the function $L(t)$ satisfies the conditions (7.2.51) and (7.2.52). By Theorem 7.2.1 the claim is clear. ☐

7.3 The cut loci of a von Mangoldt surface

Definition 7.3.1. A surface of revolution M is called a *von Mangoldt surface* if $G(x)$ is not greater than $G(y)$, for the Gaussian curvatures $G(x)$, $G(y)$ of any two points x, y on M with $d(p, x) \geq d(p, y)$, where p denotes the vertex of M.

Lemma 7.3.1. *Let M be a von Mangoldt surface with vertex p. If on the geodesic τ_q emanating from a point q there is no conjugate point of q then q is a pole.*

Remark 7.3.1. The lemma above was proved by von Mangoldt [**59**] in the case where M is a two-sheeted hyperboloid of revolution.

Proof. Suppose that q is not a pole. Take an endpoint x of the cut locus of q. Since the cut point x admits a unique sector, q is conjugate to x along any minimal geodesic joining q to x. Let $\tau : [0, d(q, x)] \to M$ be a minimal geodesic joining q to x. Since p is a pole,

$$d(p, \tau_q(t)) = |\, d(p, q) - d(q, \tau(t))\,|$$

on $[0, d(q, x)]$. Thus it follows from the triangle inequality that

$$d(p, \tau_q(t)) \leq d(p, \tau(t))$$

on $[0, d(q, x)]$. This inequality implies that

$$G(\tau_q(t)) \geq G(\tau(t))$$

on $[0, d(q, x)]$, since M is a von Mangoldt surface. By the Sturm comparison theorem (Lemma 4.2.4), there exists a conjugate point of q along τ_q. This contradicts the assumption on τ_q. ☐

Lemma 7.3.2. *Let M be a surface of revolution with vertex at p and let q be any point distinct from p. Let (r, θ) be geodesic polar coordinates around p such that $\theta(q) = 0$. If two points q_1, q_2 satisfy*

$$r(q_1) = r(q_2), \qquad 0 \leq \theta(q_1) < \theta(q_2) \leq \pi$$

then $d(q, q_1) < d(q, q_2)$.

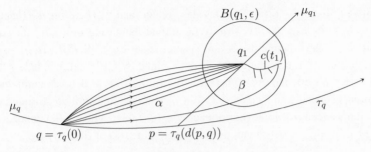

Figure 7.3.1

Proof. We may assume that

$$0 < \theta(q_1) < \theta(q_2) < \pi, \qquad (7.3.1)$$

because the remaining case reduces to this one. Let $c : [0, d(q, q_2)] \to M$ be a minimizing geodesic joining q to q_2. By the assumption (7.3.1), c intersects μ_{q_1} at a unique point x. Then it is trivial that

$$d(q, x) + d(x, q_2) = d(q, q_2). \qquad (7.3.2)$$

It follows from the assumption $r(q_1) = r(q_2)$ and the triangle inequality that

$$d(x, q_1) < d(x, q_2). \qquad (7.3.3)$$

Hence it follows from the triangle inequality that

$$d(q, q_1) \le d(q, x) + d(x, q_1) < d(q, x) + d(x, q_2) = d(q, q_2).$$

Thus the proof is complete. $\qquad\qquad\qquad\qquad\qquad\qquad\qquad\square$

Theorem 7.3.1. *If M is a von Mangoldt surface with vertex p then, for any point $q \in M \setminus \{p\}$, the cut locus C_q of q equals $\tau_q[t_0, \infty)$, where $\tau_q(t_0)$ denotes the first conjugate point of q along τ_q.*

Proof. First, we will prove that $C_q \subset \tau_q(d(p, q), \infty)$. Supposing the existence of a cut point q_1 of q such that $q_1 \notin \tau_q(d(p, q), \infty)$, we shall obtain a contradiction. If we note that p is a pole, there exists an endpoint q_1 of C_q such that $q_1 \notin \tau_q[d(p, q), \infty) \cup \mu_q[0, \infty)$ (see Figure 7.3.1). Since q_1 admits a unique sector $\Sigma_\epsilon(q_1)$, it is conjugate to q along any minimal geodesic joining q to q_1. Since any minimizing geodesics joining q and q_1 do not intersect the meridian μ_{q_1} within its interior, the inner angle of $\Sigma_\epsilon(q_1)$ at q_1 is also greater than π.

Let $c : [0, 1] \to C_q$ be a Jordan arc emanating from $q_1 = c(0)$ and for each $t \in (0, 1)$ let $\Sigma_\epsilon^+(c(t))$ be the unique sector at $c(t)$ containing $c(t, 1]$. Since the inner angle of $\Sigma_\epsilon(q_1)$ at q_1 is also greater than π, for any sufficiently small positive t, that of $\Sigma_\epsilon^+(c(t))$ at $c(t)$ is also greater than π. Therefore the function $[0, 1] \ni t \mapsto d(q, c(t))$ is strictly monotone increasing on $[0, t_1]$ for some positive t_1. Thus there exists a unit-speed geodesic $\beta : [0, d(q, c(t_1))] \to M$ joining q to $c(t_1)$ such that the angle determined by $\dot{t}_q(0)$ and $\dot{\beta}(0)$ is less than that between $\dot{t}_q(0)$ and $\dot{\alpha}(0)$, where $\alpha : [0, d(q, q_1)] \to M$ denotes a unit-speed minimal geodesic joining q to q_1. It follows from the first variation formula that β is longer than α. Moreover, for each $s \in (0, L(\alpha))$, where $L(\alpha)$ denotes the length of α, there exists a unique parameter value $t(s)$ of β satisfying

$$\theta(\alpha(s)) = \theta(\beta(t(s))).$$

Note that both the functions $\theta \circ \alpha$, $\theta \circ \beta$ are strictly monotone increasing. Here we assume the geodesic polar coordinates (r, θ) are chosen in such a way that $\theta(q) = 0 < \theta(q_1) < \pi$. Fix $s \in (0, L(\alpha))$. Let (a, b) be the connected component containing $t(s)$ of the set $\{t \in (0, L(\beta)); d(p, \beta(t)) < d(p, \alpha(s))\}$. If $a = 0$ (resp. $b \geq L(\alpha)$) then $s > a$ (resp. $s < b$). Hence we may assume that $a > 0$ (resp. $b < L(\alpha)$) in order to prove $s > a$ (resp. $s < b$). From the definition of (a, b) we have

$$d(p, \alpha(s)) = d(p, \beta(a)) = d(p, \beta(b)).$$

Since $0 < \theta(\beta(a)) < \theta(\alpha(s)) = \theta(\beta(t(s))) < \theta(\beta(b)) < \pi$, it follows from Lemma 7.3.2 that

$$a = d(q, \beta(a)) < s = d(q, \alpha(s)) < d(q, \beta(b)) = b,$$

which implies that $d(p, \beta(s)) < d(p, \alpha(s))$ for any $s \in (0, L(\alpha))$. Since M is a von Mangoldt surface,

$$G(\alpha(s)) \leq G(\beta(s))$$

for any $s \in [0, L(\alpha)]$. As noted above, q is conjugate to q_1 along α. Thus, by the Sturm comparison theorem, q is conjugate to $\beta(s_1)$ along β for some $s_1 \in (0, L(\alpha)] \subset (0, L(\beta))$. This is a contradiction. Thus C_q is contained in $\tau_q(d(p, q), \infty)$. If C_q is nonempty then it follows from Lemma 7.3.1 that the first conjugate point of q along τ_q is $\tau_q(t_0)$. For any $t > t_0$, $\tau_q|_{[0,t]}$ is not a minimizing geodesic. Thus there exist at least two minimizing geodesics joining q to $\tau_q(t)$, each of which is distinct from τ_q. Hence $\tau_q(t)$ is a cut point of q for any $t > t_0$. Thus we have proved that if C_q is nonempty then

$$\tau_q[t_0, \infty) \subset C_q.$$

Suppose that $\tau_q|_{[0,t_0]}$ is not minimizing. Then there exists a cut point $\tau_q(t_1)$ of q with $0 < t_1 < t_0$ such that $\tau_q|_{[0,t_1]}$ is minimizing. Since $\tau_q(t_1)$ is not conjugate to q along τ_q, there exists a minimizing geodesic γ joining q to $\tau_q(t_1)$ that is distinct from τ_q. Then the two minimizing geodesics $\tau_q|_{[0,t_1]}$ and $\gamma|_{[0,t_1]}$ bound a relatively compact domain D. Take any geodesic $c : [0, \infty) \to M$ emanating from q such that $c|_{[0,t]}$ lies in D for sufficiently small positive t. If the geodesic c does not meet γ or τ_q except at q then the geodesic is a ray. Note that $C_q \subset \tau_q(d(p, q), \infty)$. But this is a contradiction, since the image of c is bounded. Thus c meets γ or τ_q again. Since $\gamma|_{[0,t_1]}$, $\tau_q|_{[0,t_1]}$ are minimizing, c meets them at $\gamma(t_1) = \tau_q(t_1)$. This fact means that $\tau_q(t_1)$ is conjugate to q along τ_q. This is a contradiction. Therefore we have proved that $C_q = \tau_q[t_0, \infty)$. $\square$

Proposition 7.3.1. *Let q be a point on a von Mangoldt surface M. Suppose that there exist two distinct unit-speed minimizing geodesics α, β emanating from q that meet at $\tau_q(s_0)$ for some positive s_0. If the relatively compact domain D bounded by α, β does not contain the vertex of M then the Gaussian curvature G of M is constant on the set $\{x \in M; r_1 \le d(p, x) \le r_2\}$, where*

$$r_1 := \min\{d(p, \alpha(t)), d(p, \beta(t)); 0 \le t \le L(\alpha) = L(\beta)\}$$

and

$$r_2 := \max\{d(p, \alpha(t)), d(p, \beta(t)); 0 \le t \le L(\alpha) = L(\beta)\}.$$

Proof. Without loss of generality, we may assume that for each $t \in (0, L(\alpha))$, the subarc $\mu_{\alpha(t)}|_{[0,d(p,\alpha(t))]}$ of the meridian $\mu_{\alpha(t)}$ meets β. Then, by the proof of Theorem 7.3.1, the inequalities

$$d(p, \beta(t)) \le d(p, \alpha(t)), \qquad G(\alpha(t)) \le G(\beta(t))$$

hold on $[0, L(\alpha)]$. Furthermore, it follows from Theorem 7.3.1 that there is no cut point of q in the domain D. Therefore q is conjugate to $\tau_q(s_0)$ along α and β respectively. It follows from the Sturm comparison theorem that

$$G(\alpha(t)) = G(\beta(t))$$

for any $t \in [0, L(\alpha)]$. Since M is a von Mangoldt surface, the Gaussian curvature of M is constant on the set

$$A(t) := \{x \in M; d(p, \beta(t)) \le d(p, x) \le d(p, \alpha(t))\}$$

for each $t \in (0, L(\alpha))$. Since two sets $A(t_1)$, $A(t_2)$ have a common point if $|t_1 - t_2|$ is sufficiently small, G is constant on the set $\{x \in M; r_1 \le d(p, x) \le r_2\}$. $\square$

Definition 7.3.2. For each smooth function $f : [0, \infty) \to \mathbf{R}$ that is extensible to a smooth even function around 0, let $M(f)$ denote the surface of revolution in the three-dimensional Euclidean space E^3 defined by the equation $z = f(\sqrt{x^2 + y^2})$.

Note that the Gaussian curvature of the surface $M(f)$ at a point $(x, y, f(\sqrt{x^2 + y^2}))$, $x > 0$, is given by

$$\frac{f'f''}{x(1 + f'(x)^2)^2}$$

and at the vertex it is $f''(0)^2$.

Remark 7.3.2. It follows from Lemma 7.1.3 that $M(f)$ is a smooth Riemannian manifold. Such an $M(f)$ that is also a von Mangoldt surface was called a *flattening surface* by Elerath. Theorem 7.3.1 was proved for a flattening surface by him in [**22**].

Example 7.3.1. Let f_1 be the function defined by

$$f_1(x) := a\sqrt{x^2 + b},$$

where a, b are positive constants. Then the surface $M(f_1)$ is a two-sheeted hyperboloid of revolution. Moreover the surface is a von Mangoldt surface, since the Gaussian curvature at $(x, y, f_1(\sqrt{x^2 + y^2}))$ equals

$$\frac{a^2 b}{((1 + a^2) x^2 + b)^2}.$$

Let $x = x_1(t) : [0, \infty) \to \mathbf{R}$ be the inverse function of

$$t = \int_0^x \sqrt{1 + f_1'(x)^2} \, dx.$$

Thus

$$x_1'(t) = \frac{1}{\sqrt{1 + f_1'(x(t))^2}}. \tag{7.3.4}$$

If $L(t)$ denotes the length of the parallel on $M(f_1)$ with radius t then

$$L(t) = 2\pi x_1(t). \tag{7.3.5}$$

By l'Hôpital's theorem and (7.3.4), we get

$$\lim_{t \to \infty} \frac{L(t)}{t} = 2\pi \lim_{t \to \infty} x_1'(t) = \frac{2\pi}{\sqrt{a^2 + 1}}. \tag{7.3.6}$$

Therefore it follows from Theorem 7.2.1 that the set of poles on $M(f_1)$ forms a nontrivial closed ball centered at the vertex of $M(f_1)$.

Example 7.3.2. Let f_2 be the smooth function defined by $f_2(x) = ax^2$, where a is a positive constant. Then $M(f_2)$ is a paraboloid of revolution and a von

Mangoldt surface. Let $x = x_2(t) : [0, \infty) \to \mathbf{R}$ be the inverse function of

$$t = \int_0^x \sqrt{1 + f_2'(x)^2} \, dx.$$

If $L(t)$ denotes the length of the parallel on $M(f_2)$ with radius t then we get

$$\int_1^\infty \frac{1}{L^2(t)} dt = \frac{1}{(2\pi)^2} \int_1^\infty \frac{1}{x_2(t)^2} \, dt = \frac{1}{(2\pi)^2} \int_{x_2(1)}^\infty \frac{1}{x^2} \sqrt{1 + f_2'(x)^2} \, dx.$$

$$(7.3.7)$$

Thus

$$\int_1^\infty \frac{1}{L^2(t)} \, dt \geq \frac{a}{2\pi^2} \int_{x_2(1)}^\infty \frac{1}{x} \, dx = \infty. \qquad (7.3.8)$$

This inequality implies that the vertex is a unique pole on $M(f_2)$, by Theorem 7.2.1.

Example 7.3.3. Let $\phi : [0, \infty) \to [0, 1]$ be a smooth monotone nondecreasing function such that $\phi \equiv 0$ on $[0, 2]$ and $\phi \equiv 1$ on $[4, \infty)$. Then the function $f_3 : [0, \infty) \to \mathbf{R}$ is defined by

$$f_3''(x) = 2(1 - \phi(x)) + \frac{\phi(x)}{x},$$

with $f_3 = f_3' = 0$ at $x = 0$. In order to check that the surface $M(f_3)$ is a von Mangoldt surface, we shall introduce a function k defined by

$$k(t) = \frac{t}{(1 + t^2)^2}.$$

Since the function k is monotone decreasing on $[1/\sqrt{3}, \infty)$, the function $k(f_3'(x))$ is monotone nonincreasing on $[1/2\sqrt{3}, \infty)$. Thus on this interval the function

$$\frac{f_3' f_3''}{x (1 + f_3'^2)^2} = \frac{f_3'' k(f_3'(x))}{x}$$

is also monotone nonincreasing. Since $f_3(x) = x^2$ on $[0, 2]$,

$$\frac{f_3' f_3''}{x (1 + f_3'^2)^2} = \frac{4}{1 + 4x^2}$$

is monotone decreasing on $[0, 2]$. Thus $M(f_3)$ is a von Mangoldt surface. It is easy to check that the conditions (7.2.51) and (7.2.52) hold and that

$$\lim_{t \to \infty} \frac{L(t)}{t} = 0$$

for the surface $M(f_3)$. Thus the total curvature of the surface is 2π and the set of poles on the surface forms a nontrivial closed ball.

In [56], Maeda found an upper bound for the diameter of the set of poles on a nonnegatively curved Riemannian manifold M, as follows. Let p be a point on M. For each positive t, let $D(t)$ denote the diameter of $S_p(t)$. Then he and Sugahara proved in [100]:

Theorem 7.3.2. *Let M be a complete connected Riemannian manifold with nonnegative sectional curvature. Then the number*

$$d_0(M) := \limsup_{t \to \infty} \frac{D(t)^2}{t}$$

is independent of the choice of the point p, and the diameter of the set of poles on M is not greater than than $d_0(M)/8$.

As a remark on this theorem the following lemma implies that the vertex of a surface of revolution M with nonnegative Gaussian curvature is a unique pole if $d_0(M)$ is finite.

Lemma 7.3.3. *Let (M, g) be a surface of revolution with vertex p. If the Gaussian curvature of M is nonnegative then $\lim_{t \to \infty} D(t)/L(t)$ is positive.*

Proof. Since the claim is trivial in the case where the Gaussian curvature is identically zero, we omit this case. Let $\gamma : \mathbf{R} \to M$ be a unit-speed geodesic with $\gamma(0) = p$. Then, by Lemma 7.3.2,

$$D(t) = d(\gamma(t), \gamma(-t))$$

holds on $[0, \infty)$. Since $D(t)$ is Lipschitz continuous, its derivative exists almost everywhere, and

$$D(t) = \int_0^t D'(t) \, \mathrm{d}t \tag{7.3.9}$$

holds for any positive t. By (2.2.5), if $D(t)$ is differentiable at $t > 0$ then there exists a unique minimizing geodesic α_t joining $\gamma(t)$ and $\gamma(-t)$ in each closed half-plane cut off by γ, and

$$D'(t) = \cos \theta(t) + \cos \theta(-t) \tag{7.3.10}$$

holds, where $\theta(t)$ and $\theta(-t)$ denote the inner angles at $\gamma(t)$ and $\gamma(-t)$ of the domains bounded by α_t and $\gamma|_{[-t,t]}$ respectively. Note that $\theta(t) = \theta(-t)$ by the symmetry of M. Since the Gaussian curvature of M is not identically zero, γ is not a straight line. Thus, for sufficiently large t, the geodesic α_t and that given by the reflection with respect to γ bound a compact domain $\Delta(t)$. It follows from the Gauss–Bonnet theorem and Theorem 2.2.1 that

$$c(\Delta(t)) = 4\theta(t), \qquad c(M) \le 2\pi. \tag{7.3.11}$$

Since the Gaussian curvature of M is nonnegative, it follows from (7.3.11) that $\theta(t) \le \pi/2$ for almost all t. Therefore $D(t)$ is monotone nondecreasing by (7.3.9) and (7.3.10). Let $B(t)$ denote the open ball with center p and radius t. Since $L(t) = 2\pi m(t)$,

$$L'(t) = 2\pi - c(B(t))$$

for any positive t. Therefore $L(t)$ is also monotone nondecreasing, since the Gaussian curvature of M is nonnegative. If $L(t)$ is bounded then it is trivial that

$$\lim_{t \to \infty} \frac{D(t)}{L(t)}$$

is positive. By the Gauss–Bonnet theorem,

$$D'(t) = 2\cos\frac{c(\Delta(t))}{4} \ge \sin\left(\frac{\pi}{2} - \frac{c(B(t))}{4}\right) \tag{7.3.12}$$

holds for almost all sufficiently large t. By l'Hôpital's theorem, we get

$$\lim_{t \to \infty} \frac{L(t)}{t} = \lim_{t \to \infty} L'(t) = 2\pi - c(M),$$

$$\lim_{t \to \infty} \frac{D(t)}{t} = \lim_{t \to \infty} D'(t) = 2\cos\frac{c(M)}{4}. \tag{7.3.13}$$

Hence, if the total curvature $c(M)$ is less than 2π then

$$\lim_{t \to \infty} \frac{D(t)}{L(t)} = \frac{2\cos\frac{1}{4}c(M)}{2\pi - c(M)} \tag{7.3.14}$$

holds. On the one hand, if $c(M)$ equals 2π and $\lim_{t \to \infty} L(t) = \infty$ then by Lemma 5.2.1 we get

$$\liminf_{t \to \infty} \frac{D(t)}{L(t)} \ge \liminf_{t \to \infty} \frac{D'(t)}{L'(t)} \ge \lim_{t \to \infty} \frac{\sin t}{2t} = \frac{1}{2}. \tag{7.3.15}$$

On the other hand, the inequality

$$\limsup_{t \to \infty} \frac{D(t)}{L(t)} \le \frac{1}{2} \tag{7.3.16}$$

holds, since $2D(t) \le L(t)$ for any positive t. By (7.3.15) and (7.3.16), we have $\lim_{t \to \infty} D(t)/L(t) = 1/2$. $\qquad\square$

For a von Mangoldt surface M, let us calculate the number $r(M)$ (see Definition 7.2.2).

Theorem 7.3.3. *Let M be a von Mangoldt surface with*

$$\int_1^\infty \frac{1}{L^2(t)}\, dt < \infty.$$

If the bounded number $c(L)$ defined by

$$c(L) := 4 \int_0^\infty \frac{L(t) - tL'(t)}{L^3(t)}\, dt \tag{7.3.17}$$

is nonpositive then $r(M) = \infty$, and if $c(L)$ is positive then $r(M)$ equals the unique zero point of the function $F : (0, \infty) \to \mathbf{R}$ defined by

$$F(r) := c(L) - \int_r^\infty \frac{1}{L^2(t)}\, dt. \tag{7.3.18}$$

Proof. Let q be a point distinct from the vertex of M. It follows from (7.2.30) that the Jacobi field Y_0 along τ_q is given by

$$Y_0(t) = \left(-\frac{2}{m(a_1)} + 2m(a_1)h(m(a_1)) - 2\int_0^{m(a_1)} th'(t)\, dt \right.$$

$$+ 2\int_{a_1}^\infty \frac{1}{m^2(r)}\, dr - \int_\rho^\infty \frac{1}{m^2(r)}\, dr$$

$$\left. - \int_{t-\rho}^\infty \frac{1}{m^2(r)}\, dr \right) \left(\frac{\partial}{\partial \theta} \right)_{\tau_q(t)} \tag{7.3.19}$$

on (ρ, ∞). Since $Y_0(t)$ is independent of the choice of a_1,

$$C := -\frac{2}{m(s)} + 2m(s)h(m(s)) - 2\int_0^{m(s)} th'(t)\, dt + 2\int_s^\infty \frac{1}{m^2(r)}\, dr \tag{7.3.20}$$

is constant on $(0, a_1]$. It follows from integrating by parts that

$$4 \int_t^\infty \frac{m - rm'}{m^3}\, dr = 2\int_t^\infty \frac{1}{m^2}\, dr - \frac{2t}{m^2(t)} \tag{7.3.21}$$

holds for any positive t. Thus we have

$$C = 4 \int_s^\infty \frac{m - rm'}{m^3}\, dr + \frac{2s}{m^2(s)} - \frac{2}{m(s)}$$

$$+ 2m(s)h(m(s)) - 2\int_0^{m(s)} th'(t)\, dt,$$

for any $s \in (0, a_1]$. From l'Hôpital's theorem and Corollary 7.1.1,

$$\lim_{s \to +0} \left(\frac{s}{m^2(s)} - \frac{1}{m(s)} \right) = \lim_{s \to +0} \frac{s - m(s)}{m^2(s)} = 0.$$

Since C is independent of s, we get

$$C = \lim_{s \to +0} C = 4 \int_0^\infty \frac{m - rm'}{m^3} \, dr = (2\pi)^2 c(L). \qquad (7.3.22)$$

From (7.3.19) and (7.3.22),

$$Y_0(t) = (2\pi)^2 \left\{ F(\rho) - \int_{t-\rho}^\infty \frac{1}{L(r)^2} \, dr \right\} \left(\frac{\partial}{\partial \theta} \right)_{\tau_q(t)} \qquad (7.3.23)$$

for any $t > \rho$. If $c(L)$ is nonpositive then $Y_0(t)$ is nonzero for any $t > \rho$. This implies that there is no conjugate point of q along τ_q. Hence by Lemma 7.3.1 q is a pole. Since q is arbitrary, $r(M) = \infty$. Suppose that $c(L)$ is positive. Let s_0 be the unique zero point of the function F. Then it is trivial that if ρ is greater (resp. less) than s_0 then $Y_0(t)$ has a zero point (resp. no zero point) in (ρ, ∞). Hence $r(M)$ is equal to the zero point of F by Lemma 7.3.1. $\qquad \square$

As an application of Theorem 7.3.3, we will construct a von Mangoldt surface that has a positive constant Gaussian curvature around the vertex, but each point of which is a pole. That is, we will construct a Riemannian manifold with sectional curvature of both signs but without conjugate points. Such a manifold was first constructed by R. Gulliver in [33]. Let $m_0 : (0, \infty) \to (0, \infty)$ be a smooth function defined by

$$m_0(r) = (1 - \phi(r)) \sin r + \phi(r)(r - c + \tan 2c), \qquad (7.3.24)$$

where c is a positive constant less than $\pi/4$ and $\phi \colon \mathbf{R} \to \mathbf{R}$ denotes a smooth nondecreasing function such that $\phi(r) = 0$ for $r \leq c$ and $\phi(r) = 1$ for $r \geq 2c$.

Lemma 7.3.4. *The value*

$$\int_0^\infty \frac{m_0(r) - rm_0'(r)}{m_0^3(r)} \, dr$$

is negative.

Proof. Fix a positive constant $\epsilon \, (< c)$. Then, integrating by parts, we have

$$\int_\epsilon^\infty \frac{m_0(r) - rm_0'(r)}{m_0^3(r)} \, dr = \int_\epsilon^\infty \frac{1}{m_0^2} \, dr + \frac{1}{2} \int_\epsilon^\infty r \left(\frac{1}{m_0^2} \right)' \, dr$$

$$= \frac{1}{2} \int_\epsilon^\infty \frac{1}{m_0^2} \, dr - \frac{\epsilon}{2 \sin^2 \epsilon}. \qquad (7.3.25)$$

Since $m_0(r) \geq \sin r$ on $[c, 2c]$,

$$\int_\epsilon^\infty \frac{1}{m_0^2} \, dr \leq \int_\epsilon^{2c} \frac{1}{\sin^2 r} \, dr + \int_{2c}^\infty \frac{dr}{(r - c + \tan 2c)^2}$$

$$= -\cot 2c + \cot \epsilon + \frac{1}{c + \tan 2c}. \qquad (7.3.26)$$

By l'Hôpital's theorem,

$$\lim_{\epsilon \to +0} \left(\cot \epsilon - \frac{\epsilon}{\sin^2 \epsilon} \right) = \lim_{\epsilon \to +0} \frac{\sin \epsilon \cos \epsilon - \epsilon}{\sin^2 \epsilon} = 0.$$

It follows from (7.3.25) and (7.3.26) that

$$\int_0^\infty \frac{m_0(r) - r m_0'(r)}{m_0^3(r)} \, dr \leq -\frac{\cot 2c}{2} + \frac{1}{2(c + \tan 2c)} < 0. \quad (7.3.27)$$

$\square$

By (7.2.22), it is clear that $-m_0''/m_0 \equiv 1$ on $(0, c)$ and $-m_0''/m_0 \equiv 0$ on $(2c, \infty)$. Thus there exists a smooth nonincreasing function $K : [0, \infty) \to \mathbf{R}$ such that $K \leq -m_0''/m_0$ on $(0, \infty)$, $K \equiv 1$ on $[0, c]$ and $K \equiv$ constant on $[2c, \infty)$. If m denotes the solution of the following differential equation,

$$m'' + Km = 0,$$

with initial conditions $m(0) = 0$, $m'(0) = 1$ then $m(r)$ is not less than $m_0(r)$ for any nonnegative r, by the Sturm comparison theorem. If (r_0, θ_0) denotes geodesic polar coordinates around the origin for the Euclidean plane $(\mathbf{R}^2, g_0)$ then it follows from Corollary 7.1.1 and Theorem 7.1.1 that $(\mathbf{R}^2, dr_0^2 + m(r_0)^2 \, d\theta_0^2)$ is a von Mangoldt surface whose vertex is the origin and whose Gaussian curvature equals the function K.

Theorem 7.3.4. *The von Mangoldt surface constructed above has a positive constant Gaussian curvature around the vertex (and a negative constant Gaussian curvature outside a compact subset) but no conjugate points.*

Proof. Since $m(r) \geq m_0(r)$ on $[0, \infty)$, it follows from (7.3.25) that

$$\int_0^\infty \frac{m - r m'}{m^3} \, dr \leq \int_0^\infty \frac{m_0 - r m_0'}{m_0^3} \, dr.$$

Therefore, from Theorem 7.3.3 and Lemma 7.3.4, each point on the surface $(\mathbf{R}^2, dr_0^2 + m^2(r_0) d\theta_0^2)$ is a pole and, in particular, has no conjugate points. The claim regarding the Gaussian curvature is clear from the property of K, since the Gaussian curvature equals the function.

$\square$

We shall now prove the two analytic inequalities below ((7.3.28) and (7.3.29)).

Proposition 7.3.2. *Let* $f : (0, \pi/2] \rightarrow (0, \infty)$ *be a smooth function that is extensible to a smooth odd function around* 0 *with* $f'(0) = 1$. *Suppose that the function* f''/f *is monotone nondecreasing on* $(0, \pi/2]$ *and that* $\pi/2$ *is the least positive zero point of the function* f'. *Then the following inequalities,*

$$f\left(\frac{\pi}{2}\right) \leq 1, \qquad (7.3.28)$$

$$4 \int_0^{\pi/2} \frac{f(t) - tf'(t)}{f^3(t)} \, dt \geq \frac{\pi}{f^2(\pi/2)}, \qquad (7.3.29)$$

hold and each equality holds iff $f(t) = \sin t$.

Proof. Extend the function f to a smooth positive-valued function F on $(0, \infty)$. Let $G : [0, \infty) \rightarrow \mathbf{R}$ be a smooth monotone nonincreasing function such that $G \leq -F''/F$ on $(0, \infty)$ and $G \equiv -F''/F$ on $(0, \pi/2]$. Let m be the solution of the differential equation $m'' + Gm = 0$ with initial conditions $m(0) = 0$, $m'(0) = 1$. Then, by the Sturm comparison theorem, $m \geq F > 0$ on $(0, \infty)$. Thus it follows from Corollary 7.1.1 and Theorem 7.1.1 that we arrive at a von Mangoldt surface $(M, g) := (\mathbf{R}^2, dr^2 + m^2(r)d\theta^2)$, where (r, θ) denotes geodesic polar coordinates around the origin for the Euclidean plane $(\mathbf{R}^2, g_0)$. It follows from Lemma 7.1.4 that the parallel $S_p(\pi/2)$ is a geodesic, where p denotes the vertex of M. Let q be any point on $S_p(\pi/2)$. By Theorem 7.3.1, the semicircle of $S_p(\pi/2)$ emanating from q is minimizing. Thus the length of the semicircle is not greater than that of the geodesic segment $\tau_q|_{[0,\pi]}$. Since the length of the semicircle is $\pi m(\pi/2)$, we obtain (7.2.35). By (7.2.30), the Jacobi field $Y_0(t)$ along τ_q is given by

$$Y_0(t) = \left(-\frac{2}{m(a_1)} + 2m(a_1)h(m(a_1)) - 2\int_0^{m(a_1)} th'(t)\,dt \right.$$
$$\left. + 2\int_{a_1}^{\pi/2} \frac{1}{m^2}\,dr + \int_{\pi/2}^{t-\pi/2} \frac{1}{m^2}\,dr\right)\left(\frac{\partial}{\partial\theta}\right)_{\tau_q(t)} \qquad (7.3.30)$$

on $(\pi/2, \infty)$. Since $Y_0(t)$ is independent of a_1,

$$C := -\frac{2}{m(s)} + 2m(s)h(m(s)) - 2\int_0^{m(s)} th'(t)\,dt + 2\int_s^{\pi/2} \frac{1}{m^2}\,dr \quad (7.3.31)$$

is constant on $(0, a_1]$. It follows from integrating by parts that

$$4\int_s^{\pi/2} \frac{m - rm'}{m^3}\,dr = 2\int_s^{\pi/2} \frac{1}{m^2}\,dr + \frac{\pi}{m^2(\pi/2)} - \frac{2s}{m^2(s)} \qquad (7.3.32)$$

Thus by (7.3.31)

$$C = -\frac{2}{m(s)} + 2m(s)h(m(s)) - 2\int_0^{m(s)} th'(t)\,dt + \frac{2s}{m^2(s)}$$

$$-\frac{\pi}{m^2(\pi/2)} + 4\int_s^{\pi/2} \frac{m - rm'}{m^3}\,dr \qquad (7.3.33)$$

holds for any $s \in (0, a_1]$. It follows from l'Hôpital's theorem and Corollary 7.1.1 that

$$\lim_{s \to +0} \left(\frac{s}{m^2(s)} - \frac{1}{m(s)} \right) = 0.$$

Thus from (7.3.33) we get

$$C = \lim_{s \to +0} C = 4\int_0^{\pi/2} \frac{m - rm'}{m^3}\,dr - \frac{\pi}{m^2(\pi/2)}. \qquad (7.3.34)$$

By (7.3.30),

$$Y_0(t) = \left\{ 4\int_0^{\pi/2} \frac{m - rm'}{m^3}\,dr - \frac{\pi}{m^2(\pi/2)} + \int_{\pi/2}^{t-\pi/2} \frac{1}{m^2}\,dr \right\} \left(\frac{\partial}{\partial\theta} \right)_{\tau_q(t)}$$

$$(7.3.35)$$

for any $t \in (\pi/2, \infty)$. Since $\tau_q(\pi)$ is a cut point of q, it follows from Theorem 7.3.1 that there exists a $t_0 \in (\pi/2, \pi]$ such that $Y_0(t_0) = 0$. By (7.3.35),

$$4\int_0^{\pi/2} \frac{m - rm'}{m^3}\,dr - \frac{\pi}{m^2(\pi/2)} = \int_{t_0-\pi/2}^{\pi/2} \frac{1}{m^2}\,dr \geq 0.$$

Therefore we get (7.3.29). Suppose that the equality in (7.3.28) or in (7.3.29) holds. Then $\tau_q|_{[0,\pi]}$ is also minimizing. It follows from Proposition 7.3.1 that the Gaussian curvature of M is constant on the closed ball $\overline{B(p, \pi/2)}$. Therefore, by Corollary 7.1.1, $f(t) = m(t) = \sin t$ on $[0, \pi/2]$. Conversely, if $f(t) = \sin t$ then the Gaussian curvature of M is constant on the closed ball $\overline{B(p, \pi/2)}$. Thus it is trivial that the equalities hold. $\qquad \square$

8

The behavior of geodesics

In this chapter, we shall discuss the behavior of geodesics in Riemannian planes admitting a total curvature. Our main purpose is to study the global topological shapes of complete geodesics sufficiently close to infinity, originally studied in [93] and [95]. In particular, we give some explicit estimates of the Whitney rotation number and, in some cases, even the number of self-intersections of such geodesics in terms of the curvature at infinity (or the total curvature). Also, the existence of such a geodesic is discussed.

8.1 The shape of plane curves

In this section we introduce some preliminary notions needed in order to state the main theorems of this chapter.

Let M be a two-dimensional differentiable manifold without boundary.

Definition 8.1.1 (Transversal immersion). A C^∞-map $\alpha : I \to M$ for an interval $I \subset \mathbf{R}$ is said to be a *weakly transversal immersion* (resp. *transversal immersion*) if it satisfies the following conditions, (1) and (2) (resp. (1), (2) and (3)).

(1) (Immersibility condition) $\dot{\alpha}(t) \neq 0$ for any $t \in I$.
(2) (Source transversality condition) Whenever $\alpha(a) = \alpha(b) =: p$ for $a \neq b$, the tangent vectors $\dot{\alpha}(a)$ and $\dot{\alpha}(b)$ are linearly independent in $T_p M$.
(3) (Target transversality requirement) The mapping α has no triple points, i.e., there exist no different $a, b, c \in I$ such that $\alpha(a) = \alpha(b) = \alpha(c)$.

Note that in the case where α is a geodesic in a Riemannian manifold, α is weakly transversal iff it is nonclosed.

Lemma 8.1.1. *If* $\alpha : I \to M$ *is a proper weakly transversal immersion then the set of crossing points of* α *is a discrete subset of* M. *In particular, if* I *is compact then* α *has at most finitely many double points.*

Proof. Suppose that the set of crossing points of a proper weakly transversal immersion $\alpha : I \to M$ is not discrete. Then there is a sequence of crossing points $p_i = \alpha(a_i) = \beta(b_i)$, $i = 1, 2 \ldots$, converging to a point $p \in M$. Since α is proper, there are subsequences $a_{j(i)}, b_{j(i)}$ of a_i, b_i tending respectively to some numbers $a, b \in I$. Then it follows that $p = \alpha(a) = \beta(b)$. Recall now that for $\alpha : I \to M$ to be differentiable we require that α can be differentiably extended on an open interval containing I. The differentiability of α shows that there exists an $\epsilon > 0$ such that $\alpha|_{(t-a, t+a)}$ has no self-intersection points, so that we have $a \neq b$. Moreover, $\alpha(a_{j(i)}) = \beta(b_{j(i)})$ implies that $\dot{\alpha}(a) = \pm \dot{\beta}(b)$. This contradicts the source transversality condition of α. $\qquad\square$

Assume from now on that M is diffeomorphic to $\mathbf{R}^2$ and that an orientation of M is fixed.

Definition 8.1.2 (Rotation number). Let $\alpha : I \to M$ be a proper weakly transversal immersion of a closed interval $I \subset \mathbf{R}$. A double point $p = \alpha(a) = \alpha(b)$, $a < b$, of α is said to have a *positive sign* (sgn $p := 1$) if the basis $(\dot{\alpha}(a), \dot{\alpha}(b))$ has positive orientation and a *negative sign* (sgn $p := -1$) otherwise. Here, any n-ple point $p = \alpha(a_1) = \cdots = \alpha(a_n)$ such that $a_i \neq a_j$ for $i \neq j$ is interpreted as consisting of $\binom{n}{2}$ different double points $\alpha(a_i) = \alpha(a_j)$ for all $i \neq j$. When I is a compact interval, the *rotation number* rot α *of* α is defined by

$$\text{rot } \alpha := |n_+ - n_-|,$$

where n_+ (resp. n_-) is the number of positive (resp. negative) double points of α. When $I = \mathbf{R}$, the rotation number of α is defined by

$$\text{rot } \alpha := \limsup_{\substack{s \to -\infty \\ t \to +\infty}} \text{rot } \alpha|_{[s,t]} \qquad (\in \{0, 1, \ldots, \infty\}).$$

The *maximal subrotation number* maxrot α of a proper weakly transversal immersion $\alpha : \mathbf{R} \to M$ is defined by

$$\text{maxrot} \, \alpha := \sup\{\text{rot} \, \hat{\alpha}; \, \hat{\alpha} \text{ is a subarc of } \alpha\}.$$

It is clear that rot $\alpha \leq$ maxrot α. Note that rot α is an invariant of the compactly supported regular homotopy class of α. Here, two proper transversal immersions $\alpha, \beta : \mathbf{R} \to M$ are said to be *compact supported regular homotopic* if there exist two numbers $a < b$ such that $\alpha(t) = \beta(t)$ for any $t \in (-\infty, a] \cup [b, +\infty)$ and such that there exists a regular homotopy between $\alpha|_{[a,b]}$ and

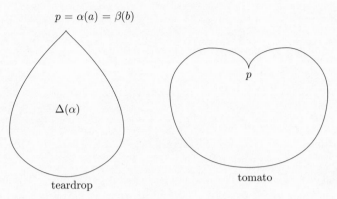

Figure 8.1.1

$\beta|_{[a,b]}$ fixing $\dot{\alpha}(a)$ and $\dot{\beta}(b)$. However, the maximal subrotation number is not an invariant up to a compactly supported regular homotopy.

Definition 8.1.3 (Semi-regular curves). A proper transversal immersion α: $\mathbf{R} \to M$ is called a *semi-regular curve* if there exists a (finite or infinite) sequence of numbers $\cdots < a_2 < a_1 < b_1 < b_2 < \cdots$ such that $\alpha(a_i) = \alpha(b_i)$ for all $i = 1, 2, \ldots$, are the double points of α. A *semi-regular arc* is defined to be a subarc of some semi-regular curve.

In order to describe semi-regular curves we now classify loops and biangles up to diffeomorphisms.

Recall that a *loop* in M is a curve $\alpha \colon [a, b] \to M$ such that $\alpha(a) = \alpha(b) =: p$, where p is called the *base point* of the loop α. A loop $\alpha \colon [a, b] \to M$ is said to be *simple* if $\alpha|_{[a,b)}$ has no crossing point. Since M is diffeomorphic to $\mathbf{R}^2$, any simple loop α bounds a compact disk domain $\Delta(\alpha)$ in M. In the absence of a Riemannian structure on M, we cannot measure the inner angle θ of $\Delta(\alpha)$ at the base point; nevertheless, for any Riemannian metric on M, we can distinguish (independently of the given Riemannian metric) which of the relations $\theta < \pi$, $\theta = \pi$, $\theta > \pi$ holds. The disk domain $\Delta(\alpha)$ is called a *teardrop* if $\theta < \pi$ and a *tomato* if $\theta > \pi$ (see Figure 8.1.1).

A biangle in M is defined to be the union of two curves $\alpha \colon [a, a'] \to M$ and $\beta \colon [b, b'] \to M$ such that $\alpha(a) = \beta(b)$ and $\alpha(a') = \beta(b')$. A biangle $\alpha \cup \beta$ is said to be *simple* if the join of the two curves α and β, which forms a piecewise-closed curve, is simple. Any simple biangle $\alpha \cup \beta$ in M bounds a compact disk domain $\Delta(\alpha \cup \beta)$ in M. When the two inner angles θ_1 and θ_2 of such a domain $\Delta(\alpha \cup \beta)$ are both not equal to π, one of the three following possibilities holds (see Figure 8.1.2).

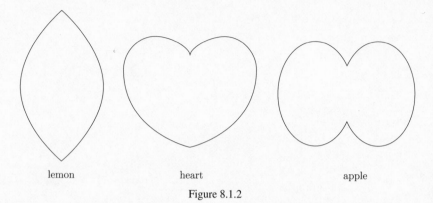

lemon   heart   apple

Figure 8.1.2

(1) $\theta_1, \theta_2 < \pi$, in which case $\Delta(\alpha \cup \beta)$ is called a *lemon*.

(2) $\theta_i < \pi$ and $\theta_{\bar{i}} > \pi$ for some $i = 1, 2$ (where $\bar{i}$ denotes the number in $\{1, 2\}$ different from i), in which case $\Delta(\alpha \cup \beta)$ is called a *heart*.

(3) $\theta_1, \theta_2 > \pi$, in which case $\Delta(\alpha \cup \beta)$ is called an *apple*.

Let α be a nonsimple semi-regular curve and let $p_i = \alpha(a_i) = \alpha(b_i)$, $i = 1, 2, \ldots$, be the double points of α such that $\cdots < a_2 < a_1 < b_1 < b_2 < \cdots$. The subarc $\alpha|_{[a_1,b_1]}$ is a simple loop bounding a disk domain, say B_1, that must be a teardrop because α is proper. The union of the two subarcs $\alpha|_{[a_i,a_{i-1}]} \cup \alpha|_{[b_{i-1},b_i]}$ is a simple biangle bounding a disk domain, say B_i, that satisfies one of the two sets of equivalent conditions:

(1) sgn $p_{i-1} \neq$ sgn $p_i \iff B_i \cap \bigcup_{j=1}^{i-1} B_j = \{p_i\} \iff B_i$ is a lemon.

(2) sgn $p_{i-1} =$ sgn $p_i \iff B_i \cup \{p_i\} \supset \bigcup_{j=1}^{i-1} B_j \iff B_i$ is a heart.

Definition 8.1.4 (Semi-regularity index). The *semi-regularity index* (or simply *index*) ind $\alpha \in \{0, 1, 2, \ldots, \infty\}$ of a semi-regular curve α in M is defined to be the number of disks B_i associated with α that are not lemons (see Figure 8.1.3).

The following proposition is easily proved.

Proposition 8.1.1. *For a semi-regular curve α we have*:

(1) rot $\alpha \leq$ maxrot $\alpha \leq$ ind $\alpha \leq \#\{p_i\}$;

(2) α *is simple iff* ind $\alpha = 0$.

Proposition 8.1.2. *Let α be a semi-regular curve in M and $f : \mathbf{N} \cup \{0\} \to \mathbf{Z}$ a recursive function defined as follows*:

$$f(0) := 0;$$

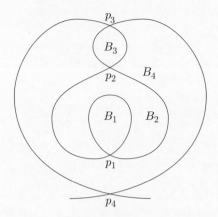

Figure 8.1.3 A semi-regular curve of index 3.

for $i \geq 1$,

$$f(i) := \begin{cases} 1 + f(i-1) & \textit{if } B_i \textit{ is not a lemon,} \\ 1 - f(i-1) & \textit{otherwise;} \end{cases}$$

for $i \geq \#\{p_i\}$,

$$f(i) := f(i-1).$$

Then

$$\operatorname{rot} \alpha = \limsup_{i \to \infty} |f(i)|.$$

Proof. Eliminate the lemons by using compactly supported regular homotopies. $\square$

Definition 8.1.5 (Regular and almost regular curve). A semi-regular curve α is said to be *almost regular* if the largest heart (if any) contains no lemons. A semi-regular curve α is said to be *regular* if it has no lemons.

Define a sequence of integers $k(j)$ for $j = 0, 1, \ldots,$ ind α associated with a semi-regular curve α by the following:

$$k(0) := 0;$$
$$k(1) := 1 \qquad \text{if } \alpha \text{ is nonsimple;}$$
$$k(j+1) := \min\{i \in \mathbf{N}; \ i \geq k(j) + 1 \text{ and } B_i \text{ is a heart}\}.$$

The following are obvious.

Proposition 8.1.3

(1) *If α is a regular curve then* rot α = maxrot α = ind α = #$\{p_i\}$.
   Given that a semi-regular curve α has a finite index, we have:
(2) α *is regular iff* ind α = #$\{p_i\}$;
(3) α *is almost regular iff* ind α = k(ind α);
(4) *if α is almost regular, then*

$$\text{rot } \alpha = \begin{cases} \text{ind } \alpha - 1 & \text{if } \#\{p_i\} - \text{ind } \alpha \text{ is odd,} \\ \text{ind } \alpha & \text{if } \#\{p_i\} - \text{ind } \alpha \text{ is even or infinite.} \end{cases}$$

Proposition 8.1.4. *Let K be a compact disk domain in M, and let α be a semi-regular curve in M outside K such that $K \subset$ int B_1 if α is nonsimple. Then any lift $\tilde{\alpha}$ of α to the universal covering space $\tilde{N}$ of $N := \overline{M \setminus K}$ is a semi-regular curve in the differentiable plane $\tilde{N} \setminus \partial\tilde{N}$. Moreover, if α is almost regular then the lift $\tilde{\alpha}$ is simple.*

Exercise 8.1.1. Prove Propositions 8.1.3 and 8.1.4.

Exercise 8.1.2. A point p in a Riemannian manifold M is called a *pole* if the exponential map $\exp_p : T_pM \to M$ is a local diffeomorphism. In particular, if M is simply connected, this condition is equivalent to the condition that the exponential map $\exp_p : T_pM \to M$ is a diffeomorphism. Prove that in a Riemannian plane admitting a pole any semi-regular complete geodesic becomes regular.

Definition 8.1.6 (Reverse semi-regular arc). A *reverse semi-regular arc* in M is defined to be a transversal immersion α of a compact interval $I \subset \mathbf{R}$ into M satisfying the following condition: there exists a diffeomorphism $\varphi : M \to \mathbf{C}$ such that $0 \notin \varphi \circ \alpha(I)$ and the curve $I \ni t \mapsto 1/\varphi \circ \alpha(t) \in \mathbf{C}$ (where $\mathbf{C}$ is the complex plane) is a semi-regular arc with an associated teardrop containing $0 \in \mathbf{C}$.

8.2 Main theorems and examples

In this section we shall present the main results of this chapter; proofs will be given in the later sections.

Riemannian planes M admitting a total curvature are classified into three classes, contracting Riemannian planes, expanding Riemannian planes and building Riemannian planes (see Figure 8.2.1 and Definition 8.3.1 below).

For example, if $c(M) < 2\pi$ then M is expanding (see Proposition 8.3.2). The *positive* (*resp. negative*) *curvature locus* M^+ (resp. M^-) of M is defined to be the subset of M where the Gaussian curvature is positive (resp. negative).

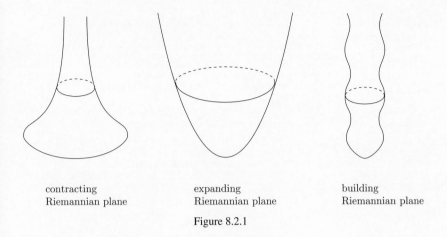

contracting   expanding   building
Riemannian plane   Riemannian plane   Riemannian plane

Figure 8.2.1

In the case $c(M) = 2\pi$, if M^+ is bounded and M^- is unbounded then M is contracting; if M^+ is unbounded and M^- is bounded then M is expanding; if M is flat outside a compact subset then M is building. It holds that M is building iff arbitrarily close enough to infinity there exists a simple closed geodesic (see Proposition 8.3.1 below).

Theorem 8.2.1. *If M is an expanding Riemannian plane, any complete geodesic in M close enough to infinity is semi-regular.*

Under the more strict assumption that $c(M) < 2\pi$, we have the (almost) regularity of a complete geodesic and the control of its index. Let us first see the following simple (but typical) example.

Example 8.2.1

(1) In a flat cone $C := \text{cone } S^1(\theta/2\pi)$, $\theta > 0$, any geodesic γ (not passing through the vertex o) is regular. Let $\tilde{C}$ be the universal covering space of $C \setminus \{o\}$, and let $\tilde{\gamma}$ be a lift of γ (see Figure 8.2.2). There is a ray σ in C from the vertex such that a lift $\tilde{\sigma}$ of σ is parallel to $\tilde{\gamma}$. The index of γ coincides with the number of all lifts of σ intersecting γ, which is equal to $n(\theta) := \max\{k \in \mathbf{Z}; k\theta < \pi\}$.

(2) Let M be a Riemannian plane that contains a compact subset K outside which the Gaussian curvature is zero. Then there is a compact disk domain K' of M containing K such that $M \setminus K'$ is isometrically embedded into the cone over $S^1(\lambda_\infty(M)/2\pi)$. Therefore, any complete geodesic γ in M close enough to infinity is regular and with index ind $\gamma = n(\lambda_\infty(M))$.

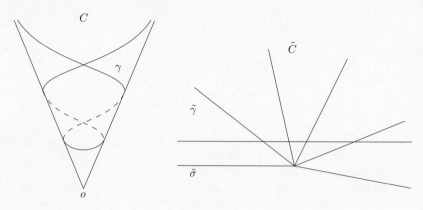

Figure 8.2.2

In a general Riemannian plane M with $c(M) < 2\pi$, all geodesics close enough to infinity almost behave as in the above examples. Indeed, we have:

Theorem 8.2.2. *Let M be a Riemannian plane with $\lambda_\infty(M) > 0$ such that either $\lambda_\infty(M) = +\infty$ or $\pi/\lambda_\infty(M)$ is not an integer. Then any complete geodesic sufficiently close to infinity is regular and its index is $[\pi/\lambda_\infty(M)]$.*

In particular, if $c(M) < \pi$, the above theorem implies that any complete geodesic sufficiently close to infinity is proper and simple.

Theorem 8.2.3. *Let M be a Riemannian plane such that $\pi/\lambda_\infty(M)$ is an integer. For a complete geodesic γ, consider the following three conditions*:

(i) *γ is regular and*

$$\text{ind } \gamma = \pi/\lambda_\infty(M) - 1;$$

(ii) *γ is regular and*

$$\text{ind } \gamma = \begin{cases} \pi/\lambda_\infty(M) & \text{if } c(M) \neq \pi, \\ 0 \text{ or } 1 & \text{if } c(M) = \pi; \end{cases}$$

(iii) *γ is not regular but almost regular (with possibly infinitely many double points) and*

$$\text{ind } \gamma = \pi/\lambda_\infty(M).$$

Then, for a complete geodesic γ sufficiently close to infinity we have:

(1) *if M^- is bounded then all γ satisfy condition (i)*;
(2) *if M^- is unbounded and M^+ is bounded then all γ satisfy condition (ii)*;
(3) *if both M^+ and M^- are unbounded then all γ satisfy (i), (ii) or (iii)*.

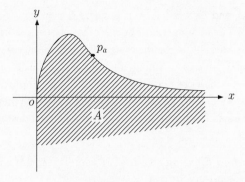

Figure 8.2.3

Remark 8.2.1. As shown by the next example, the condition M^- is unbounded, M^+ is bounded and $c(M) = \pi$ does not determine the index of a complete geodesic γ in M suffciently close to infinity; the index may then satisfy ind $\gamma = 0$ or 1. Nevertheless, if there is a compact subset of M outside which M is negatively curved then any such geodesic γ satisfies ind $\gamma = 1$.

Example 8.2.2. Let $f \colon [0, +\infty) \to \mathbf{R}$ be a unique continuous function such that $f(t) := 1/(t - \ln t)$ for all $t \in (0, +\infty)$, and set $A := \{(x, y) \in \mathbf{R}^2; \ x \geq 0, \ y \leq f(x)\}$; see Figure 8.2.3. The curve $c := \partial A$ is the image of a smooth proper curve defined on $\mathbf{R}$ containing the negative y-axis, asymptotic to the positive x-axis and with only one inflection point, $p_a := (a, f(a)), 1 < a < 2$. Therefore, considered as a function with values in $[0, +\infty)$, the radius of curvature of c is bounded away from 0 by a positive number r_0.

For any $p = (x_p, y_p) \in \mathbf{R}^2$ let

$$B_p := \{(x, y, z) \in \mathbf{R}^3; \ |z| \leq 1/2,$$

$$(x - x_p)^2 + (y - y_p)^2 \leq (\epsilon \ln 4)^2 g(z)^2\},$$

where $0 < \epsilon < r_0$ and where $g \colon [-1/2, 1/2] \to \mathbf{R}$ is a unique continuous function such that $g(t) := (\ln(1/4 - t^2))^{-1}$ for all $t \in (-1/2, 1/2)$. Note that B_p is a smooth revolution ball of axis $\{p\} \times \mathbf{R}$ with north pole $(x_p, y_p, 1/2)$ and south pole $(x_p, y_p, -1/2)$. The boundary $M := \partial B$ of $B := \bigcup_{p \in A} B_p$ is a smooth Riemannian plane embedded in $\mathbf{R}^3$ with $c(M) = \pi$. The surface M is flat on $(A \times \{1/2, -1/2\}) \cup D$, where $D := \{(x, y, z) \in M; \ x \leq 0, \ y \leq 0\}$. Note that the meridian $\mu_p := M \cap B_p$ for any $p \in c$ is a geodesic arc of M and satisfies $\bigcup_{p \in c} \mu_p = M \setminus \text{int} (A \times \{-1/2, 1/2\})$. Decompose $c - \{p_a\}$ into three subarcs, c_0, c_+ and c_-, where c_0 is the closed negative y-axis, c_+ is the open bounded subarc between the origin and p_a and c_- is the open unbounded subarc starting at p_a asymptotic to the positive x-axis. Now, it follows that $D = \bigcup_{p \in c_0} \mu_p$,

$M^+ = \bigcup_{p \in c_+} \mu'_p$ and $M^- = \bigcup_{p \in c_-} \mu'_p$, where $\mu'_p := \mu_p - \{$poles$\}$ and $M^\pm$ is the positive or negative curvature locus of M. Let $\hat{c} := M \cap (\mathbf{R}^2 \times \{0\})$, and let $\hat{c}_*$, with $* = 0, +, -$, be the image of the mapping of c_* into $\hat{c}$ by the normal projection from c to $\hat{c}$. Among all complete geodesics of M, it is easily verified that those which orthogonally intersect $\hat{c}_0$ are simple and those which orthogonally intersect $\hat{c}_-$ are regular and of index unity with their double points on $\hat{c}_0$. Arbitrary close to infinity there exist complete geodesics in M of these two types.

Theorem 8.2.4. *In a Riemannian plane M with $\int_M G_+ \, dM < 2\pi$, any complete geodesic is semi-regular and*

$$\text{ind } \gamma < \frac{\pi}{2\pi - \int_M G_+ \, dM}.$$

Theorem 8.2.5. *Let M be a Riemannian plane with $c(M) = 2\pi$. For any $n \geq 1$ there exists a compact subset K_n of M such that any proper complete geodesic in M outside K_n satisfies*

$$\text{maxrot } \gamma \geq n.$$

The following corollary is a direct consequence of Theorems 8.2.1 and 8.2.5.

Corollary 8.2.1. *Let M be an expanding Riemannian plane with $c(M) = 2\pi$. Then for any $n \geq 1$ there exists a compact subset K_n of M such that any complete geodesic γ in M outside K_n is semi-regular and of index ind $\gamma \geq n$.*

Regarding the existence of a complete geodesic arbitrarily close to infinity, we have:

Theorem 8.2.6

(1) *If M is a contracting Riemannian plane then there exists a compact subset of M having an intersection with any complete geodesic in M.*
(2) *If M is an expanding Riemannian plane then for any subset K of M there exists a complete geodesic in M not intersecting K.*

8.3 The semi-regularity of geodesics

Throughout this section, let M be a Riemannian plane for which a total curvature exists. The purpose of this section is to prove the semi-regularity of geodesics (see Theorem 8.3.3 below).

Lemma 8.3.1. *For any compact subset K of M there exists a compact disk domain C in M containing K and which is either locally convex or locally concave.*

Proof. To prove the lemma, it may be assumed that K is a compact disk domain. Let $\{p_i\}$ be a sequence of points in $M \setminus K$ such that $d(p_i, K)$ tends to $+\infty$ as $i \to \infty$. Among all loops in $\overline{M \setminus K}$ with base point p_i that are not homotopic to zero, we can find a minimal loop c_i for each i. Then each c_i is a simple closed curve bounding a compact disk domain D_i containing K that is locally convex except possibly at p_i. If c_{i_0} for an i_0 does not intersect K then it must be a geodesic loop, and consequently $D_{i_0} =: C$ is either locally convex or locally concave (in fact, D_{i_0} is a teardrop, a tomato or a disk domain bounded by a simple closed geodesic). Assume that every c_i intersects K. Find lifts $\tilde{p}_i$ of p_i for all i in a common fundamental domain $D(\sigma^0, \sigma^1)$ of the universal covering space $\tilde{U}$ of $U := \overline{M \setminus K}$, where σ is a ray in M from K. For each i, there exist different two subarcs α_i and β_i of $pr^{-1}(c_i)$ connecting $\tilde{p}_i$ to $\partial \tilde{U}$, both of which are contained in $D(\sigma^{-1}, \sigma^2)$ because of the minimal property of c_i. Applying Lemma 3.2.2 yields that the inner angle of D_i at p_i tends to zero, so that D_i is locally convex for sufficiently large i. □

Definition 8.3.1 (Classification of Riemannian planes). A *locally convex* (resp. *concave*) *filling of M* is defined to be a family $\mathcal{F}$ of compact locally convex (resp. concave) contractible subsets of M such that for any compact subset K of M there exists a $C \in \mathcal{F}$ with $K \subset C$. By Lemma 8.3.1, Riemannian planes are classified into the following three cases.

(1) (Contracting case) M has a locally concave filling and no locally convex filling, in which case it is said to be *contracting*.
(2) (Expanding case) M has a locally convex filling and no locally concave filling, in which case it is said to be *expanding*.
(3) (Building case) M has both a locally concave filling and a locally convex filling, in which case it is said to be *building*.

Proposition 8.3.1. *The Riemannian plane M is building iff M has a sequence $\{\gamma_i\}$ of simple closed geodesics such that $d(p, \gamma_i) \to +\infty$ as $i \to \infty$ for a fixed point $p \in M$.*

Proof. Assume that M is building. Then there exists a monotone increasing sequence $\{C_i\}$ of locally concave disk domains in M with $\bigcup_i C_i = M$. For each i, we can find a locally concave disk domain $C'_i \supset C_i$. Since each $C'_i \setminus C_i$ is locally convex and homotopy-equivalent to S^1, if we can find a minimal closed curve γ_i in $C_i' \setminus C_i$ not homotopic to zero, we see that it must be a simple closed geodesic. It is clear that $d(p, \gamma_i) \to +\infty$.

Let us prove the converse. Assume that there exists such a sequence $\{\gamma_i\}$ of simple closed geodesics. Each γ_i surrounds a compact disk domain $\Delta(\gamma_i)$. Choosing any compact disk domain K in such a way that $\int_{M \setminus K} G_+ \, dM < 2\pi$, we can exclude the case where $D_i \cap K = \emptyset$, because $c(D_i) = 2\pi$. If i is sufficiently large then we have $\gamma_i \cap K = \emptyset$, so that $\Delta(\gamma_i)$ must contain K. The arbitrariness of K shows that $\mathcal{F} := \{\Delta(\gamma_i)\}$ is a locally convex and locally concave filling. $\qquad\square$

For a compact subset K of M let

$$\delta(K) := 2\pi - \int_{M \setminus K} G_+ \, dM - \max\{0, c(K)\}$$

$$= \min\left\{2\pi - \int_{M \setminus K} G_+ \, dM, \ \ 2\pi - \int_M G_+ \, dM + \int_K G_- \, dM\right\}.$$

It follows that

$$\delta(K) = \min\left\{2\pi - \int_{M \setminus K} G_+ \, dM, \ \ \lambda_\infty(M) - \int_{M \setminus K} G_- \, dM\right\}$$

provided $c(M) > -\infty$.

We observe the following easy lemma.

Lemma 8.3.2. *Let K and K' be two compact subsets of M.*

(1) *If $K \subset K'$ then $\delta(K) \leq \delta(K')$.*
(2) *If either $K \subset K'$ or $K \cap K' = \emptyset$ holds then $2\pi - c(K') \geq \delta(K)$.*
(3) *As K tends to M, $\delta(K)$ tends to $\lambda_\infty(M)$.*
(4) *$\delta(\emptyset) = 2\pi - \int_M G_+ \, dM$.*

Proposition 8.3.2. *If $c(M) < 2\pi$ then M is expanding.*

Proof. Since $\lambda_\infty(M) > 0$, Lemma 8.3.2(3) implies that there exists a contractable compact subset K of M such that $\delta(K) > 0$. By Lemma 8.3.2(2) and the Gauss–Bonnet theorem, any compact contractable subset K' of M containing K is not locally concave, which implies that M is expanding. $\qquad\square$

Let A be a Riemannian manifold diffeomorphic to the annulus $S^1 \times [0, 1]$, and let A_n, A_w be the two connected components of ∂A. For an arc α in A, denote by $\Delta(\alpha)$ the union of α and of all connected components of $A \setminus \alpha$ not intersecting A_w. Let $\iota_* : (A, A_*) \to (D^2, \partial D^2)$, with $* = n, w$, be an embedding; D^2 is the two-dimensional unit disk.

Lemma 8.3.3. *Let γ be a geodesic arc in* int *A such that*

(1) *if γ is nonsimple, it has a subarc that is a simple geodesic loop not homotopic to zero;*

(2) *for any subarc $\hat{\gamma}$ of γ, the region $\Delta(\hat{\gamma})$ is not locally concave.*

Then, $\iota_w \circ \gamma$ and $\iota_n \circ \gamma$ are respectively semi-regular and reverse semi-regular arcs in D^2.

Proof. Let $\gamma : [a, b] \to A$ be a geodesic arc. The lemma is trivial if γ is simple, so we will assume that γ is nonsimple. Then, we can find a subarc $\gamma|_{[a_1, b_1]}$ that is a simple geodesic loop. It follows from condition (2) that $\angle_{p_1} \Delta(\gamma|_{[a_1, b_1]}) < \pi$, where $p_1 := \gamma(a_1) = \gamma(b_1)$, and hence $\iota_w \circ \gamma|_{[a_1, b_1]}$ is semi-regular. Assume now that a subarc $\iota_w \circ \gamma|_{[a_k, b_k]}$ is semi-regular and that $\gamma(a_i) = \gamma(b_i) =: p_i$ for $i = 1, \ldots, k$ are the double points of $\gamma|_{[a_k, b_k]}$, where $a_k < \cdots < a_1 < b_1 < \cdots < b_k$. Since γ has at most a finite number of crossing points, it suffices to prove that if there is a self-intersection point of γ other than $p_1, \ldots, p_k$ then we can find a double point $p_{k+1} = \gamma(a_{k+1}) = \gamma(b_{k+1})$ such that $a_{k+1} < a_k$, $b_k < b_{k+1}$ and such that $\iota_w \circ \gamma|_{[a_{k+1}, b_{k+1}]}$ is semi-regular. Let us prove the following:

Sublemma 8.3.1. *The subarc $\gamma|_{(b_k, b]}$ does not intersect $\gamma|_{[a_k, b_k]}$ and has no self-intersection points.*

Proof. Suppose the contrary and let

$$t_0 := \inf \{t \in (b_k, b]; \ \gamma(t) \in \gamma[a_k, t)\};$$

see Figure 8.3.1. We will show that $D_0 := \Delta(\gamma|_{[a_k, t_0]})$ is locally concave, which is contrary to condition (2). Note in general that for a geodesic arc σ in A, $\Delta(\sigma)$ is locally concave iff for any $p \in \partial\Delta(\sigma)$ there exist $v \in S_p M$ and $\epsilon > 0$ such that $\exp_p tv \in \Delta(\sigma)$ for any $t \in \mathbf{R}$ with $|t| < \epsilon$. Since $\Delta(\gamma|_{[a_k, b_k]})$ is locally concave except at p_k, it suffices to verify that D_0 is locally concave at p_k and $\gamma(t_0)$. Since $\gamma(b_k - \epsilon, b_k + \epsilon) \subset D_0$ for a sufficiently small $\epsilon > 0$, the region D_0 is locally concave at p_k. Moreover, we can find a number $s_0 \in (a_k, t_0)$ such that $\gamma(s_0) = \gamma(t_0)$. Then we have $\gamma(s_0 - \epsilon, s_0 + \epsilon) \subset D_0$ for a sufficiently small $\epsilon > 0$. Thus D_0 is locally concave, which leads to a contradiction. $\qquad\square$

As well as Sublemma 8.3.1, we have that $\gamma|_{[a, a_k)}$ does not intersect $\gamma|_{[a_k, b_k]}$ and has no self-intersection points. Suppose that there exists a self-intersection point of γ other than $p_1, \ldots, p_k$. Then $\gamma[a, a_k) \cap \gamma(b_k, b]$ is nonempty, and hence there exist $a_{k+1} \in [a, a_k)$ and $b_{k+1} \in (b_k, b]$ such that $\gamma(a_{k+1}) = \gamma(b_{k+1})$ and $\gamma(a_{k+1}, a_k) \cap \gamma(b_k, b_{k+1}) = \emptyset$. Since, by condition (2), $\Delta(\gamma|_{[a_{k+1}, b_{k+1}]})$ is

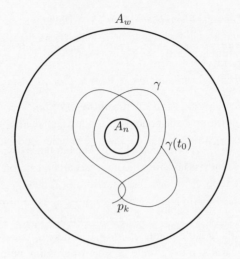

Figure 8.3.1

not locally concave, it must be convex at p_{k+1}. Therefore $\iota_w \circ \gamma|_{[a_{k+1}, b_{k+1}]}$ is a semi-regular arc, which completes the proof of Lemma 8.3.3. □

Theorem 8.3.1 (Semi-regularity of geodesic arcs). *For a Riemannian plane M we have the following.*

(1) *If M is expanding, any geodesic arc in M close enough to infinity is semi-regular.*
(2) *If M is contracting, any geodesic arc in M close enough to infinity is reverse semi-regular.*
(3) *If $\int_M G_+ \, dM \leq 2\pi$, any nonclosed geodesic arc is semi-regular.*

Proof. (1), (2): Let M be a Riemannian plane that is either expanding or contracting. Then there exists a compact disk domain K with smooth boundary such that

(i) $\int_{M \setminus K} G_+ \, dM \leq \pi$;
(ii) if M is expanding (resp. contracting), there are no contractible compact locally concave (resp. convex) subsets containing K.

Let γ be a geodesic arc in M outside K and find a compact disk domain K' in M, with smooth boundary, such that $\gamma \cup K \subset \text{int } K'$. Set $A := \overline{K' \setminus K}$ and

$$A_n := \partial K, \quad A_w := \partial K' \quad \text{if } M \text{ is expanding,}$$
$$A_n := \partial K', \quad A_w := \partial K \quad \text{if } M \text{ is contracting.}$$

To verify condition (1) of Lemma 8.3.3, we suppose that there is a subarc of γ that is a simple geodesic loop homotopic to zero; then it bounds a disk domain whose total curvature is $> \pi$ by the Gauss–Bonnet theorem, which contradicts (i). Condition (2) of Lemma 8.3.3 is implied by (ii). Note that the definition of $\Delta(\hat{\gamma})$ in Lemma 8.3.3 depends on which side of the components of ∂K A_n lies. The lemma leads to (1) and (2) of the theorem.

(3): Let γ be a nonsimple and nonclosed geodesic arc in M with $\int_M G_+ \, dM \leq 2\pi$. Then there exists a subarc $\gamma|_{[a_1,b_1]}$ that is a simple geodesic loop. Find two compact disk domains K and K' in M with smooth boundary in such a way that $K \subset \text{int} \, \Delta(\gamma|_{[a_1,b_1]})$ and $\gamma \cup K \subset \text{int} \, K'$. We apply Lemma 8.3.3 to $A := \overline{K' \setminus K}$, $A_n := \partial K$, $A_w := \partial K'$ and γ, where condition (1) of Lemma 8.3.3 is trivial, and, with regard to condition (2), the Gauss–Bonnet theorem and $\int_M G_+ \, dM \leq 2\pi$ together show that any locally concave domain must be bounded by a simple closed geodesic; this implies condition (2), on recalling that γ is nonsimple. $\qquad\square$

Remark 8.3.1. Let γ be a geodesic in M such that any subarc of γ is semi-regular. Then, we cannot conclude that γ is semi-regular if γ is improper, and Theorem 8.3.1 does not directly imply Theorem 8.2.1. Nevertheless there exists a sequence $\cdots < a_2 < a_1 < b_1 < b_2 \cdots$ such that $\gamma(a_i) = \gamma(b_i), i = 1, 2, \ldots,$ are the double points of γ.

Proposition 8.3.3. *For a Riemannian plane M we have the following.*

(1) *If $\int_M G_+ \, dM \leq 2\pi$, any closed geodesic is simple.*
(2) *If $\int_M G_+ \, dM < 2\pi$, there exist no closed geodesics in M.*

Proof. Apply the Gauss–Bonnet theorem to the region $\Delta(\gamma)$ for any closed geodesic γ. $\qquad\square$

Exercise 8.3.1. For a contractible compact subset K of M, prove the following.

(1) If $\delta(K) \geq 0$ then
   (i) any closed geodesic in M outside K is simple;
   (ii) any nonclosed geodesic arc γ in M outside K is semi-regular.
(2) If $\delta(K) > 0$, there exist no closed geodesics in M outside K.

Lemma 8.3.4

(1) *Let K be a compact disk domain in M such that*
   $K^\rho := \{x \in K; d(x, \partial K) \geq \rho\}$ *for some $\rho > 0$ is contractible and*
   $\delta(K^\rho) > 0$. *Then any half-geodesic in M outside K is proper.*
(2) *If $\int_M G_+ \, dM < 2\pi$, any half-geodesic in M is proper.*

Note that the disk domain K as defined in (1) necessarily exists if $c(M) < 2\pi$.

Proof. Let K be the empty set if $\int_M G_+ \, dM < 2\pi$ and let it be as defined in (1) otherwise. We set $\rho := +\infty$ if $K = \emptyset$. In either case we have $\delta(K) > 0$. In order to prove (1) and (2), it suffices to show that any half-geodesic γ with $\gamma \cap K = \emptyset$ is proper. Suppose there exists an improper half-geodesic $\gamma : [0, +\infty) \to M$ such that $\gamma \cap K = \emptyset$. Then there is a sequence $t_i \nearrow +\infty$ such that $\dot{\gamma}(t_i)$ tends to some nonzero vector $v \in T_p M$, $p \in \overline{M \setminus K}$. For $r := \min\{\rho, \text{conv } p\}/2$, there exists an i_0 such that $\gamma(t_i) \in B(p, r)$ and $t_i + 4r < t_{i+1}$ for any $i \geq i_0$. For each $i \geq i_0$, we can find a unique geodesic segment $\sigma_i : [0, s_i] \to B(p, 2r)$ from $\gamma(t_{i+1})$ to $\gamma(t_i + r)$ whose image is different from that of $\gamma_i := \gamma|_{[t_i+r, t_{i+1}]}$. Each $\Delta(\gamma_i \cup \sigma_i)$ has at most two positive exterior angles, possibly at $\gamma(t_i + r)$ and $\gamma(t_{i+1})$, both tending to zero as $i \to +\infty$. Hence, by the Gauss–Bonnet theorem,

$$\liminf_{i \to \infty} c(\Delta(\gamma_i \cup \sigma_i)) \geq 2\pi.$$

No geodesic σ_i, for $i \geq i_0$, intersects K^ρ and neither does $\gamma_i \cup \sigma_i$, where we agree that $K^\rho = \emptyset$ if $K = \emptyset$. Since $\Delta(\gamma_i \cup \sigma_i)$ either contains K or does not intersect K, Lemma 8.3.2(2) implies that

$$c(\Delta(\gamma_i \cup \sigma_i)) \leq 2\pi - \delta(K^\rho) \qquad \text{for any } i \geq i_0,$$

which is a contradiction. □

The following simple lemma will be used frequently later in the chapter.

Lemma 8.3.5. *If $K \subset D \subset M$ then*

$$c(M) - \int_{M \setminus K} G_+ \, dM \leq c(D) \leq c(M) + \int_{M \setminus K} G_- \, dM.$$

Proof. The lemma follows from

$$c(D) = c(M) - c(M \setminus D),$$

$$-\int_{M \setminus K} G_- \, dM \leq c(M \setminus D) \leq \int_{M \setminus K} G_+ \, dM. \qquad □$$

Lemma 8.3.6. *Assume that a contractible subset K of M satisfies $\int_{M \setminus K} G_+ \, dM$ $< c(M) - \pi$. Then M contains no simple proper complete geodesic outside K.*

Note that such a K exists if $c(M) > \pi$.

Proof. Suppose there exists a simple proper complete geodesic in M outside K. Such a geodesic will split M into two Riemannian half-planes, one of which, say H, contains K. The Cohn-Vossen theorem implies that $c(H) \leq \pi$. On the

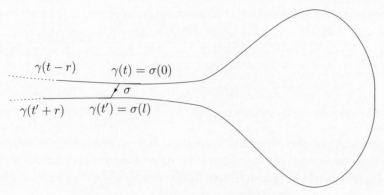

Figure 8.3.2

contrary, however, by Lemma 8.3.5

$$c(H) \geq c(M) - \int_{M \setminus K} G_+ \, dM > \pi.$$ □

Lemma 8.3.7. *Assume that $c(M) = 2\pi$ and that there exists a compact disk domain K in M such that $\overline{M \setminus K}$ contains no simple closed geodesic, K^ρ for some $\rho > 0$ is contractible and $\int_{M \setminus K^\rho} G_+ \, dM < \pi$. Then any simple half-geodesic outside K is proper.*

Note that if M is either contracting or expanding then such a K exists.

Proof. For a compact subset C of M, a number $0 < \delta < \min\{\rho, r\}$ with $r := \operatorname{conv} C$ and a simple geodesic arc $\gamma : [t - r, \ t' + r] \to M$ outside K, $t < t'$, we consider the following condition:

(i) $\gamma(t), \gamma(t') \in C, \qquad t' - t > r, \qquad d(\gamma(t), \gamma(t')) < \delta.$

Under condition (i), for a unique minimal geodesic $\sigma : [0, \ell] \to M$ from $\gamma(t)$ to $\gamma(t')$ we describe the second condition:

(ii) The two bases $(\dot{\gamma}(t), \dot{\sigma}(0))$ and $(\dot{\gamma}(t'), \dot{\sigma}(\ell))$ have orientations opposite to each other (see Figure 8.3.2).

Sublemma 8.3.2. *For any compact subset C of M there exists a number $0 < \delta < \min\{\rho, r\}$ such that no simple geodesic arc $\gamma : [t - r, \ t' + r] \to M$ outside K satisfies conditions (i) and (ii) simultaneously.*

Proof. Suppose the contrary, that there exist a compact subset C of M and simple geodesic arcs $\gamma_\delta : [t_\delta - r, \ t'_\delta + r] \to M$ outside K for all $\delta \in (0, \min\{\rho, r\})$

that satisfy conditions (i) and (ii). Taking a subarc of γ_δ for each δ if necessary, we may assume that γ_δ intersects σ_δ only at $\gamma_\delta(t_\delta) = \sigma_\delta(0)$ and $\gamma_\delta(t'_\delta) = \sigma_\delta(\ell)$, where σ_δ is as in condition (ii), so that $\gamma_\delta|_{[t_\delta, t'_\delta]} \cup \sigma_\delta$ forms a simple closed curve. Since γ_δ is simple, an easy discussion using the triangle comparison theorem (Theorem 1.7.4) implies that

$$d(\gamma_\delta(t_\delta - s), \ \gamma_\delta(t'_\delta + s)) < \omega(\delta) \qquad \text{for any } s \in [-r/2, \ r/2],$$

where $\omega(\delta)$ is a function tending to zero as $\delta \to 0$ and depending only on r and the supremum on C of the absolute value of the Gaussian curvature of M. By using the triangle comparison theorem (Theorem 1.7.4) together with condition (ii), it can be proved easily that

$$|\angle(\dot{\gamma}_\delta(t_\delta), \ \dot{\sigma}_\delta(0)) - \angle(-\dot{\gamma}_\delta(t'_\delta), \ \dot{\sigma}_\delta(\ell))| < \omega(\delta).$$

Denote by D_δ the disk domain bounded by $\gamma|_{[t_\delta, t'_\delta]} \cup \sigma$. The above inequality and the Gauss–Bonnet theorem imply that

$$|c(D_\delta) - \pi| < \omega(\delta).$$

Note that $\partial D_\delta = \gamma_\delta|_{[t_\delta, t'_\delta]} \cup \sigma_\delta$ possibly intersects K but does not intersect K^ρ. If D_δ contains K^ρ, we have $2\pi - c(D_\delta) = c(M \setminus D_\delta) \leq \int_{M \setminus K^\rho} G_+ \, dM$, which together with the above inequality implies that $0 < \pi - \int_{M \setminus K^\rho} G_+ \, dM < \omega(\delta)$; this is contrary to the arbitrariness of δ. Hence D_δ is contained in $\overline{M \setminus K^\rho}$ and so $c(D_\delta) \leq \int_{M \setminus K^\rho} G_+ \, dM$, which leads to a contradiction. $\square$

To prove Lemma 8.3.7, we suppose that there exists an improper and simple half-geodesic γ outside K. Then there exists a sequence $a_i \to +\infty$ of numbers such that $\dot{\gamma}(a_i)$ tends to a vector $v \in T_p M$, $p \in \overline{M \setminus K}$. Let $\{C_i\}_{i=1,2,\ldots}$ be a monotone increasing sequence of compact subsets of M such that $\bigcup_i C_i = M$ and $p \in \text{int } C_1$, and let δ_i be associated with C_i as in Sublemma 8.3.2. For a vector $w \in T_p M$ perpendicular to v we define a geodesic $\sigma : (-\delta/2, \ \delta/2) \to M$ by $\sigma(t) := \exp_p tw$, $t \in (-\delta/2, \ \delta/2)$, where δ is a small number such that $\delta \leq \delta_1$ and $\sigma \subset C_1$. Let s, s', t and t' be any four numbers such that $s \neq s'$, $t \neq t'$, $\sigma(s) = \gamma(t)$ and $\sigma(s') = \gamma(t')$ and such that $\sigma_{s,s'} := \sigma|_{(\min\{s,s'\}, \ \max\{s,s'\})}$ does not intersect $\gamma_{t,t'} := \gamma|_{(\min\{t,t'\}, \ \max\{t,t'\})}$ (see Figure 8.3.3). Sublemma 8.3.2 shows that both $(\dot{\sigma}(s), \dot{\gamma}(t))$ and $(\dot{\sigma}(s'), \dot{\gamma}(t'))$ induce the same orientation of M and that $\gamma - \gamma_{t,t'}$ does not intersect $\sigma_{s,s'}$. Therefore, the intersection points of σ and γ are expressed as $\sigma(s_j) = \gamma(t_j)$, $j = 1, 2, \ldots$, where $s_j \to 0$ and $t_j \nearrow +\infty$ are monotone sequences. By reversing the parameter of σ if necessary, it may be assumed that $\{s_j\}$ is monotone increasing. We obtain:

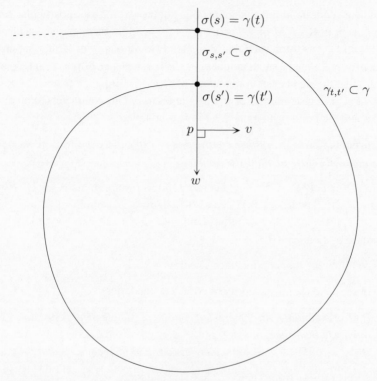

Figure 8.3.3

Sublemma 8.3.3. *For any i, j, t and t' such that $t_j \leq t < t' \leq t_{j+1}$, $d(\gamma(t), \gamma(t')) < \delta_i$ and $\gamma(t), \gamma(t') \in C_i$, we have*

$$\min\{t' - t, \quad t - t_j + t_{j+1} - t' - (s_{j+1} - s_j)\} < \delta_i.$$

Proof. Observe that $\sigma|_{[s_j, s_{j+1}]} \cup \gamma|_{[t_j, t_{j+1}]}$ forms a simple closed curve with two break points and apply Sublemma 8.3.2 to $\gamma|_{[t_j, t_{j+1}]}$. $\square$

We denote by D_j the disk domain bounded by $\sigma|_{[s_j, s_{j+1}]} \cup \gamma|_{[t_j, t_{j+1}]}$. Note that $\{D_j\}$ is a monotone sequence. If γ is bounded, then by Sublemma 8.3.3 the limit set $D_\infty := \lim_j D_j$ is a disk domain and its boundary is the simple closed geodesic $\gamma_\infty := \lim_j \gamma|_{[t_j, t_{j+1}]}$, which does not intersect int K. The existence of the closed geodesic γ_∞ contradicts the assumption, so that γ is unbounded. Thus, $\{D_j\}$ is monotone increasing and the limit set $D_\infty = \bigcup_j D_j$ is bounded by the limit $\gamma_\infty = \lim_j \gamma|_{[t_j, t_{j+1}]}$, which consists of disjoint simple complete geodesics not intersecting int K. Now, by Lemma 8.3.6, any simple complete geodesic in M outside int K is not proper, so that in particular each component $\tau : \mathbf{R} \to M$ of γ_∞ is not proper. Therefore there exists a sequence $\{b_k\}$ with $|b_k| \to +\infty$ such that $\tau(b_k)$ tends to a point $q \in M$. Taking an i_0 with $q \in$ int C_{i_0} and a small number $r \in (0, \delta_{i_0}]$ such that $B(q, r) \subset$ int C_{i_0}, we obtain that the number of components of $B(q, r) \cap \gamma|_{[t_j, t_{j+1}]}$ tends to $+\infty$ as $j \to +\infty$, because a subarc of $\gamma|_{[t_j, t_{j+1}]}$ tends to τ. This contradicts Sublemma 8.3.3, and the proof of Lemma 8.3.7 is now complete. $\qquad\square$

Theorem 8.3.2. *If M is either contracting or expanding, then any geodesic in M sufficiently close to infinity is proper.*

Proof. When $c(M) < 2\pi$, the theorem follows from Lemma 8.3.4(1). When $c(M) = 2\pi$, Theorem 8.3.1 (see also Remark 8.3.1) implies that any half-geodesic sufficiently close to infinity contains a simple sub-half-geodesic, which is proper by Lemma 8.3.7. $\qquad\square$

The following theorem includes Theorem 8.2.1.

Theorem 8.3.3 (Semi-regularity of complete geodesics).

(1) *If M is expanding, any complete geodesic in M sufficiently close to infinity is semi-regular.*

(2) *If $\int_M G_+ \, dM < 2\pi$, any complete geodesic in M is semi-regular.*

Proof. (1) follows from Theorems 8.3.1(1) and 8.3.2.

(2) follows from Theorem 8.3.1(3) and Lemma 8.3.4(2) (see also the proof of Theorem 8.3.2). $\qquad\square$

8.4 Almost-regularity of geodesics; estimate of index

Let $\gamma : \mathbf{R} \to M$ be a semi-regular geodesic in a Riemannian plane M. With the notation of Section 8.1, let $\{B_i\}$ be the sequence of disk domains in M associated with γ, and set $D_j := B_{k(j)} \cup \cdots \cup B_{k(j+1)-1}$ for $j = 1, \ldots,$ ind $\alpha - 1$. Moreover, when ind $\alpha < \infty$, set $D_{\text{ind }\alpha} := B_{k(\text{ind }\alpha)}$. Set $\theta_0 := \pi$, $\theta_j := \angle_{p_{k(j+1)-1}} B_{k(j+1)-1}$ and $\varphi_j := \angle_{p_{k(j)}} B_{k(j)}$, for $j = 1, \ldots,$ ind $\gamma - 1$. When ind $\gamma < \infty$, set also $\varphi_{\text{ind }\gamma} := \theta_{\text{ind }\gamma} := \angle_{p_{k(\text{ind }\gamma)}} D_{\text{ind }\gamma}$.

Lemma 8.4.1. *If there exists a constant $c < 2\pi$ such that $c(D_j) \le c$ for each $j = 1, \ldots,$ ind γ then*

$$\text{ind } \gamma < \frac{\pi}{2\pi - c}.$$

Proof. Applying the Gauss–Bonnet theorem to each D_j yields

$$2\pi + \theta_j - \theta_{j-1} \leq c(D_j) \leq c.$$

Adding up this for $j = 1, \ldots,$ ind γ completes the proof. $\qquad\square$

Proof of Theorem 8.2.4. By Theorem 8.3.3(2), any complete geodesic γ in M is semi-regular. For $c := \int_M G_+ \, dM$ we apply Lemma 8.4.1 and obtain an estimate of the index of γ. $\qquad\square$

Lemma 8.4.2. *Let γ be a semi-regular geodesic for which $k(j) = j$ for some $j \geq 1$. If there exists a constant c' such that $c(B_i) \geq c'$ for each $i = 1, \ldots, j$ then*

$$\varphi_j \geq \pi - j(2\pi - c'),$$

and if in addition $c(B_i) > c'$ for some $i = 1, \ldots, j$ then

$$\varphi_j > \pi - j(2\pi - c').$$

Proof. The Gauss–Bonnet theorem implies that

$$2\pi + \varphi_i - \varphi_{i-1} = c(B_i) \qquad \text{for each } i = 1, \ldots, j,$$

which leads to the lemma. $\qquad\square$

Let us here demonstrate a method of proving the regularity of a geodesic and of estimating its index (see Theorems 8.2.2 and 8.2.3). Assume that $-\infty < c(M) < 2\pi$. Choose a sufficiently large compact disk domain K in M. Let γ be a complete geodesic in M outside K, which is semi-regular by Theorem 8.3.3, and let j_0 be the maximal number of j-values for which $k(j) = j$. Then, since each D_i contains K, each $c(D_i)$ is nearly equal to $c(M)$. Assume for simplicity that $c(D_i) = c(M)$ for any $i = 1, \ldots, j_0$. By Lemma 8.4.1, we have an upper estimate of the index of γ. If $a := \pi - j_0(2\pi - c(M)) > 0$ then Lemma 8.4.2 implies that $\varphi_{j_0} \geq a$, and we can assume that $\int_{M \setminus K} G_+ \, dM < a$, so that if B_{j_0+1} exists and is a lemon then $\int_{M \setminus K} G_+ \, dM \geq c(B_{j_0+1}) \geq \varphi_{j_0} \geq a$, which is a contradiction. In this way we obtain the almost-regularity of γ. Finally, we will prove that $\varphi_{\text{ind } \gamma}$ is so small that the inequality $\varphi_{\text{ind } \gamma} \geq \pi - \text{ind } \gamma \, (2\pi - c(M))$ implies the lower estimate of the index.

Lemma 8.4.3. *Let K be a compact contractible subset of M such that $\delta(K) > \pi$. Then any geodesic outside K is simple.*

Note that, by Lemma 8.3.2(3), if $c(M) < \pi$ then a sufficiently large compact subset K of M satisfies $\delta(K) > \pi$.

Proof. Suppose that a nonsimple geodesic outside K exists. Then it has a subarc that is a simple geodesic loop. The disk domain bounded by this geodesic loop, say D, satisfies $c(D) > \pi$ by the Gauss–Bonnet theorem. If $D \cap K = \emptyset$ then $\pi < c(D) \leq \int_{M \setminus K} G_+ \, dM \leq 2\pi - \delta(K)$, contrary to $\delta(K) > \pi$. If $D \supset K$ then $\pi \geq \pi - \int_{M \setminus D} G_+ \, dM = \delta(D) \geq \delta(K)$. $\qquad\square$

For $\theta > 0$, we set

$$n(\theta) := \max\{k \in \mathbf{Z}; \, k\theta < \pi\}, \qquad n'(\theta) := [\pi/\theta],$$

$$\epsilon_+(\theta) := \frac{\pi - n(\theta)\theta}{n'(\theta) + 1}, \qquad \epsilon_-(\theta) := \frac{(n'(\theta) + 1)\theta - \pi}{n'(\theta) + 1}.$$

It follows that

$$n'(\theta) - n(\theta) = \begin{cases} 0 & \text{if } \pi/\theta \notin \mathbf{Z}, \\ 1 & \text{if } \pi/\theta \in \mathbf{Z}, \end{cases}$$

$$\pi - n(\theta) = \min\{\pi - k\theta; \, k \in \mathbf{Z}, \, 0 \leq k\theta < \pi\},$$

$$(n'(\theta) + 1)\theta - \pi = \min\{k\theta - \pi; \, k \in \mathbf{Z}, \, k\theta > \pi\}$$

and in particular that $\epsilon_\pm(\theta)$ are positive for any $\theta > 0$.

From now on, assume that $-\infty < c(M) < 2\pi$ and that a compact contractible subset K of M satisfies

$$\int_{M \setminus K} G_+ \, dM < \epsilon_+(\lambda_\infty(M)), \qquad \int_{M \setminus K} G_- \, dM < \epsilon_-(\lambda_\infty(M)).$$

Note that such a K exists and satisfies $\delta(K) > 0$. For simplicity we set

$$\epsilon_\pm := \epsilon_\pm(\lambda_\infty(M)), \qquad n := n(\lambda_\infty(M)), \qquad n' := n'(\lambda_\infty(M)).$$

Proposition 8.4.1 (Upper estimate of the index). *For any semi-regular complete geodesic γ in M outside K, the index satisfies*

$$\operatorname{ind} \gamma \leq n',$$

and if $M^- \subset K$ then

$$\operatorname{ind} \gamma \leq n.$$

Proof. Let $c := c(M) + \int_{M \setminus K} G_- \, dM$. It follows that $0 < \delta(K) \leq 2\pi - c$ and hence $c < 2\pi$. By Lemma 8.3.5, we have $c(D_j) \leq c$ for any $j = 1, \ldots, \operatorname{ind} \gamma$. Applying Lemma 8.4.1 yields

$$\operatorname{ind} \gamma < \frac{\pi}{\lambda_\infty(M) - \int_{M \setminus K} G_- \, dM} < \frac{\pi}{\lambda_\infty(M) - \epsilon_-} = n' + 1.$$

If $M^- \subset K$, since $\int_{M\backslash K} G_- \, dM = 0$ we have ind $\gamma < \pi/\lambda_\infty(M)$, which means that ind $\gamma \leq n$. $\qquad\square$

Lemma 8.4.4 (Almost-regularity of complete geodesics). *If γ is a complete geodesic in M outside K that is semi-regular but not regular then $\pi/\lambda_\infty(M)$ is an integer ≥ 1 and γ is almost regular with index* ind $\gamma = n' = n + 1$.

Proof. Let j_0 be the smallest integer j such that the disk domain B_{j+1} associated with γ is a lemon, in other words, the greatest integer j such that $k(j) = j$. Let $c' := c(M) - \int_{M\backslash K} G_+ \, dM$. Since $c(D_j) \geq c'$ for any $j = 1, \ldots, j_0$ (see Lemma 8.3.5), applying Lemma 8.4.2 yields

$$\varphi_{j_0} \geq \pi - j_0 \left(\lambda_\infty(M) + \int_{M\backslash K} G_+ \, dM \right).$$

By the Gauss–Bonnet theorem, we have $\varphi_{j_0} < c(B_{j_0+1}) \leq \int_{M\backslash K} G_+ \, dM$, which together with the above implies that

$$\pi - j_0 \lambda_\infty(M) \leq (j_0 + 1) \int_{M\backslash K} G_+ \, dM < (j_0 + 1)\epsilon_+.$$

Proposition 8.4.1 implies that $j_0 \leq$ ind $\gamma \leq n'$ and hence that

$$(j_0 + 1)\epsilon_+ \leq (n' + 1)\epsilon_+ = \pi - n\lambda_\infty(M).$$

Therefore, we obtain

$$n + 1 \leq j_0 \leq n',$$

which completes the proof. $\qquad\square$

Proposition 8.4.2. *If $M^+ \subset K$, any semi-regular geodesic outside K is regular.*

Proof. The Gauss–Bonnet theorem proves the nonexistence of associated lemons. $\qquad\square$

Proposition 8.4.3 (Lower estimate of the index). *If γ is a regular complete geodesic in M outside K then*

$$\text{ind } \gamma \geq n.$$

Proof. When $n = 0$, the lemma is trivial. Assume that $n \geq 1$ or equivalently $c(M) > \pi$. Since $\int_{M\backslash K} G_+ \, dM < \epsilon_+ \leq \pi - \lambda_\infty(M) = c(M) - \pi$, Lemma 8.3.6 implies that γ is nonsimple. With the notation of Section 8.1, let $D_{\text{ind } \gamma+1}$ denote the Riemannian half-plane bounded by $\gamma(-\infty, a_{\text{ind } \gamma}] \cup \gamma[b_{\text{ind } \gamma}, +\infty)$ and containing $B_{\text{ind } \gamma} = D_{\text{ind } \gamma}$ (where ind γ is finite, by Proposition 8.4.1). Note that $D_{\text{ind } \gamma+1}$ has a unique inner angle equal to $2\pi - \theta_{\text{ind } \gamma}$. It follows from

the Cohn-Vossen theorem that $c(D_{\text{ind } \gamma+1}) \leq 2\pi - \theta_{\text{ind } \gamma}$, which together with
$c(D_{\text{ind } \gamma+1}) \geq c(M) - \int_{M \setminus K} G_+ \, dM$ (see Lemma 8.3.5) implies that

$$\theta_{\text{ind } \gamma} \leq \lambda_\infty(M) + \int_{M \setminus K} G_+ \, dM. \tag{8.4.1}$$

However, applying Lemma 8.4.1 to $c' := c(M) - \int_{M \setminus K} G_+ \, dM$ yields

$$\theta_{\text{ind } \gamma} \geq \pi - \text{ind } \gamma \left(\lambda_\infty(M) + \int_{M \setminus K} G_+ \, dM \right).$$

Therefore

$$\text{ind } \gamma + 1 \geq \frac{\pi}{\lambda_\infty(M) + \int_{M \setminus K} G_+ \, dM} > \frac{\pi}{\lambda_\infty(M) + \epsilon_+} \geq n(M). \qquad \square$$

Lemma 8.4.5. *Assume that M^+ is bounded, M^- is unbounded and $\pi/\lambda_\infty(M)$ is an integer ≥ 2. If $M^+ \subset K$ then the index of any regular complete geodesic γ in M outside K satisfies*

$$\text{ind } \gamma \geq n' = n + 1.$$

Proof. Since $\pi/\lambda_\infty(M) \geq 2$ we have $c(M) \geq 3\pi/2 > \pi$, so that by Lemma 8.3.6, γ is nonsimple. Since M^- is unbounded and $M^+ \subset K \subset B_1$, we have $0 > c(M \setminus B_1) = c(M) - c(B_1)$ and so $c(B_1) < c' := c(M)$. Applying Lemma 8.4.1 yields

$$\theta_{\text{ind } \gamma} > \pi - \text{ind } \gamma \, \lambda_\infty(M).$$

However, remarking that $\int_{M \setminus K} G_+ \, dM = 0$ and using (8.4.1) we have $\theta_{\text{ind } \gamma} \leq \lambda_\infty(M)$. Thus

$$\text{ind } \gamma > \frac{\pi}{\lambda_\infty(M)} - 1 = n' - 1 = n. \qquad \square$$

Proof of Theorems 8.2.2 and 8.2.3. Theorems 8.2.2 and 8.2.3 (1), (3) follow from Propositions 8.4.1 and 8.4.3 and Lemma 8.4.4.

(2) of Theorem 8.2.3 follows from Propositions 8.4.1, 8.4.3 and 8.4.2 and Lemma 8.4.5. $\qquad \square$

8.5 The rotation numbers of proper complete geodesics

The purpose of this section is to prove Theorem 8.2.5.

Proof of Theorem 8.2.5. For an arbitrarily given number $n \geq 1$ we take a compact disk domain C in M with smooth boundary in such a way that

$\int_{M\setminus C} |G|\, dM < 1/(100n)$. Moreover, we can find such a C that will satisfy also

$$\lambda_{\mathrm{abs}}(\partial C) \le \int_{M\setminus C} |G|\, dM. \tag{8.5.1}$$

In fact, we can find a sufficiently large compact disk domain that is either locally convex or locally concave (see Lemma 8.3.1) and deform its boundary to obtain a smooth compact disk C that is also either locally convex or locally concave. Since $|\lambda(\partial C)| = |2\pi - c(C)| = |c(M\setminus C)| \le \int_{M\setminus C} |G|\, dM$, this C is the one desired.

Theorem 6.1.3 implies the nonexistence of a critical point in $M\setminus B(C, R)$ of the distance function to C for large $R > 0$. Hence, by Lemma 4.4.3, for any two numbers r_1 and r_2 with $R \le r_1 < r_2$ the annulus $A(C, r_1, r_2)$ is homeomorphic to $S^1 \times [r_1, r_2]$. Set $K_n := \overline{B(C, R)}$ and $N := \overline{M\setminus C}$ and let $pr : \tilde{N} \to N$ be the universal covering. Find any proper complete geodesic γ in M outside K_n and a lift $\tilde\gamma$ of γ in $\tilde N$. Since γ is proper, the function $\mathbf{R} \ni t \mapsto d(\gamma(t), C) = d(\tilde\gamma(t), \partial\tilde N)$ attains its minimum at some number $t_{\min}$. Let σ be a ray in M from C (i.e., $d(\sigma(s), C) = s$ for any $s > 0$) and let σ^i for $i \in \mathbf{Z}$ be the lifts in $\tilde N$ of σ such that $\cdots < \sigma^{-1} < \sigma^0 < \sigma^1 < \cdots$ and $\tilde\gamma(t_{\min}) \in D(\sigma^0, \sigma^1)$, where $D(\sigma^i, \sigma^j)$ for $i < j$ denotes the Riemannian half-plane in $\tilde N$ bounded by σ^i, σ^j and the subarc of $\partial\tilde N$ from $\sigma^i(0)$ to $\sigma^j(0)$. For $t \in \mathbf{R}$ denote by σ_t a minimal segment from $\partial\tilde N$ to $\tilde\gamma(t)$. Let $\partial\tilde N : \mathbf{R} \to \partial\tilde N$ be a positive parameterization relative to the orientation of $\tilde N$ and let $s(t) := \partial\tilde N^{-1}(\sigma_t(0))$. It follows from the first variation formula that

$$\angle(\dot\sigma_t(0),\ \partial\dot{\tilde N}(s(t))) = \angle(\dot\sigma_{t_{\min}}(L(\sigma_{t_{\min}})),\ \dot\gamma(t_{\min})) = \frac{\pi}{2}$$

for any $t \in \mathbf{R}$, where $L(\cdot)$ denotes the length of a curve; note that $s(t)$ is strictly monotonic in t near $t_{\min}$. We then assume that $s(t)$ is monotone increasing in t near $t_{\min}$, reversing the parameterization of γ if necessary. Define two minimal segments σ_t^- and σ_t^+ from $\partial\tilde N$ to $\tilde\gamma(t)$ as follows:

$$\sigma_t^-(0) = \partial\tilde N(\min\{s \in \mathbf{R};\ \partial\tilde N(s) \text{ is a foot point on } \partial\tilde N \text{ of } \tilde\gamma(t)\})$$

and

$$\sigma_t^+(0) = \partial\tilde N(\max\{s \in \mathbf{R};\ \partial\tilde N(s) \text{ is a foot point on } \partial\tilde N \text{ of } \tilde\gamma(t)\}).$$

Let I_t be the subarc of $\partial\tilde N$ from $\sigma_t^-(0)$ to $\sigma_t^+(0)$ and let D_t be the region bounded by σ_t^-, σ_t^+ and I_t (see Figure 8.5.1). The Gauss–Bonnet theorem implies that

$$c(D_t) = \angle_{\tilde\gamma(t)} D_t - \lambda(I_t).$$

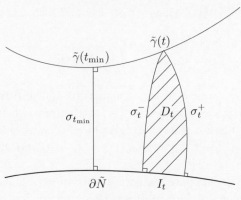

Figure 8.5.1

Since D_t is contained in $D(\sigma^i, \sigma^{i+1})$ for some i, and since by (8.5.1), we have $|c(D_t)|$, $|\lambda(I_t)| \leq \int_{M\setminus C} |G|\, dM$. Hence

$$\angle_{\tilde{\gamma}(t)} D_t \leq 2 \int_{M\setminus C} |G|\, dM < \frac{1}{50}. \tag{8.5.2}$$

Denote by J_t the subarc of $\partial \tilde{N}$ joining $\sigma_{t_{\min}}(0)$ to $\sigma_t(0)$. In the case where $\tilde{\gamma}|_{[t_{\min},t)}$ does not intersect σ_t for some $t > t_{\min}$, the four arcs $\sigma_{t_{\min}}$, J_t, σ_t and $\tilde{\gamma}|_{[t_{\min},t]}$ together bound a disk domain E_t. Note that if t is close enough to $t_{\min}$ then E_t is defined. For any $t > t_{\min}$ such that E_t is defined, we have

$$c(E_t) = \angle_{\tilde{\gamma}(t)} E_t - \frac{\pi}{2} - \lambda(J_t),$$

and hence, if E_t is contained in $D(\sigma^0, \sigma^n)$,

$$\left| \angle_{\tilde{\gamma}(t)} E_t - \frac{\pi}{2} \right| \leq 2n \int_{M\setminus C} |G|\, dM < \frac{1}{50}. \tag{8.5.3}$$

We will prove the following:

Sublemma 8.5.1. *If $\tilde{\gamma}|_{[t_{\min},t)}$ for some $t > t_{\min}$ is entirely contained in $D(\sigma^0, \sigma^n)$ then $\tilde{\gamma}|_{[t_{\min},t)}$ does not intersect σ_t.*

Proof. Suppose the contrary. Then, setting

$$t_0 := \inf\{t > t_{\min};\ \tilde{\gamma}|_{[t_{\min},t)} \subset D(\sigma^0, \sigma^n),\ \tilde{\gamma}|_{[t_{\min},t)} \cap \sigma_t \neq \emptyset\},$$

we have $t_{\min} < t_0 < +\infty$. It follows from (8.5.2) and (8.5.3) that $s(t)$ is strictly monotone increasing in $t \in [t_{\min}, t_0)$. Since σ_t does not intersect $\tilde{\gamma}|_{[t_{\min},t)}$ for any $t \in (t_{\min}, t_0)$, it follows that $\sigma_{t_0}^- = \lim_{t \nearrow t_0} \sigma_t$ does not intersect $\tilde{\gamma}|_{[t_{\min},t_0)}$. However, there exists a sequence $\{t_i \searrow t_0\}_{i \geq 1}$ such that $\tilde{\gamma}|_{[t_{\min},t_i)}$ intersects σ_{t_i}

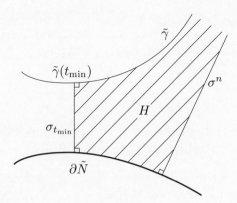

Figure 8.5.2

for any $i \geq 1$. Since any sequence $\{s_i\}$ with $\tilde{\gamma}(s_i) \in \sigma_{t_i}$ and $s_i \in [t_{\min}, t_i)$ does not accumulate at t_0, a limiting minimal geodesic τ of $\{\sigma_{t_i}\}$ intersects $\tilde{\gamma}|_{[t_{\min}, t_0)}$ and satisfies $\partial \tilde{N}^{-1}(\sigma_{t_0}^{-}(0)) < \partial \tilde{N}^{-1}(\tau(0))$. Let $\tilde{\gamma}(a)$, $t_{\min} \leq a < t_0$, be a point in $\tau \cap \tilde{\gamma}|_{[t_{\min}, t_0)}$. Then we must have $s(a) = \partial \tilde{N}^{-1}(\tau(0))$, which contradicts the monotonicity of $[t_{\min}, t_0) \ni t \mapsto s(t)$. This completes the proof of Sublemma 8.5.1. $\qquad\square$

Let $t_+ := \sup\{t > t_{\min}; \tilde{\gamma}|_{[t_{\min}, t)} \subset D(\sigma^0, \sigma^n)\}$. Sublemma 8.5.1 implies that $[t_{\min}, t_+) \ni t \mapsto s(t)$ is a strictly monotone increasing function and in particular that $\tilde{\gamma}|_{[t_{\min}, t_+)}$ is simple. We next prove the following:

Sublemma 8.5.2. $\tilde{\gamma}|_{[t_{\min}, +\infty)}$ *intersects* σ^n.

Proof. Suppose the contrary, that $\tilde{\gamma}|_{[t_{\min}, +\infty)}$ is contained in $D(\sigma^0, \sigma^n)$. We denote by H the Riemannian half-plane bounded by $\tilde{\gamma}|_{[t_{\min}, +\infty)}$, $\sigma_{t_{\min}}$, the subarc of $\partial \tilde{N}$ from $\sigma_{t_{\min}}(0)$ to $\sigma^n(0)$ and σ^n (see Figure 8.5.2). Since $H \subset D(\sigma^0, \sigma^n)$, we have $c(H) \geq -n \int_{M \backslash C} |G|\, dM$. However, it follows that

$$\lambda(\partial H) \geq \frac{3\pi}{2} - n\, \lambda_{\mathrm{abs}}(\partial C),$$

which together with $\lambda_\infty(H) \geq 0$ implies that

$$c(H) = \pi - \lambda_\infty(H) - \lambda(\partial H) \leq -\frac{\pi}{2} + n \int_{M \backslash C} |G|\, dM.$$

This is a contradiction. $\qquad\square$

In the same way as above, we can prove that $\tilde{\gamma}|_{(-\infty, t_{\min}]}$ intersects σ^{-n-1}. Set $t_- := \inf\{t < t_{\min}; \tilde{\gamma}|_{(t, t_{\min}]} \subset D(\sigma^{-n-1}, \sigma^1)\}$ and let $t_{\max}$ be one of the t-values for which the function $[t_-, t_+] \ni t \mapsto d(\tilde{\gamma}(t), \partial \tilde{N})$ attains its maximum. Set

$r_{\min} := d(\gamma(t_{\min}), C)$ and $r_{\max} := d(\gamma(t_{\max}), C)$. By remarking that $A(C, r_{\max}, r_{\min})$ is homeomorphic to $S^1 \times [r_{\max}, r_{\min}]$, we deduce that if $t_{\max} \leq t_{\min}$ then the number of intersection points of $\gamma|_{[t_{\min}, t_+]}$ and $\gamma|_{[t_{\max}, t_{\min}]}$ is at least n, and $\mathrm{rot}(\gamma|_{[t_-, t_+]}) \geq n$. When $t_{\max} > t_{\min}$, we also have $\mathrm{rot}(\gamma|_{[t_-, t_+]}) \geq n$. $\qquad \square$

8.6 The existence of complete geodesics arbitrarily close to infinity

Proof of Theorem 8.2.6(1). Suppose that there is a sequence $\{\gamma_i\}$ of complete geodesics tending to infinity. By Theorem 8.3.3, for every sufficiently large i we have that γ_i is semi-regular. Since the family of teardrops associated with the semi-regular geodesics γ_i is a locally convex filling, M cannot be contracting. $\qquad \square$

A main purpose of this section is to prove Theorem 8.2.6(2).

Lemma 8.6.1. *If $c(M) > \pi$ then for any compact subset K of M there exists a simple geodesic loop that bounds a disk domain containing K.*

Proof. Let ϵ be a fixed positive number such that $5\epsilon < c(M) - \pi$ and $\epsilon < \pi/2$. Find a core C_ϵ of M such that $\int_{M \setminus C_\epsilon} |G|\, dM \leq \epsilon$ and C_ϵ contains a given compact subset K of M. Then, for a ray σ from C_ϵ, we have $\lambda_\infty(D(\sigma^0, \sigma^1)) = 2\pi - c(M) < \pi - 5\epsilon$. Applying Lemma 3.7.2 yields that there exist a number $T > 0$ and a minimal segment γ_T in $D(\sigma^0, \sigma^1)$ joining $\sigma^0(T)$ to $\sigma^1(T)$ that does not intersect ∂C_ϵ (in particular, γ_T is a geodesic segment). The projection of γ_T into $M \setminus C_\epsilon$ is a simple geodesic loop with base point $\sigma(T)$ that bounds a disk domain containing K. $\qquad \square$

The following lemma is part of Theorem 8.2.6(2).

Lemma 8.6.2. *If M is expanding and $c(M) > \pi$ then for any compact subset K of M there exists a complete geodesic in M outside K.*

Proof. Let K be a given compact subset of M. Since M is expanding and, by Theorem 8.3.1(1), there exists a compact contractible subset K' of M such that the interior $\mathrm{int}\, K'$ of K' contains K and any geodesic arc outside $\mathrm{int}\, K$ is semi-regular. By Lemma 8.6.1, there exists a simple geodesic loop γ that bounds a disk domain D containing K'. Since D is not locally concave, it is a teardrop. Let γ' be the geodesic that is maximal, so that $\gamma \subset \gamma' \subset M \setminus \mathrm{int}\, K'$. Then, since γ' is semi-regular, $\gamma' \setminus \gamma$ cannot intersect D, so that γ' is a complete geodesic in M not intersecting K. $\qquad \square$

Definition 8.6.1 (Visual image and diameter). For a point $p \in M$ and a subset $A \subset M$, let

$$\Gamma_p^+(A) := \{v \in S_p M; \ \exp_p tv \in A \text{ for some } t \geq 0\}.$$

We call $\Gamma_p^+(A)$ the *visual image of A at p*. The *visual diameter of A at p* is defined to be the diameter of $\Gamma_p(A)$ with respect to the canonical angle–distance function on $S_p M$.

If A is a fixed closed subset of M then the visual diameter diam $\Gamma_p^+(K)$ is upper semicontinuous in $p \in M$.

Theorem 8.6.1. *Let M be a finitely connected complete Riemannian 2-manifold possibly with boundary, and let C be a core of M. If a component V of $\overline{M \setminus C}$ satisfies $\lambda_\infty(V) > 0$ then the visual diameter* diam $\Gamma_p^+(C)$ *tends to zero as $p \in V$ tends to the end corresponding to V.*

Proof. Since C is compact, so is the visual image $\Gamma_p(C)$. Hence there are two vectors $u_p, v_p \in \Gamma_p(C)$ such that $\angle(u_p, v_p) = \text{diam} \, \Gamma_p(C)$. We can find lifts $\tilde{u}_p, \tilde{v}_p \in S_{\tilde{p}} \tilde{V}$, $\tilde{p} \in D(\sigma^0, \sigma^1)$ for a ray σ from C. Applying Lemma 3.3.2 to any geodesic segment γ emanating from $\tilde{p}$ yields that γ is contained in $D(\sigma^n, \sigma^m) \setminus (\sigma^n \cup \sigma^m)$ for some $m < 0$ and $n > 1$ independent of p and γ. Thus, the two geodesics α_p and β_p emanating from $\tilde{p}$ and respectively having directions $u_{\tilde{p}}$ and $v_{\tilde{p}}$ are both contained in $D(\sigma^m, \sigma^n)$ and connect $\tilde{p}$ to $I(\sigma^m, \sigma^n) = \partial \tilde{V} \cap D(\sigma^m, \sigma^n)$, because $u, v \in \Gamma_p(C)$. As in the proof of Proposition 3.2.3, we can find two curves α_p' and β_p' connecting a point $\tilde{p}' \in \partial \Delta(\alpha_p \cup \beta_p)$ to $I(\sigma^m, \sigma^n)$. Lemma 3.2.2 shows that if p is close enough to infinity then $\tilde{p}' = \tilde{p}$ and the angle $\angle(u_p, v_p)$ tends to zero as p tends to infinity. $\qquad \square$

Theorem 8.6.1 implies the following:

Corollary 8.6.1. *Let M be a Riemannian plane with $c(M) < 2\pi$. Then for any compact subset K of M there exists a number $R > 0$ such that through any point $p \in M \setminus B(K, R)$ there passes a complete geodesic outside K.*

Proof of Theorem 8.2.6(2). The theorem now follows from Lemma 8.6.2 and Corollary 8.6.1. $\qquad \square$

Let Γ denote the set of unit-speed geodesics $\gamma : \mathbf{R} \to M$. Recall that there exists the canonical Lebesgue measure on the unit tangent bundle SM with Sasaki metric (see Example 1.3.2). The bijective map from the unit tangent bundle SM to Γ that assigns $\gamma_v \in \Gamma$ to each $v \in SM$, where $\gamma_v(t) := \exp tv$, $t \in \mathbf{R}$, induces a measure m on Γ. Let C be a compact subset of M with nonzero volume. On the set $\Gamma_C := \{ \gamma \in \Gamma; \gamma(0) \in C \}$, the probability measure prob_{Γ_C} is defined naturally by

$$\text{prob}_{\Gamma_C} A := \frac{m(A)}{m(\Gamma_C)} \qquad \text{for measurable } A \subset \Gamma_C.$$

For a measurable subset K of M let

$$E(K) := \{ \gamma \in \Gamma; \gamma \cap K \neq \emptyset \}.$$

Then $E(K) \cap \Gamma_C$ is a probability event with respect to the probability space $(\Gamma_C, \text{prob}_{\Gamma_C})$. We define Γ_C^+, $\text{prob}_{\Gamma_C^+}$ and $E^+(K)$ in the same manner, replacing Γ with the set Γ^+ of half-geodesics $\gamma : [0, +\infty) \to M$.

Corollary 8.6.2. *Let M be a Riemannian plane with $c(M) < 2\pi$, let K be a compact subset of M and let $\{C_i\}$ be a sequence of compact subsets of M with nonzero volume such that $\bigcup_i C_i = M$. Setting $\Gamma_i := \Gamma_{C_i}$ and $\Gamma_i^+ := \Gamma_{C_i}^+$, we have*

$$\lim_{i \to \infty} \text{prob}_{\Gamma_i}(E(K) \cap \Gamma_i) = \lim_{i \to \infty} \text{prob}_{\Gamma_i^+} \big(E^+(K) \cap \Gamma_i^+ \big) = 0.$$

Proof. Let $\epsilon > 0$ be fixed. By Theorem 8.6.1, there exists a compact subset C_ϵ of M such that $\operatorname{diam} \Gamma_p^+(K) < \epsilon$ for any $p \in M \setminus C_\epsilon$. Therefore

$$\text{prob}_{\Gamma_i^+} \big(E^+(K) \cap \Gamma_i^+ \setminus \Gamma_{C_\epsilon}^+ \big) < \frac{\epsilon}{2\pi} \qquad \text{for each } i.$$

It follows that

$$\text{prob}_{\Gamma_i^+}(E^+(K) \cap \Gamma_i^+) = \frac{m\big(\Gamma_{C_\epsilon}^+\big)}{m\big(\Gamma_i^+\big)} \text{prob}_{\Gamma_i^+} \big(E^+(K) \cap \Gamma_i^+ \cap \Gamma_{C_\epsilon}^+ \big)$$

$$+ \frac{m\big(\Gamma_i^+ \setminus \Gamma_{C_\epsilon}^+\big)}{m\big(\Gamma_i^+\big)} \text{prob}_{\Gamma_i^+} \big(E^+(K) \cap \Gamma_i^+ \setminus \Gamma_{C_\epsilon}^+ \big).$$

Theorem 5.2.1 implies that M has infinite volume and hence that $m(\Gamma_i^+)$ tends to infinity as $i \to \infty$. Consequently, we have

$$\lim_{i \to \infty} \text{prob}_{\Gamma_i^+} \big(E^+(K) \cap \Gamma_i^+ \big) = \lim_{i \to \infty} \text{prob}_{\Gamma_i^+} \big(E^+(K) \cap \Gamma_i^+ \setminus \Gamma_{C_\epsilon}^+ \big) \leq \frac{\epsilon}{2\pi},$$

which completes the proof.   □

Conjecture 8.6.1. *Let M be a Riemannian plane with $c(M) = 2\pi$. Then we have*

$$\liminf_{M \ni p \to \infty} \mu\big(\Gamma_p^+(K)\big) \geq \pi,$$

where μ denotes the canonical Lebesgue measure on $S_p M$ and $M \ni p \to \infty$ means that the distance between $p \in M$ and a fixed point in M tends to infinity. In addition, if M is expanding then

$$\lim_{p \to \infty} \mu\big(\Gamma_p^+(K)\big) = \pi.$$

If the conjecture were true we would obtain the following statements in the same way, as in the proof of Corollary 8.6.2. With the same notation as in Corollary 8.6.2, if M is a Riemannian plane with $c(M) = 2\pi$ then

$$\lim_{i \to \infty} \text{prob}_{\Gamma_i}(E(K) \cap \Gamma_i) = 1,$$

$$\liminf_{i \to \infty} \text{prob}_{\Gamma_i^+}\big(E^+(K) \cap \Gamma_i^+\big) \geq \frac{1}{2}.$$

In addition, if M is expanding, then

$$\lim_{i \to \infty} \text{prob}_{\Gamma_i^+}\big(E^+(K) \cap \Gamma_i^+\big) = \frac{1}{2}.$$

References

[1] A. D. Alexandrov, A theorem on triangles in a metric space and some of its applications, *Trudy Mat. Inst. Steklov.* **38** (1951), 5–23.

[2] *Die Innere Geometrie der Konvexen Flächen*, Akademie-Verlag, Berlin, 1955.

[3] Über eine Verallgemeinerung der Riemannschen Geometrie, *Schriften Forschungsinst. Math.* **1** (1957), 33–84.

[4] A. D. Alexandrov, V. N. Berestovskii and I. G. Nikolaev, Generalized Riemannian spaces, *Russian Math. Surveys* **41** (1986), no. 3, 1–54.

[5] A. D. Alexandrov and V. A. Zalgaller, Two-dimensional manifolds of bounded curvature, *Trudy Mat. Inst. Steklov.* **63** (1962), 33–84.

[6] *Intrinsic Geometry of Surfaces*, Translations of Mathematical Monographs, vol. 15, Amer. Math. Soc., Providence, RI, 1967.

[7] W. Ballmann, M. Gromov and V. Schroeder, *Manifolds of Nonpositive Curvature*, Progress in Math., vol. 61, Birkhäuser, Boston-Basel Stuttgart, 1985.

[8] S. Banach, Sur les linges rectifiables et les surfaces dont l'aire est finie, *Fundamenta Mathematicae* **7** (1925), 225–236.

[9] V. Bangert, Geodesics and totally convex sets on surfaces, *Invent. Math.* **63** (1981), 507–517.

[10] On the existence of escaping geodesics, *Comment. Math. Helv.* **56** (1981), 59–65.

[11] Yu. D. Burago, Closure of the class of manifolds of bounded curvature, *Proc. Steklov Inst. Math.* **76** (1967), 175–183.

[12] H. Busemann, *The Geometry of Geodesics*, Academic Press, 1955.

[13] *Convex Surfaces*, Interscience Tracts in Pure and Applied Mathematics, no. 6, Interscience Publishers, New York, London, 1958.

[14] M. Cassorla, Approximating compact inner metric spaces by surfaces, *Indiana Univ. Math. J.* **41** (1992), 505–513.

[15] J. Cheeger, Critical points of distance functions and applications to geometry, in *Geometric Topology: Recent Developments*, Lecture Notes in Math., no. 1504, Springer-Verlag, 1991, pp. 1–38.

[16] J. Cheeger and D. G. Ebin, *Comparison Theorems in Riemannian Geometry*, North-Holland Mathematical Library, vol. 9, North-Holland, Amsterdam, 1975.

[17] J. Cheeger and D. Gromoll, On the structure of complete manifolds of nonnegative curvature, *Ann. Math. (2)* **96** (1972), 413–443.

[18] X. Chen, Weak limits of Riemannian metrics in surfaces with integral curvature bound, *Calc. Var. Partial Differential Equations* **6** (1998), no. 3, 189–226.

[19] S. Cohn-Vossen, Kürzeste Wege und Totalkrümmung auf Flächen, *Compositio Math.* **2** (1935), 63–113.

[20]   Totalkrümmung und geodätische Linien auf einfach zusammenhängenden offenen volständigen Flächenstücken, *Recueil Math. Moscow* **43** (1936), 139–163.

[21] P. Eberlein and B. O'Neill, Visibility manifolds, *Pacific J. Math.* **46** (1973), 45–109.

[22] D. Elerath, An improved Toponogov comparison theorem for non-negatively curved manifolds, *J. Differential Geom.* **15** (1980), 187–216.

[23] L. C. Evans and R. F. Gariepy, *Measure Theory and Fine Properties of Functions*, Studies in Advanced Math., CRC Press, Boca Raton, FL, 1992.

[24] K. J. Falconer, *The Geometry of Fractal Sets*, New York, New Rochelle, Melbourne, Sydney, 1985.

[25] H. Federer, *Geometric Measure Theory*, Springer, Berlin, 1969.

[26] F. Fiala, Le problème isopérimetres sur les surfaces ouvertes à courbure positive, *Comment. Math. Helv.* **13** (1941), 293–346.

[27] H. Gluck and D. Singer, Scattering of geodesic fields. I, *Ann. Math. (2)* **108** (1978), no. 2, 347–372.

[28]   Scattering of geodesic fields. II, *Ann. Math. (2)* **110** (1979), no. 2, 205–225.

[29] D. Gromoll, W. Klingenberg and W. Meyer, *Riemannsche Geometrie im Großen*, Lecture Notes in Math., vol. 55, Springer-Verlag, 1968.

[30] D. Gromoll and W. Meyer, On complete manifolds of positive curvature, *Ann. Math. (2)* **75** (1969), 75–90.

[31] M. Gromov, *Volume and Bounded Cohomology*, Inst. Hautes Études Sci. Publ. Math. (1982), no. 56, 5–99 (1983).

[32]   *Metric Structures for Riemannian and non-Riemannian Spaces*, Birkhäuser Boston Inc., Boston, MA, 1999. Based on the 1981 French original (MR 85e:53051), with appendices by M. Katz, P. Pansu and S. Semmes; translated from the French by Sean Michael Bates.

[33] R. Gulliver, On the variety of manifolds without conjugate points, *Trans. Amer. Math. Soc.* **210** (1975), 185–201.

[34] P. Hartman, Geodesic parallel coordinates in the large, *Amer. J. Math.* **86** (1964), 705–727.

[35] J. J. Hebda, Parallel translation of curvature along geodesics, *Trans. Amer. Math. Soc.* **299** (1987), no. 2, 559–572.

[36]   Metric structure of cut loci in surfaces and Ambrose's problem, *J. Differential Geom.* **40** (1994), no. 3, 621–642.

[37] E. Heintze and H.-C. Imhof, Geometry of horospheres, *J. Differential Geom.* **12** (1977), no. 4, 481–491.

[38] H. Hopf and W. Rinow, Über den Begriff der volständigen differentialgeometrischen Fläche, *Comment. Math. Helv.* **3** (1931), 209–225.

[39] A. Huber, On subharmonic functions and differential geometry in the large, *Comment. Math. Helv.* **32** (1952), 13–72.

[40] M. Igarashi, K. Kiyohara and K. Sugahara, Noncompact Liouville surfaces, *J. Math. Soc. Japan* **45** (1993), 459–480.

[41] J. Itoh, The length of a cut locus on a surface and Ambrose's problem, *J. Differential Geom.* **43** (1996), 642–651.

[42] A. Kasue, A compactification of a manifold with asymptotically nonnegative curvature, *Ann. Sci. École Norm. Sup. (4)* **21** (1988), no. 4, 593–622.

[43] ——— A convergence theorem for Riemannian manifolds and some applications, *Nagoya Math. J.* **114** (1989), 21–51.

[44] K. Kiyohara, Compact Liouville surfaces, *J. Math. Soc. Japan* **43** (1991), 555–591.

[45] Y. Kubo and V. K. Senanayake, The Busemann total excess of complete open Alexandrov surfaces, *Kyushu J. Math.* **49** (1995), no. 1, 135–142.

[46] D. P. Ling, Geodesics on surfaces of revolution, *Trans. Amer. Math. Soc.* **59** (1946), 415–429.

[47] Y. Machigashira, The Gaussian curvature of Alexandrov surfaces, *J. Math. Soc. Japan* **50** (1998), no. 4, 859–878.

[48] Y. Machigashira and F. Ohtsuka, Total excess on length surfaces, Math. Ann. **319** (2001), 675–706.

[49] M. Maeda, On the existence of the rays, *Sci. Rep. Yokohama Nat. Univ. Sect. I* (1979), no. 26, 1–4.

[50] ——— Remarks on the distribution of rays, *Sci. Rep. Yokohama Nat. Univ. Sect. I* (1981), no. 28, 15–21.

[51] ——— A geometric significance of total curvature on complete open surfaces, in *Geometry of Geodesics and Related Topics* (*Tokyo, 1982*), North-Holland, Amsterdam, 1984, pp. 451–458.

[52] ——— A note on the set of points which are poles, *Sci. Rep. Yokohama Nat. Univ. Sect. I* (1985), no. 32, 1–5.

[53] ——— On the total curvature of noncompact Riemannian manifolds. II, *Yokohama Math. J.* **33** (1985), no. 1–2, 93–101.

[54] ——— A note on the geodesic circles and total curvature, *Sci. Rep. Yokohama Nat. Univ. Sect. I* (1988), no. 35, 1–9.

[55] ——— Geodesic circles and total curvature, *Yokohama Math. J.* **36** (1989), no. 2, 91–103.

[56] ——— Geodesic spheres and poles, in *Geometry of Manifolds* (*Matsumoto, 1988*), Academic Press, Boston, MA, 1989, pp. 281–293.

[57] ——— On the diameter of geodesic circles, *Sci. Rep. Yokohama Nat. Univ. Sect. I* (1989), no. 36, 1–5.

[58] ——— Geodesic sphere and poles. II, *Sci. Rep. Yokohama Nat. Univ. Sect. I* (1990), no. 37, 19–23.

[59] H. von Mangoldt, Über die jenigen Punkte auf positive gekrümmten Flächen, welche die Eigenschaft haben, dass die von ihnen ausgehenden geodätischen Linien nie aufhöven, pürzeste Linien zu sein, *J. Reine Angew. Math.* **91** (1881), 23–53.

[60] S. B. Myers, Connections between differential geometry and topology I, *Duke Math. J.* **1** (1935), 376–391.

[61] ——— Connections between differential geometry and topology II, *Duke Math. J.* **2** (1935), 95–102.

[62] T. Oguchi, Total curvature and measure of rays, *Proc. Fac. Sci. Tokai Univ.* **21** (1986), 1–4.

[63] F. Ohtsuka, On a relation between total curvature and Tits metric, *Bull. Fac. Sci. Ibaraki Univ. Ser. A* (1988), no. 20, 5–8.

[64] On the existence of a straight line, *Tsukuba J. Math.* **12** (1988), no. 1, 269–272.

[65] P. Petersen, S. D. Scheingold, and G. Wei, Comparison geometry with integral curvature bounds, *Geom. Funct. Anal.* **7** (1997), no. 6, 1011–1030.

[66] P. Petersen and C. Sprouse, Integral curvature bounds, distance estimates and applications, *J. Differential Geom.* **50** (1998), no. 2, 269–298.

[67] P. Petersen and G. Wei, Relative volume comparison with integral curvature bounds, *Geom. Funct. Anal.* **7** (1997), no. 6, 1031–1045.

[68] H. Poincaré, Sur les linges gésiques de surfaces convexes, *Trans. Amer. Math. Soc.* **6** (1905), 237–274.

[69] W. Poor, Some results on nonnegatively curved manifolds, *J. Differential Geom.* **9** (1974), 583–600.

[70] Yu. G. Reshetnyak (ed.), *Geometry IV, Encyclopaedia of Math. Sci.*, vol. 70, Springer-Verlag, 1993.

[71] Two-dimensional manifolds of bounded curvature, in *Encyclopaedia of Math. Sci.*, vol. 70, 1993, pp. 3–163.

[72] G. de Rham, Sur la réductibilité d'un espace de Riemann, *Comment. Math. Helv.* **26** (1952), 341.

[73] T. Sakai, *Riemannian Geometry*, Mathematical Monographs, vol. 149, Amer. Math. Soc., 1992.

[74] V. K. Senanayake, The existence of geodesic loops on Alexandrov surfaces, *Nihonkai Math. J.* **8** (1997), no. 1, 95–100.

[75] Mass of rays on Alexandrov surfaces, *Kyushu J. Math.* **54** (2000), no. 1, 139–146.

[76] K. Shiga, On a relation between the total curvature and the measure of rays, *Tsukuba J. Math.* **6** (1982), no. 1, 41–50.

[77] A relation between the total curvature and the measure of rays. II, *Tôhoku Math. J. (2)* **36** (1984), no. 1, 149–157.

[78] K. Shiohama, Busemann function and total curvature, *Invent. Math.* **53** (1979), 281–297.

[79] The role of total curvature on complete noncompact Riemannian 2-manifolds, *Illinois J. Math.* **28** (1984), 597–620.

[80] Topology of complete noncompact manifolds, in *Geometry of Geodesics and Related Topics (Tokyo, 1982)*, Advanced Studies in Pure Math., vol. 3, North-Holland, Amsterdam, 1984, pp. 423–450.

[81] Cut locus and parallel circles of a closed curve on a Riemannian plane admitting total curvature, *Comment. Math. Helv.* **60** (1985), 125–138.

[82] Total curvatures and minimal areas of complete open surfaces, *Proc. Amer. Math. Soc.* **94** (1985), 310–316.

[83] An integral formula for the measure of rays on complete open surfaces, *J. Differential Geom.* **23** (1986), 197–205.

[84] K. Shiohama, T. Shioya and M. Tanaka, Mass of rays on complete open surfaces, *Pacific J. Math.* **143** (1990), no. 2, 349–358.

[85] K. Shiohama and M. Tanaka, An isoperimetric problem for infinitely connected complete open surfaces, in *Geometry of Manifolds* (*Matsumoto, 1988*), Academic Press, Boston, MA, 1989, pp. 317–343.

[86] The length function of geodesic parallel circles, in *Progress in Differential Geometry*, Math. Soc. Japan, Tokyo, 1993, pp. 299–308.

[87] Cut loci and distance spheres on Alexandrov surfaces, in *Actes de la table ronde de géométrie différentielle* (*Luminy, 1992*), Soc. Math. France, Paris, 1996, pp. 531–559.

[88] T. Shioya, The ideal boundaries of complete open surfaces admitting total curvature $c(M) = -\infty$, in *Geometry of Manifolds* (*Matsumoto, 1988*), Perspect. Math., vol. 8, Academic Press, Boston, MA, 1989, pp. 351–364.

[89] The ideal boundaries and global geometric properties of complete open surfaces, *Nagoya Math. J.* **120** (1990), 181–204.

[90] On asymptotic behavior of the mass of rays, *Proc. Amer. Math. Soc.* **108** (1990), no. 2, 495–505.

[91] The ideal boundaries of complete open surfaces, *Tôhoku Math. J. (2)* **43** (1991), no. 1, 37–59.

[92] Diameter and area estimates for S^2 and P^2 with non-negatively curved metrics, in *Progress in Differential Geometry*, Advanced Studies in Pure Math., vol. 22, Math. Soc. Japan, Tokyo, 1993, pp. 309–319.

[93] Behavior of distant maximal geodesics in finitely connected complete 2-dimensional Riemannian manifolds, *Mem. Amer. Math. Soc.* **108** (1994), no. 517.

[94] Mass of rays in Alexandrov spaces of nonnegative curvature, *Comment. Math. Helv.* **69** (1994), no. 2, 208–228.

[95] Geometry of total curvature, in *Actes de la table ronde de géométrie différentielle* (*Luminy, 1992*), Sémin. Congr., vol. 1, Soc. Math. France, Paris, 1996, pp. 561–600.

[96] The Gromov-Hausdroff limits of two-dimensional manifolds under integral curvature bound, in *Geometry and Topology* (Y. W. Kim, S. E. Koh, Y. J. Song and Y. G. Choi, eds.), *Proceedings of Workshop in Pure Mathematics*, vol. 16, Part III, Pure Mathematics Research Association, Korean Academic Council, 1996, pp. 35–55.

[97] The limit spaces of two-dimensional manifolds with uniformly bounded integral curvature, *Trans. Amer. Math. Soc.* **351** (1999), no. 5, 1765–1801.

[98] Behavior of distant maximal geodesics in finitely connected 2-dimensional Riemannian manifolds II, to appear in *Geom. Dedicata*.

[99] T. Shioya and T. Yamaguchi, Collapsing three-manifolds under a lower curvature bound, *J. Differential Geom.* **56** (2000), no. 1, 1–66.

[100] K. Sugahara, On the poles of Riemannian manifolds of nonnegative curvature, in *Progress in Differential Geometry*, Advanced Studies in Pure Math., vol. 22 (1993), pp. 321–332.

[101] M. Tanaka, On a characterization of a surface of revolution with many poles, *Mem. Fac. Sci. Kyushu Univ. Ser. A* **46** (1992), no. 2, 251–268.

[102] On the cut loci of a von Mangoldt's surface of revolution, *J. Math. Soc. Japan* **44** (1992), no. 4, 631–641.

[103] V. A. Toponogov, Riemannian spaces containing straight lines, *Dokl. Akad. Nauk SSSR* **127** (1959), 977–979.

[104] ——— Riemannian spaces having their curvature bounded below by a positive number, *Amer. Math. Soc. Transl. Ser.* **37** (1964), 291–336.

[105] M. Trojanov, Un principe de concentration-compacité pour les suites de surfaces riemaniennes, *Ann. Inst. Henri Poincaré* **8** (1991), no. 5, 1–23.

[106] R. L. Wheeden and A. Zygmund, *Measure and Integral*, Marcel Dekker, New York, Basel, 1977.

[107] J. H. C. Whitehead, On the covering of a complete space by the geodesics through a point, *Ann. Math. (2)* **36** (1935), 679–704.

[108] J. H. C. Whitehead, Convex regions in the geometry of paths, *Quart. J. Math. Oxford Ser. (2)* **3** (1932), 33–42.

[109] H. Whitney, On regular closed curves in the plane, *Compositio Math.* **4** (1937), 276–284.

[110] D. Yang, Convergence of Riemannian manifolds with integral bounds on curvature I, *Ann. Sci. École Norm. Sup. (4)* **25** (1992), 77–105.

Index